INSECT HORMONES

H. FREDERIK NIJHOUT

INSECT HORMONES

PRINCETON UNIVERSITY PRESS

PRINCETON, NEW JERSEY

Published by Princeton University Press, 41 William Street,
Princeton, New Jersey 08540
In the United Kingdom: Princeton University Press,
Chichester, West Sussex

Library of Congress Cataloging-in-Publication Data

Nijhout, H. Frederik.
Insect Hormones / H. Frederik Nijhout.
p. cm.
Includes bibliographical references (p.) and index.
ISBN 0-691-03466-4 (cl)
1. Insects—Physiology. 2. Insect hormones.
3. Insects—Development. I. Title.
QL495.N54 1994
597.5·0142—dc20 93-42301

This book has been composed in Times Roman

Princeton University Press books are printed on
acid-free paper and meet the guidelines for
permanence and durability of the Committee on
Production Guidelines for Book Longevity
of the Council on Library Resources

Printed in the United States of America

2 4 6 8 10 9 7 5 3 1

CONTENTS

PREFACE

INSECT endocrinology is one of the oldest branches of insect physiology and the most prolific, both in terms of the number of its practitioners and the number of papers published each year. Yet, in spite of its long history and active current interest, there exists no single concise source where a student can go to get a modern overview of the field. The classical texts of Wigglesworth (1970, 1972) and the review of Doane (1973) continue to provide excellent historical synopses, but they are long out of date. The excellent set of volumes edited by Kerkut and Gilbert (1985) provide an encyclopedic view of the status of insect endocrinology in the early 1980s but are somewhat unwieldy and overwhelming as an introductory text. Modern information must be gleaned from the primary literature and topical review articles, but these tend to give little biological context and almost no sense of the cohesiveness and interconnections that define insect endocrinology as a discipline. The overarching aim of this book, then, is to provide students and researchers with a single source from which they can get a fairly complete and reasonably up to date overview of the field.

One reason for the widespread interest in insect endocrinology is that almost every aspect of an insect's life is regulated by hormones at one time or another. Molting and metamorphosis are, of course, the most obvious of the endocrine-stimulated events in the insect life cycle, and the best studied. But hormones also control such disparate physiological and developmental phenomena as metabolism, water balance, seasonal polymorphisms, caste determination, reproductive cycles, and diapause, as well as behaviors such as eclosion, pheromone production, migration, and social dominance.

Insect hormones have a pervasive role in the regulation of postembryonic development, and studies on the way in which hormones control the progressive differentiation of metamorphosis form the cornerstone of insect endocrinology. Yet—and this is a most curious case of scientific inertia—the way in which juvenile hormone and ecdysone control metamorphosis has been widely misunderstood and continues to be misinterpreted in all biological textbooks and in all but a few of the professional reviews. The metamorphic progression from larva to pupa to adult is *not* cued by a progressive decrease in juvenile hormone concentration, as the simple "standard" view would have it; the control of metamorphosis is actually quite different, and vastly more interesting. A secondary aim of this book, then, is to provide a summary account of what, at present, we understand to be the correct version of the hormonal control of metamorphosis.

The functions of insect hormones in the regulation of metabolism and homeostasis are much less well understood than their roles in development, though that is likely to change in the near future. By using readily available antibodies to vertebrate hormones, several groups of investigators have isolated a large array of insect peptides that cross-react with the vertebrate antibodies. When purified (or artificially synthesized) doses of these peptides are injected into insects, they stimulate, or inhibit, a variety of physiological and metabolic functions. Many of these peptides belong to genetically related "families," and within these families there has been an apparent diversification of physiological effects. Unfortunately, development of our knowledge of basic insect physiology has not kept pace with the advances in macromolecular technology, with the result that little is known about the normal functions of these new presumptive hormones. We appear to be at an interesting transition point in insect endocrinology. Progress in the ease with which polypeptides can be detected and analyzed suggests that the next decades of insect endocrinology research may be dominated by a plethora of hormones in search of physiological functions, in contrast to the early decades, which were dominated by a plethora of physiological functions in search of regulatory hormones.

While the primary aim of this book is to introduce advanced undergraduate and graduate students to the roles of hormones in insect development and physiology, the book will also be useful for researchers in other fields who need an overview of the role of hormones in the biology of insects. In keeping with this purpose, this book is not an encyclopedic review of the literature on insect endocrinology; rather, it is a synthesis that provides a relatively broad view of many aspects of insect biology that are regulated or affected by hormones. The focus of this book is on the biology of the organism. Accordingly, readers will find the book light on technical detail of experimental design and results, and heavier on the biological context in which insect hormones work. Readers who are interested in the cellular and subcellular aspects of insect hormone action can consult several excellent recent reviews on the subject in the series of volumes edited by Kerkut and Gilbert (1985). A sufficient number of references to the primary and review literature are provided so that the book can be used as an entry point to the detailed literature in nearly all subjects that intersect with insect endocrinology.

The book is divided into nine chapters. Chapters 1 and 2 provide general background information on the structure of the insect endocrine system and the mechanisms of action of hormones. Chapter 3 deals with the role of hormones in various homeostatic physiological functions, with particular emphasis on carbohydrate and lipid metabolism and water balance. Chapters 4 and 5 deal with the processes of growth and metamorphosis. Chapter 4 provides a rather extensive outline of the biology of growth and metamorphosis of insects and sets the context for chapter 5. Chapter 5 deals with the endocrine and physiological processes that control the orderly progression of the complex developmental processes associated with molting and metamorphosis. Chapter 6 deals with the endocrine control of reproduction. It includes a discussion of the great diversity of reproductive cycles among the insects, presents several detailed case histories, and also outlines the ways in which hormones are involved in the control of functions ancillary to reproduction such as sex determination. Chapter 7 is an outline of the mechanisms by which insect hormones control seasonal diapause at various stages in the life cycle. Chapter 8 is concerned with a unique aspect of insect development, namely polyphenism: the ability of a single individual to develop into one of several alternative phenotypes in response to token stimuli from its environment. The developmental switches that control these polyphenisms are mediated by hormones, often the same hormones that control metamorphosis. Finally, chapter 9 provides an outline of the role of hormones in the regulation of various aspects of insect behavior.

MANY friends and colleagues have contributed to my ability to understand the questions, and at least a few of the answers, about the complex processes that control insect physiology and postembryonic development. Foremost among these are Jim Truman and Lynn Riddiford, whose dedication to biology in general, and insect physiology in particular, continues to serve as a model and inspiration. Together they have provided a deeper insight into the processes that regulate insect postembryonic development than any other person or research group in the second half of the twentieth century. If the "classical" view of insect developmental endocrinology is due to Wigglesworth and Williams, then certainly the "modern" view is due to Truman and Riddiford. Others who have allowed me to use them as sounding boards, and who have helped me see things more clearly, are Skip Bollenbacher, Noelle Granger, Claire Kremen, Debbie Rountree, Wendy Smith, and Diana Wheeler. I want to thank, even though he is no longer with us, Carroll Williams, for providing me, at an early stage in my career, with an environment of total and utter freedom to explore. I would also like to thank Lynn Riddiford, Gene Robinson, Debbie Rountree, Wendy Smith, Jan Veenstra, and Diana Wheeler, for reading either all or portions of the manuscript at various stages in its genesis and for providing critical and useful comments that greatly improved the final product. Special thanks also go to Laura Grunert, who helped with the preparation of the figures, and whose incredible sense of organization allows me to do two (and sometimes three) things at the same time.

INSECT HORMONES

CHAPTER 1

ANATOMY OF THE INSECT ENDOCRINE SYSTEM

INSECTS, like vertebrates, have two very different kinds of endocrine organs. The first of these are the conventional glandular tissues that are specialized for the synthesis and internal secretion of hormones. The principal endocrine glands in insects are the prothoracic glands, which secrete ecdysteroids, and the corpora allata, which secrete the juvenile hormones. In addition, the ovaries and testes of many adult female insects also produce ecdysteroids, and, as the singular instance of a male-specific hormone in insects, the testes of the European firefly *Lampyris noctiluca* produce an androgenic hormone. No other conventional endocrine glands are known in insects.

The second kind of endocrine organ consists of groups of specialized neurons in the central nervous system, the neurosecretory cells. Neurosecretory cells are specialized neurons with unusually large cell bodies that, instead of neurotransmitters, produce small polypeptides, the neurohormones. Neurosecretory cells occur in all the ganglia of the central nervous system, but are particularly abundant in the brain. Neurosecretory cells usually do not secrete their products at synaptic endings but send their axons to specialized structures where their secretions are released directly into the hemolymph. These release sites are called neurohemal organs if they are anatomically compact and distinct, or simply neurohemal areas if they are not.

The majority of insect hormones are neurosecretory products and these control a broad diversity of physiological and developmental processes. As in the case of the vertebrate hypothalamic-pituitary axis of endocrine control, the secretory activity of the conventional endocrine glands in insects is controlled via the secretion of tropic or inhibitory neurohormones. Neurohormones are therefore either directly or indirectly responsible for all forms of endocrine control in insects. It may be said, then, since neurosecretory cells reside in the central nervous system and are controlled by physiological processes in the central nervous system, that the nervous system regulates the broad diversity of physiological and developmental events that transpire in the course of an insect's life cycle. This is an important fact to recognize, because it helps one to understand that physiological as well as developmental processes are contingent and that they may, therefore, be adjusted in time and space to achieve homeostatic or adaptive effects.

The Prothoracic Glands

The prothoracic glands are the primary source of ecdysteroids in developing insects. In spite of their common function, the prothoracic glands in the different orders of insects are morphologically quite diverse (fig. 1.1). In the Hymenoptera and Lepidoptera, for instance, they consist of loosely connected strands of very large secretory cells surrounded by a basal lamina (figs. 1.1C,D, 1.2). They are largely centered in the prothorax, but in some species extend well into the mesothorax and occasionally the metathorax. In locusts the prothoracic glands are likewise loosely organized but extend anteriorly and lie mostly within the head. In Hemiptera, by contrast, the prothoracic glands are fairly compact structures, located almost entirely within the prothorax. The ultrastructure of the prothoracic

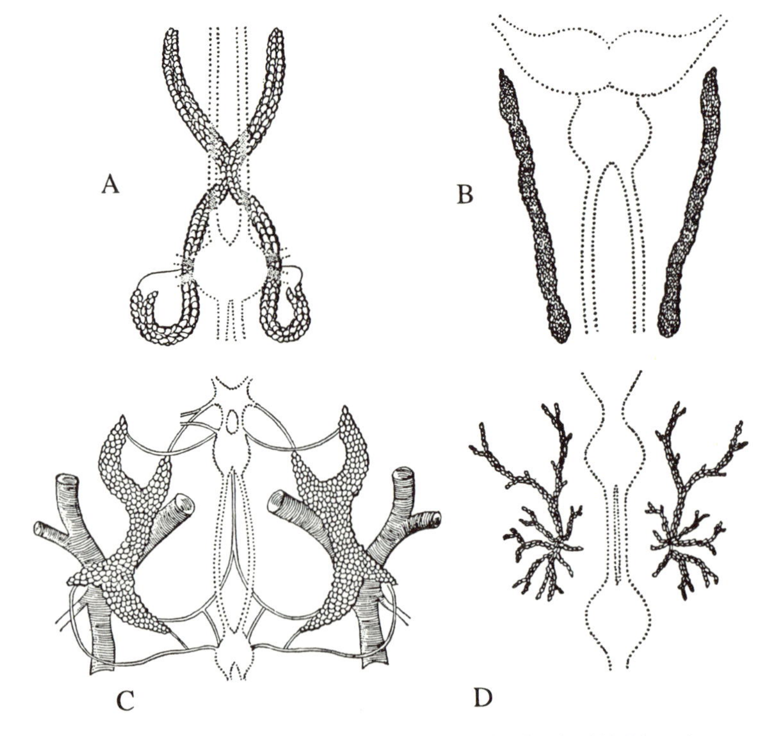

Figure 1.1. Various types of insect prothoracic glands: (A) Blattaria; (B) Hemiptera; (C) Lepidoptera; (D) Hymenoptera. (From Novak, 1975. Reprinted with permission of Chapman and Hall.)

gland cells has been reviewed by Sedlak (1985) and Beaulaton (1990). Sedlak notes that in spite of the great differences in gross morphology, the cellular structure of the prothoracic glands is remarkably uniform in different orders. One of the most notable ultrastructural features of prothoracic gland cells is the deeply infolded plasma membrane. The depth and density of these folds vary over time and appear to be correlated with the secretory activity of the glands. Prothoracic gland cells also tend to have a dense endoplasmic reticulum, as is common in steroid-secreting tissues in other

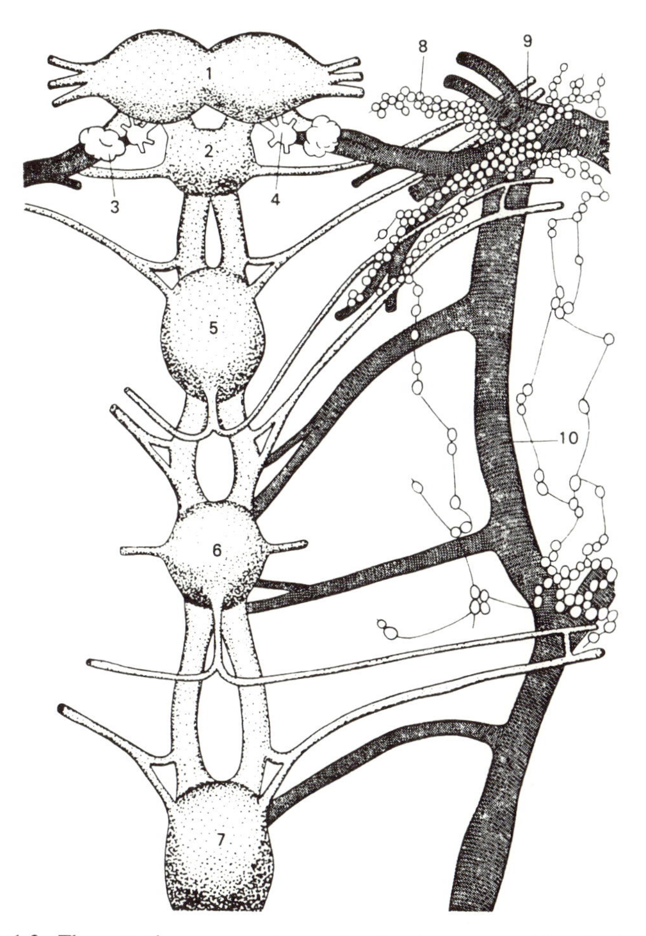

Figure 1.2. The central nervous system and endocrine system of the *Hyalophora cecropia* pupa. 1, brain; 2, subesophageal ganglion; 3, corpus allatum; 4, corpus cardiacum; 5, prothoracic ganglion; 6, mesothoracic ganglion; 7, metathoracic ganglion; 8, prothoracic gland; 9, spiracle; 10, trachea. (From Herman and Gilbert, 1966, as modified by Cymborowski, 1992. Reprinted with permission from Elsevier Science Publishers.)

animals. The extraordinarily complex and dynamic structure of the endoplasmic reticulum and the mitochondria of the prothoracic glands of *Manduca sexta* has been described by Hanton et al. (1993). The mitochondria of *Manduca* are shown to undergo elaborate changes in their ultrastructure and in their associations with the endoplasmic reticulum during periods of ecdysteroid production. The most unusual morphology of the prothoracic glands occurs in the higher Diptera (which, as we will see on various occasions in this book, have many unusual and highly derived morphological and developmental characteristics), where the prothoracic glands are fused with the corpora allata and corpora cardiaca to form a compact structure called the *ring gland*, which encircles the foregut (fig. 1.4).

The prothoracic glands are usually innervated by nerves from the subesophageal ganglion, and the first and second thoracic ganglia, or some subset of these three ganglia (Scharrer, 1964; Romer, 1971; Granger and Bollenbacher, 1981; Gersch et al., 1975). Some of these nerves contain neurosecretory granules and may play a role in regulating the secretory activity of the prothoracic glands, though it is not yet clear whether this regulation is stimulatory or inhibitory. At least one instance of inhibitory control via nerves has been demonstrated in *Mamestra brassicae*, using an in vitro culture system (Okajima and Kumagai, 1989). The prothoracic glands are not believed to store ecdysone because the hormone is generally not detectable in homogenates of the glands even during periods of high activity. Ecdysone appears to be released as soon as it is synthesized, so that the rate and timing of secretion of this hormone are determined entirely by the rate and timing of its synthesis.

In all insects except certain Apterygota, the prothoracic glands undergo programmed cell death during metamorphosis. In hemimetabolous insects this happens shortly after the molt to the adult, and in holometabolous insects in the pupal stage shortly after the initiation of adult development (Wigglesworth, 1952, 1955; Ozeki, 1968; Cassier and Fain-Maurel, 1970; Smith and Nijhout, 1982, 1983). In many insects cell death is preceded by the appearance of autophagic vacuoles, presumably lysosomes (Scharrer, 1966; Sedlak, 1985) except in *Oncopeltus fasciatus*, where cell disintegration proceeds rapidly in the absence of these organelles (Smith and Nijhout, 1982).

In adult insects ecdysteroids play an important role in the control of testicular and ovarian maturation, as will be discussed in chapter 6. In all cases studied so far, the gonads are the primary source of ecdysteroids in the adult stage. In females, the cells of the ovarian follicles secrete ecdysteroids, while in males the sheath of the testes appears to be a significant source, at least in *Heliothis virescens* (Hagedorn, 1985). Finally, several investigators have reported that ecdysteroid production can occur in isolated abdomens of larvae, though whether the developing gonads serve as the source for these hormones is not known (Delbecque and Slama, 1980). In *Tenebrio molitor*, abdominal enocytes are apparently capable of synthesizing ecdysteroids (Romer at al., 1974), but the role of this function in normal development remains to be clarified.

The Corpora Allata

The corpora allata are the glands that produce the juvenile hormones. The corpora allata are a pair of small glands that can be found in the neck region of most insects. They are attached to the brain by a nerve that also passes through the corpora cardiaca (figs. 1.3, 1.4, 5.2). The corpora allata are compact organs of tightly packed cells, surrounded by a tough membra-

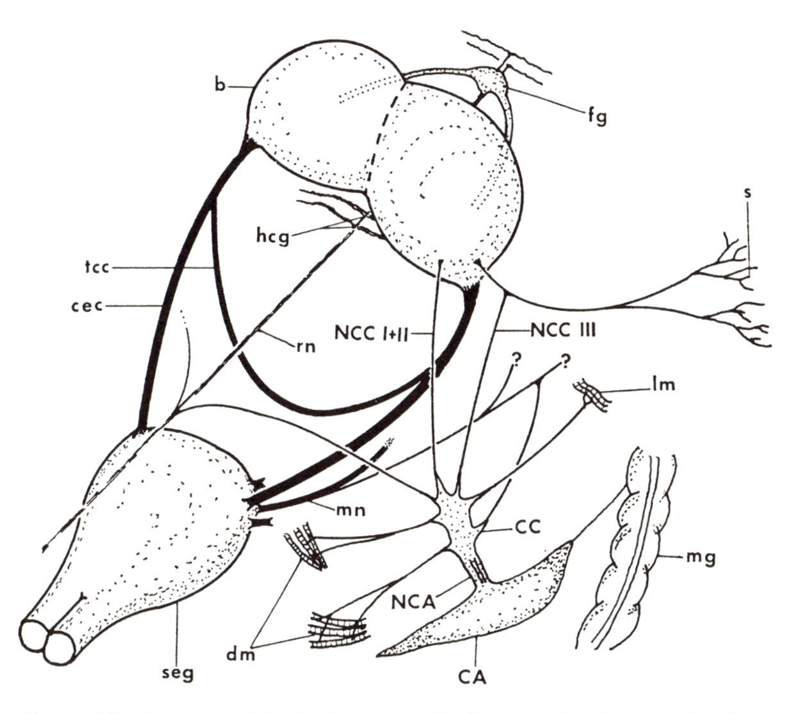

Figure 1.3. Anatomy of the brain-retrocerebral neuroendocrine complex in a final instar larva of *Manduca sexta*, showing the anatomical relations of parts and their nervous interconnections. b, brain; CA, corpus allatum; CC, corpus cardiacum; cec, circumesophageal connective; dms, dilator muscles of stomodaeum; fg, frontal ganglion; hcg, hypocerebral ganglion; lm, labial muscles; mg, mandibular gland; mn, maxillary nerve; NCA, nervus corporis allati; NCCI+II, nervi corporis cardiaci I and II; NCCIII, nervus corporis cardiaci III; rn, recurrent nerve; s, sensillae; seg, subesophageal ganglion; tcc, tritocerebral commissure. Question marks indicate nerves with unknown destination. (From Nijhout, 1975a.)

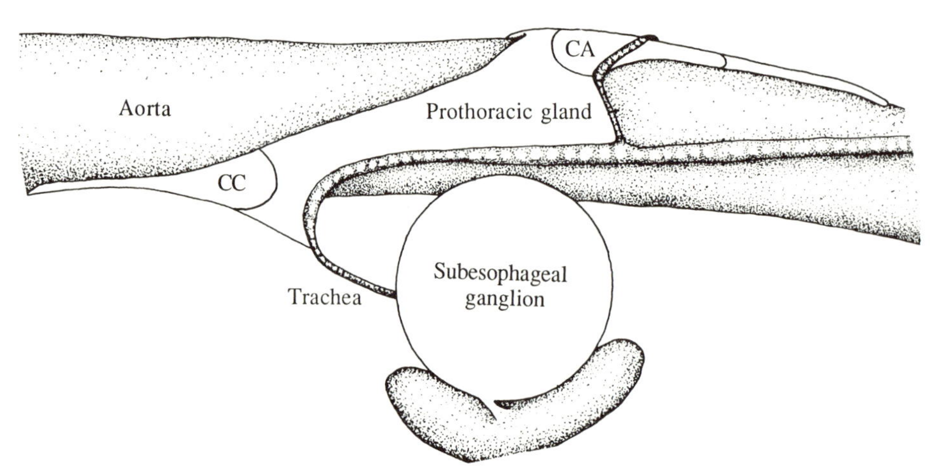

Figure 1.4. Anatomy of the retrocerebral complex (corpora allata and corpora cardiaca) and prothoracic glands of a higher dipteran, *Lucilia cuprina*. The three endocrine organs are fused into a single compact structure, the ring gland, that encircles the esophagus. (From Meurant and Sernia, 1993. Reprinted with kind permission of *Insect Biochem. & Mol. Biol.* and Pergamon Press Ltd.)

nous sheath. The cells of the corpora allata are usually heavily interdigitated, and are rich in mitochondria and endoplasmic reticulum (Sedlak, 1985; Cassier, 1990).

The innervation of the corpora allata consists of conventional as well as neurosecretory neurons that have their cell bodies in the brain, and, in some insects, possibly in the corpora cardiaca as well. Both types of neurons may be involved in the control of corpus allatum secretory activity (De Kort and Granger, 1981; Tobe and Stay, 1985). In Lepidoptera the corpora allata also act as neurohemal organs for some of the neurosecretions from the brain. In other insects only the corpora cardiaca (see below) serve this function (Nijhout, 1975a; Bollenbacher and Granger, 1985). In almost all insects the corpora allata receive their primary innervation from the brain, but in the Thysanura, Ephemeroptera, and Orthoptera, the corpora allata are also innervated from the subesophageal ganglion (Joly, 1968; Mason, 1973; see fig. 1.6C).

The corpora allata synthesize and secrete juvenile hormone but, as in the case of the prothoracic glands, are not known to store their product. The rate and timing of secretion of juvenile hormone appears to be determined entirely by the rate and timing of its synthesis. There is a correlation between cell size and secretory activity in the corpora allata of several species of insects (e.g., Scharrer, 1964; Panov and Bassurmanova, 1970; Schooneveld et al., 1979). In addition, in the roaches *Leucophaea* and *Diploptera* there is an increase in cell number during periods of high juvenile hormone production (Scharrer and Von Harnack, 1958; Szibo and Tobe, 1981). Consequently, many researchers have used total corpus allatum volume as an indicator of secretory activity, though this is not a safe assumption in all species of insects, or in all stages of their life cycle (Sedlak, 1985).

The Neuroendocrine System

Neurosecretory cells can often be seen quite readily in intact live brains because they have rather large cell bodies and because the neurosecretory granules they contain scatter light and give a nice Tyndall-blue opalescent effect. Neurosecretory cells have traditionally been identified by their characteristic staining properties with several specialized histochemical stains, such as paraldehyde-fuchsin, phloxine, or victoria blue. Each of these dyes stains a different group of neurosecretory cells, presumably because the stains have different affinities for the various kinds of neurosecretory peptides and carrier proteins they contain (Maddrell and Nordmann, 1979; Panov, 1980; Raabe, 1983; Orchard and Loughton, 1985). These histological methods have been used to characterize different general groups of neurosecretory cells, but they cannot be used to identify the product of those cells. Today, with the use of modern cellular and molecular technologies, there have been great advances in the analysis and identification of neurosecretory cells and their products. It is now possible to identify specifically the cells that produce a particular hormone by immunocytochemical methods (O'Brien et al., 1988; Westbrook et al., 1993), using antibodies to the hormone in question and reacting those with secondary antibodies that are tagged with a fluorescent molecule or with an enzyme that produces a colorimetric reaction (fig. 5.2). Once an antibody is available, it is possible to use it to detect temporal and spatial variation in the production of neurosecretory hormones throughout the central nervous system. By this method Westbrook et al. (1993) have shown that the prothoracicotropic hormone of *Manduca sexta* is produced by different groups of cells at different stages in development. In addition, it is now possible to make antibodies to the contents of single neurosecretory cells, and those antibodies can then be used as probes to identify the products of that cell and to assay the secretory behavior of the cell.

The Brain-Retrocerebral Neuroendocrine Complex

While neurosecretory cells occur in all ganglia of the central nervous system, the brain contains the greatest number and diversity of these cells and is thus the principal neuroendocrine organ of insects. The brain and its associated glands and neurohemal organs together form an integrated control, synthesis, and release system for hormones from the head region

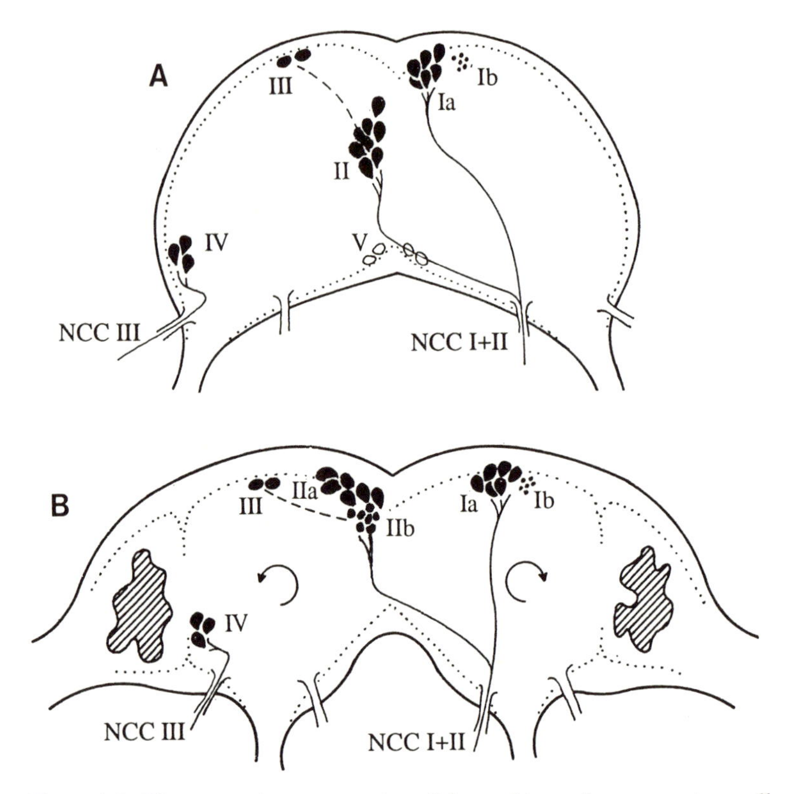

Figure 1.5. Diagrammatic representation of the positions of neurosecretory cell groups in the brains of larval (top) and pupal (bottom) *Manduca sexta*. Arrows indicate a slight rotation of the cerebral hemispheres that occurs upon metamorphosis. Roman numerals are used to designate the various neurosecretory cell groups. (Modified from Nijhout, 1975a.)

called the *brain-retrocerebral neuroendocrine complex*, consisting of the brain, corpora cardiaca, and corpora allata. The brain neurosecretory cells occur in several clusters with distinctive locations, from which they derive their names. The medial neurosecretory cells form one or two paired clusters on either side of the dorsal midline of the brain. The lateral neurosecretory cells form one or two clusters dorsally and laterally in each brain lobe. In addition, some species have ventral neurosecretory cells in the ventral and posterior part (the tritocerebrum) of each brain lobe. During metamorphosis in the moth *Manduca sexta*, the two hemispheres of the brain rotate with respect to each other so that cell groups that were medial in the larva become lateral in the pupa (fig. 1.5). A numbering system is therefore used to identify the five major clusters of neurosecretory cells in *Manduca* (Nijhout, 1975a). During metamorphosis of *Manduca*, a new group of neurosecretory cells appears next to the larval group II cells (Copenhaver and Truman, 1986b; Truman and Riddiford, 1989). In the adult moth we therefore have two groups, called IIa (the larval cells) and IIb (the new adult cells) (fig. 1.5B). The endocrine products of several of these groups of cells are now known and will be discussed in subsequent chapters.

The corpora cardiaca serve as the principal (though not the exclusive) neurohemal sites for the neurosecretory cells of the brain. The corpora cardiaca derive their name from the fact that in many species they are intimately associated with the heart (actually the anterior portion of the dorsal vessel), where they release their secretions into the hemolymph. Typically there are three main bundles of neurons, the nervi corporis cardiaci (generally abbreviated as NCC I, NCC II, and NCC III), which run between the corpora cardiaca and the brain. The NCC I carries axons from the medial neurosecretory cells of the protocerebrum, and the NCC II carries those from the lateral neurosecretory cells of the protocerebrum (Nijhout, 1975a). The NCC III carries neurons from neurosecretory cells in the tritocerebrum (Raabe, 1983). In most insects, except the Lepidoptera, all these neurons terminate in the corpora cardiaca, which serve as both the storage site as well as the release site for the neurosecretions they carry.

The corpora cardiaca also contain a number of intrinsic neurosecretory cells that produce and release their hormones locally. The adipokinetic hormone, for instance, is a product of the intrinsic neurosecretory cells of the corpora cardiaca. In the Orthoptera the neurohemal and intrinsic secretory functions of the corpora cardiaca are clearly localized to anatomically distinct regions called the secretory and glandular lobes, respectively. There is a nervous connection between the corpora cardiaca and either the recurrent nerve or the hypocerebral ganglion (both of which are part of the stomatogastric nervous system), depending on the species. The function of this innervation is not yet understood.

The corpora allata are attached to the corpora cardiaca by a bundle of neurons, the nervi corporis allati I (NCA I). In the Lepidoptera these are, at least in part, continuations of the NCC I and NCC II, but in other insects these nerves may arise in the corpora cardiaca. In locusts there is also a nervous connection between the corpora allata and the subesophageal ganglion, the NCA II (fig. 1.6; Mason, 1973). Control over the secretory activities of the corpora allata may be carried by neurosecretions or action potentials along these nerves, but their function is evidently not critical since in many insects isolated corpora allata appear to secrete normally (suggesting a humoral control pathway; see chapter 4).

The pathways of neurosecretory axons can be readily traced by cutting one of the NCCs and dipping the cut end of the nerve in a solution of cobalt chloride. Cobalt ions then diffuse up the neurosecretory axon to the cell body and can be precipitated as black cobalt sulfide, thus marking all the

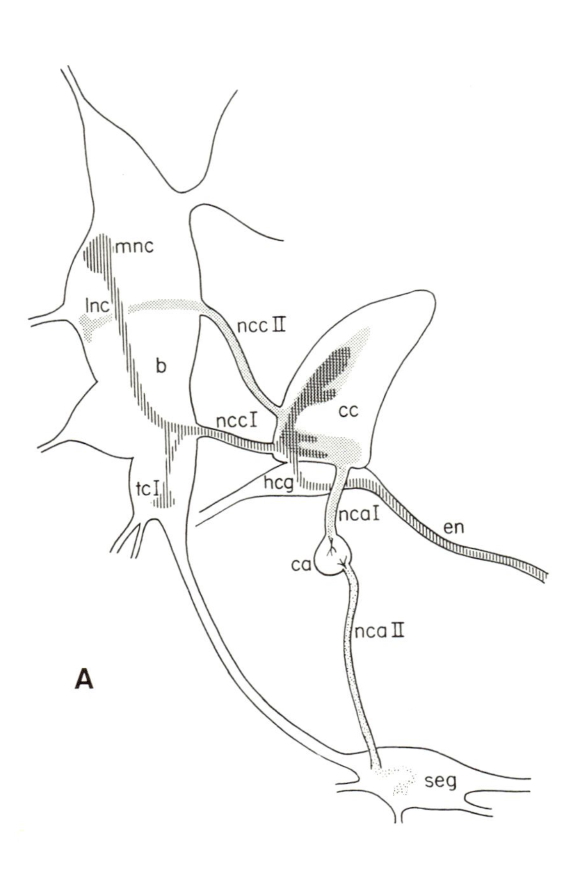

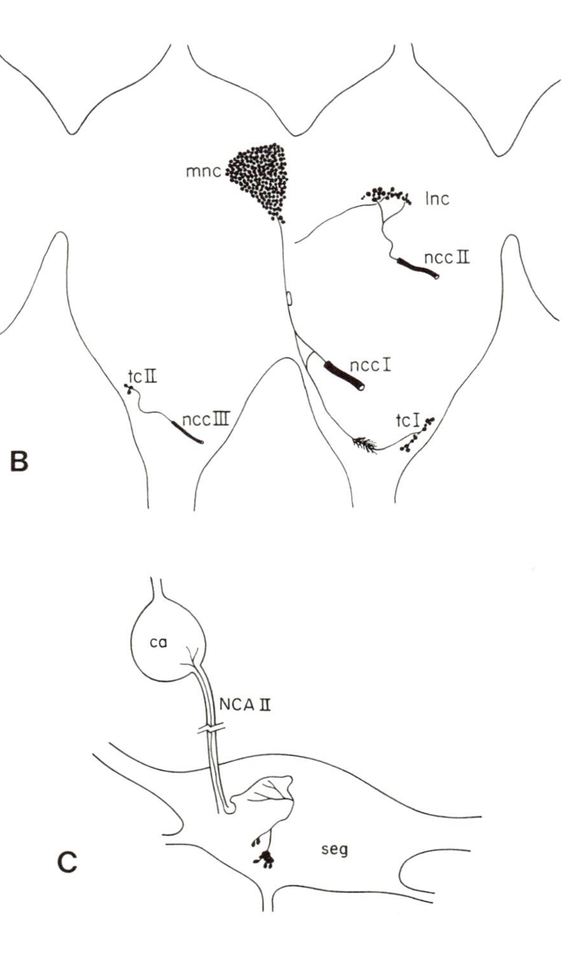

Figure 1.6. Diagram of the pattern of neurosecretory fiber projections from the brain (b) and the subesophageal ganglion (seg) into the corpora cardiaca (cc) and the corpora allata (ca) of *Schistocerca vaga*. (A): composite lateral view of entire brain retrocerebral system showing pathways of major neurosecretory cell groups; mnc, medial neurosecretory cells. (B): dorsal view of brain showing the projections of the major groups of neurosecretory cells.
(C): lateral view of subesophageal ganglion showing projections of neurosecretory cells into the corpora allata; lnc, lateral neurosecretory cells; tcI and tcII, tritocerebral neurosecretory cells group I and group II; ncc, nervi corporis cardiaci; nca, nervi corporis allati; hcg, hypocerebral ganglion; en, esophageal nerve (=recurrent nerve); other designations as in figure 1.3. (From Mason, 1973. Reprinted with permission of Springer-Verlag.)

cell bodies that send axons out that nerve, as well as the pathway of those axons through the brain (Pitman et al., 1972; Mason, 1973; Nijhout, 1975a). This so-called backfilling technique will label all the neurosecretory cells that send their axons down a particular nerve. A more precise view of neurosecretory anatomy can be obtained by injecting or iontophoresing a fluorescent dye directly into the cell body of an identified neurosecretory cell. This method will reveal the entire branching pattern of the axon and dendrites of that cell and is superior to the backfilling method because it will immediately show whether or not a cell projects down more than one nerve, and where its neurohemal area lies. The projections of the brain neurosecretory cells of *Manduca* and *Schistocerca* are shown in figures 1.5 and 1.6B. In both cases the axons from the medial neurosecretory cells decussate (cross over) and exit via the contralateral NCC I, while those of the lateral neurosecretory cells exit via the ipsilateral NCC II. Those of the tritocerebral cells exits via the ipsilateral NCC III.

A few of the neurosecretory cells of the brain do not use the corpora cardiaca as their neurohemal organ. The ventral neurosecretory cells in larvae of *Manduca sexta* (group V in fig. 1.5) do not use the NCC III but send their axons down the length of the ventral nerve cord and have their neurohemal site on the proctodeal nerve near the terminal abdominal ganglion (Truman and Copenhaver, 1989). The corpora allata also serve a significant neurohemal function. In the Lepidoptera they are the principal release sites for the prothoracicotropic hormone, which is produced by the lateral neurosecretory cells of the brain (Nijhout, 1975a; Bollenbacher and Granger, 1985), and in the cockroach *Diploptera punctata* the corpora allata appear to be the neurohemal sites for the cerebral neurosecretory cells that produce allatostatin (Stay et al., 1992).

In the higher Diptera (Cyclorrhapha), the retrocerebral complex and prothoracic glands are fused together into a single compact structure that encircles the esophagus. This endocrine complex is the ring gland, or Weismann's ring (fig. 1.3). Within the ring gland different specialized and morphological distinct cell clusters can be recognized that correspond to the corpora cardiaca, corpora allata, and prothoracic glands of other insects. The prothoracic glands form the two lateral arms of the ring and most of the large anterior lobe. The corpora allata are a compact cluster of cells within the anterior lobe, and the corpora cardiaca are a small cluster of cells at the posterior end of the ring adjacent to the brain (King et al., 1966; Aggarwal and King, 1969; Meurant and Sernia, 1993).

The Neuroendocrine System of the Ventral Nerve Cord

The general distribution of neurosecretory cells in the subesophageal ganglion and the ganglia of the ventral nerve cord of some selected insects are shown in figure 1.7. The neurohemal organs for neurosecretory cells in the

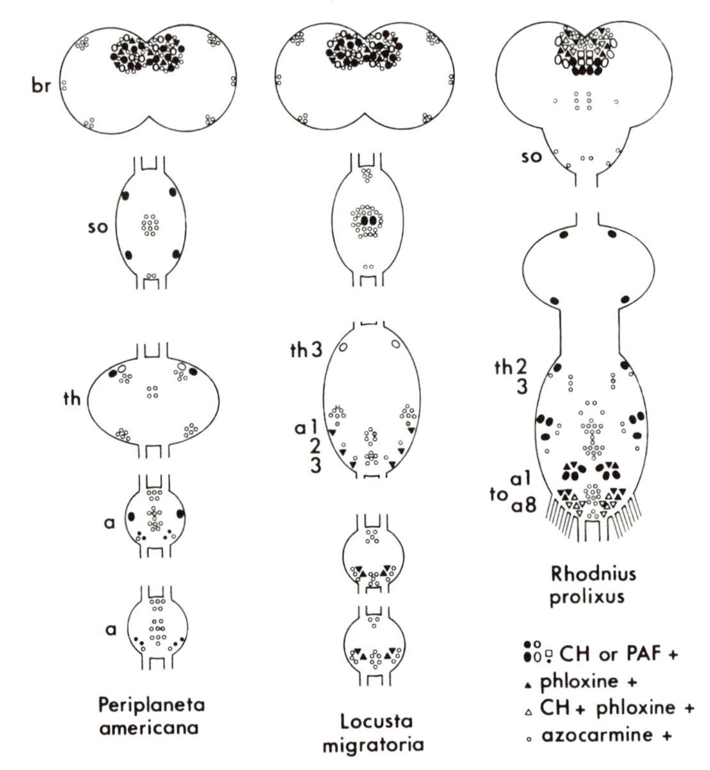

Figure 1.7. General arrangements of neurosecretory cell groups in the central nervous systems of three representative insects. Neurosecretory cells differ in their morphology and staining characteristics. Stains used in these preparations are paraldehyde fuchsin (PAF), phloxine, azocarmine, and chrome hematoxylin (CH); br, brain; so, subesophageal ganglion; th, thoracic ganglia; a, abdominal ganglia. (From Raabe, 1983. Reprinted with permission of *Adv. in Insect Physiol.* and Academic Press.)

ventral nerve cord are the perivisceral organs (also called perisympathetic organs; Raabe, 1971; Raabe et al., 1974). These are somewhat elongated structures that emerge from the ganglion or, more commonly, from the interganglionic connectives (fig. 1.8). In the thorax they usually emerge posterior to each ganglion, while in the abdomen they generally emerge anterior to the ganglion they serve. Perivisceral organs are the segmental analogs of the corpora cardiaca. In many species, however, the neurohemal areas of the abdominal neurosecretory cells occur along much of the length of one of the nerves that projects to the body wall, without much morphological specialization (Raabe, 1983), and are not always easy to detect.

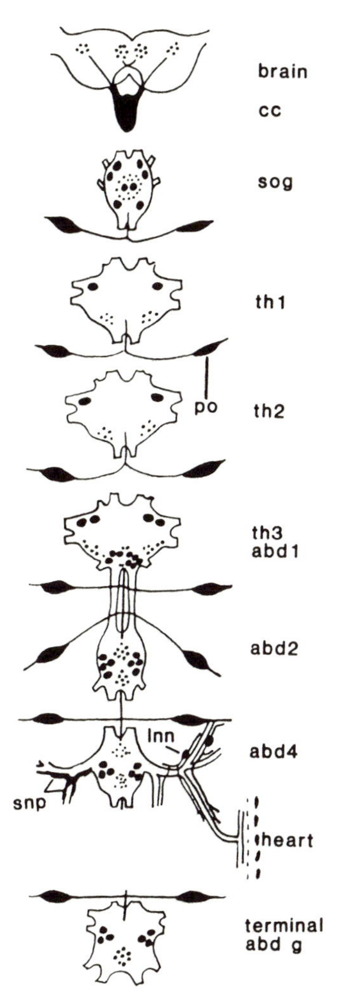

Figure 1.8. Distribution of neurosecretory cells (dots in the ganglia) and anatomical arrangement of neurohemal organs (black swellings on nerves) in a typical insect central nervous system. The neurohemal organs of the brain are the corpora cardiaca (cc) and those of the abdominal ganglia are the perivisceral organs (po); abd, abdominal ganglia; sog, subesophageal ganglion; th, thoracic ganglia. (From Orchard and Loughton, 1985. Reprinted with kind permission of Pergamon Press Ltd.)

CHAPTER 2

MECHANISMS OF HORMONE ACTION AND EXPERIMENTAL METHODS

INSECTS, like all animals, must be able to respond effectively to a great variety of stimuli and contingencies in their internal and external environments. Two systems of internal communication and integration have evolved that serve to initiate and coordinate appropriate responses to stimuli ranging from the acute dangers posed by predation to the more gradual but equally severe threats to survival posed by seasonal changes in food supply and temperature. These two systems are the nervous system and the endocrine system. As a rule, the nervous system mediates rapid short-term responses through direct and dedicated pathways between receptors, effectors, and one or more coordinating centers. The endocrine system mediates slower but more persistent long-term responses through blood-borne chemicals, the hormones. Hormones travel throughout the body and interact with any tissue or organ that bears specific receptors. In spite of substantial differences in the morphology and temporal characteristics of these two systems of internal communication, both operate on fundamentally similar principles. In both systems a message is initiated only in response to a specific stimulus; this message affects only specific target organs, and there it triggers a response that is specific to the target. Extraordinarily complex and exquisitely sensitive physiological control mechanisms can be built by linking many of these simple stimulus-response units into interacting activation, inhibition, and feedback pathways.

It is important to recognize that the characteristics of the response to a nervous or hormonal message are determined, not by the properties of the messenger, but by the properties of the target. Each target tissue or organ has a complex internal genetic and biochemical machinery that is preadapted to respond in a very specific and stereotyped way when a signal arrives. The nervous or hormonal signal simply serves to trigger this built-in response mechanism. The target-specificity of the response to a hormone accounts for the fact that in many organisms one hormone can have dramatically different effects in different target tissues. Furthermore, one hormone can have very different effects on the same target tissue at different times in the life of an individual. The juvenile hormones of insects provide an excellent illustration. In the larval stage they maintain the status quo of the epidermis in the response to ecdysteroids, but they also repress hormonal secretion by the brain in the control of metamorphosis and larval diapause, and play a role in the selection of alternate developmental pathways in insects that are potentially polyphenic. In the adult, by contrast, juvenile hormones may stimulate vitellogenin synthesis by the fat body, or its uptake by the developing egg (depending on the species), and also play a role in the regulation of adult diapause and migratory behavior.

In addition to the fact that the nervous system and endocrine system operate on fundamentally similar principles, the two are also functionally integrated. Endocrine glands produce hormones only when stimulated by an outside agent and the central nervous system is almost always involved in the stimulation or regulation of an endocrine gland. Although some endocrine glands are directly innervated, the principal pathway by which nervous and endocrine systems interact is through neurosecretions. Neurosecretory hormones, produced by specialized nerve cells (fig. 2.1), are at the basis of most endocrine regulatory pathways in insects as well as in vertebrates. Not only is the secretory activity of most endocrine glands controlled by neurosecretory hormones, the neurosecretory hormones themselves often have direct effects on a number of physiological functions. In vertebrates, neuroendocrine hormones of the hypothalamic-pituitary axis control the vast majority of hormone-mediated physiological responses. In insects, likewise, neuroendocrine hormones from the brain and segmental ganglia directly and indirectly affect virtually every aspect of physiological regulation in which hormones have been shown to play a role. Of course, to say that the brain is ultimately in control of endocrine secretion is an oversimplification. The brain does not stimulate secretion autonomously, but only in response to information it receives from the outside. This includes a variety of sensory input from the external environment as well as proprioceptive input from stretch receptors and a variety of chemical agents (substrates, metabolites, ions, and the hormones themselves). Neuroendocrine control, like much of physiological regulation, thus consists of feedback loops and complex causal chains whose origin cannot always be traced, if indeed there ever is a discrete origin.

Mechanisms of Hormone Action

The target tissues of hormones are tissues whose cells possess receptors for the hormone. Depending on the hormone, and on the time in development, one, few, or many tissues may bear such receptors, and correspondingly restricted or pervasive physiological changes ensue in response to the secretion of the hormone.

The characteristics of the cellular response to a hormone depend on the manner in which the hormone interacts with cell membranes. Two general categories of hormones are recognized whose effects at the level of the cell differ dramatically. We can classify these two categories as the lipid, or nonpolar hormones, and the polar proteinaceous, or peptide, hormones.

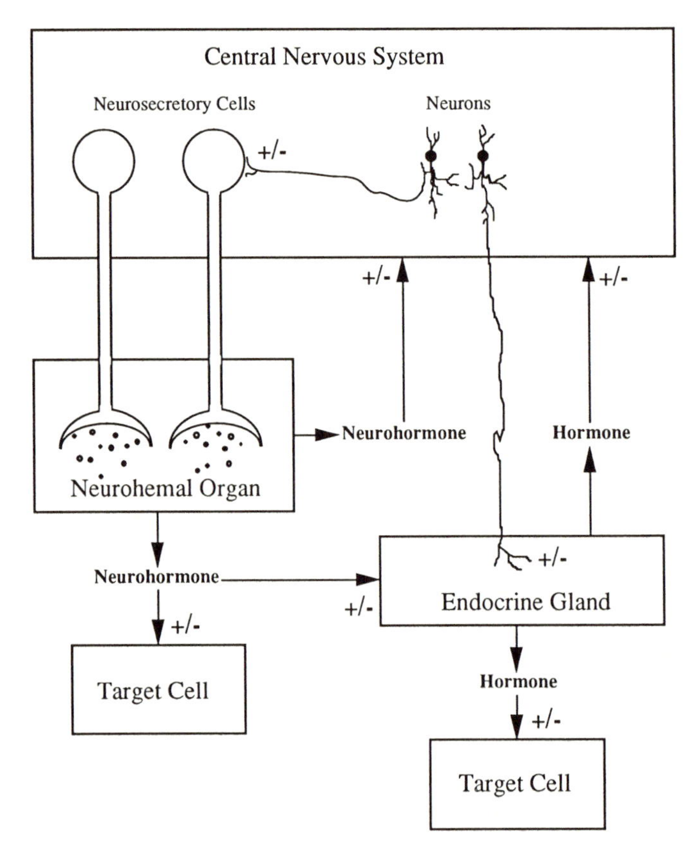

Figure 2.1. A general scheme for pathways of neuroendocrine regulation in insects. The central nervous system is the source of a large variety of neurosecretory hormones. These hormones are released from specialized neurohemal organs and may have an effect directly on a target tissue, or indirectly via the stimulation (+) or inhibition (–) of the secretory activity of conventional endocrine glands. Neurosecretory hormones can also feed back on the central nervous system and stimulate or inhibit its nervous and neuroendocrine activity. The central nervous system can also stimulate (or inhibit) neurosecretory cells and endocrine glands directly via conventional neurons.

Lipid Hormones

Lipid hormones pass readily through cell membranes and as a rule bind to receptors that occur freely in solution in the cell's cytoplasm, though most receptors appear to be restricted to the nucleus. Receptors are specialized proteins that bind to the hormone and in doing so undergo a conformational change that alters the way they interact with other constituents in the cell. In the nucleus the steroid hormone-receptor complex binds to specific base

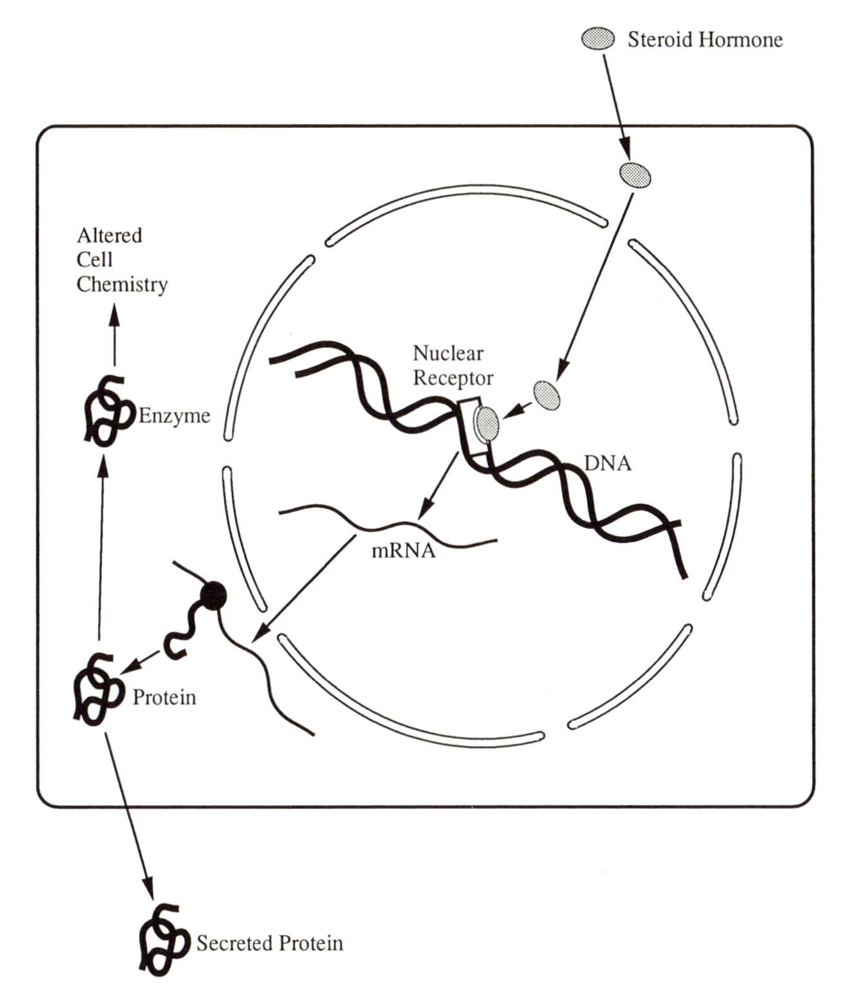

Figure 2.2. Mechanism of action of a steroid hormone. A steroid hormone passes through the cell membrane, enters the nucleus, and binds to a nuclear receptor protein. The receptor hormone complex then binds to specific regions of the DNA to control gene transcription.

sequences in the DNA (fig. 2.2). When the hormone-receptor complex binds to these specific locations in the genome, it initiates (or facilitates) or inhibits the transcription of particular genes. The appearance or disappearance of specific gene products, in turn, initiates what we recognize as the physiological response to the hormone. Different target tissue may differ in the genes that are activated or inhibited by a given hormone, presumably due to differences in the binding properties of the receptors and this accounts for the tissue-specific response to the hormone. In insects, ecdysteroids and juvenile hormones—one a group of steroids, the other a group of nonpolar terpenoids (see chapter 5)—are believed to act through such a mechanism.

The action of ecdysteroids at the cellular and molecular level has been well studied (O'Connor, 1985; Riddiford, 1985). Much of our initial knowledge of the mechanism of action of ecdysteroids came from studies on the effects of these hormones on transcriptional control in polytene chromosomes of Diptera. In fact, it was the observation that ecdysteroids could stimulate specific puffing patterns in polytene chromosomes that first prompted the theory that steroid hormones act directly on genes. Puffs in polytene chromosomes are due to a relaxation of the tight coiling of the chromatin in these chromosomes and are associated with local transcription of genes. Clever and Karlson (1960) found that injection of ecdysteroids into larvae of the midge *Chironomus tentans* caused two sequences of puffs to appear: an early set, within a few minutes of the injection, that did not require prior protein synthesis and a later set, a few hours later, that did require prior protein synthesis. The same two types of responses are seen in salivary gland chromosomes of mature larvae of *Drosophila melanogaster* after an injection of ecdysteroids, but in this species there is, in addition, an immediate regression of the puffs that are characteristic of the late larval stage (Ashburner and Berendes, 1978). The regression of late larval puffs appears to be associated with the shutdown of larval activities in preparation for pupariation and metamorphosis. The early puffs apparently respond directly to the hormone, and the proteins encoded by genes in these early puffs are believed to be involved in turning them off after some time and in activating gene transcription in the late puffs (Ashburner et al., 1974). Thus at least some of the early puffs may code for transcription factors or other regulatory proteins that control gene activity specific to the target tissue. The polytene chromosomes of Diptera therefore provide illustrations of three mechanisms of ecdysteroid action: the repression of active genes, the direct stimulation of new gene expression, and the indirect stimulation of new gene expression via the induction of regulatory proteins. Ecdysteroid-regulated gene expression in insects has been reviewed by Lepesant and Richards (1989).

Peptide Hormones

Hormones that are polar, usually polypeptides or amines, cannot pass easily through cell membranes and such hormones affect the cell's physiology in a very different way. Polar and peptide hormones bind to specific receptor molecules on the cell surface which serve as signal transducers to the cell's interior. The cell surface receptors are integral-membrane proteins. When a hormone binds to the extracellular portion of the receptor, it is believed to induce a conformational change in the receptor that causes the

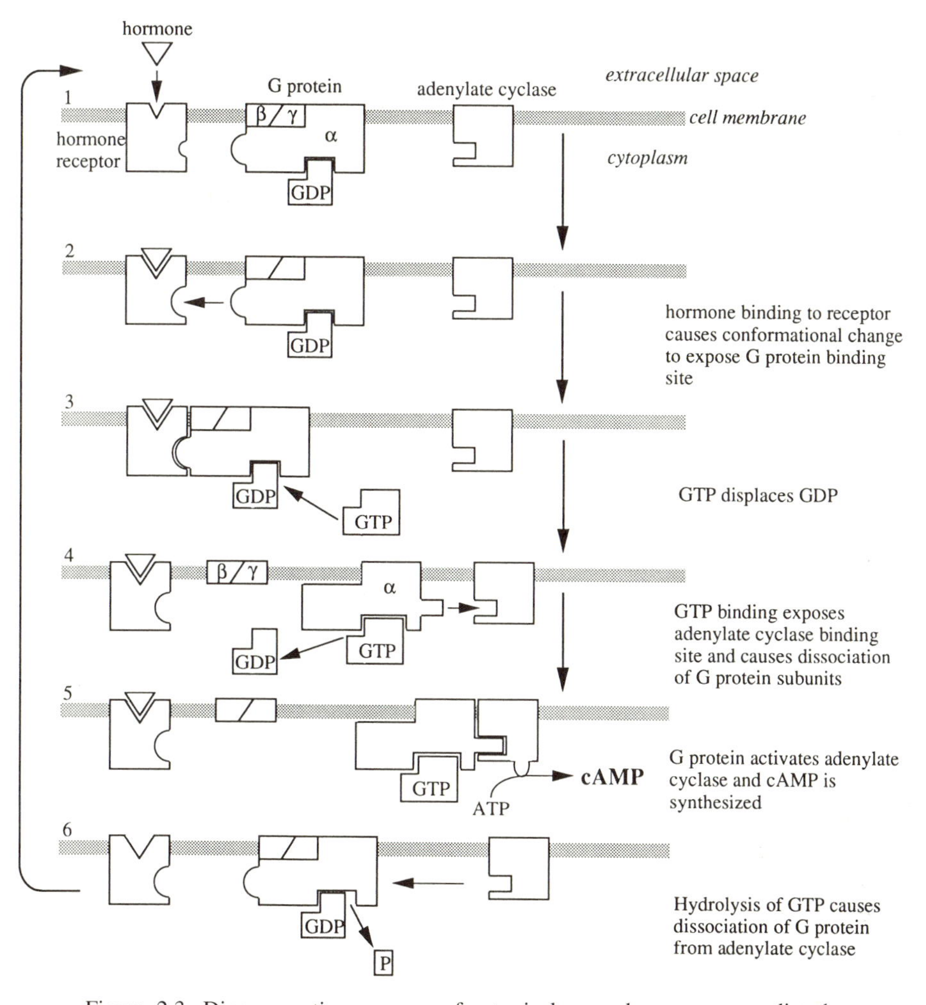

Figure 2.3. Diagrammatic summary of a typical second-messenger-mediated hormonal response pathway using cAMP. A GTP-binding membrane protein (G-protein), composed of several subunits, mediates the activation of adenylate cyclase after a hormone binds to its plasma membrane receptor. For possible intracellular effect of cAMP and other second messengers, see figure 2.5. (After Albers et al., 1989, Darnell et al., 1990, and Barritt, 1992.)

receptor-hormone complex to bind to other membrane proteins and initiate a complex sequence of reactions (fig. 2.3). In many instances, the receptor-hormone complex first binds to a GTP-dependent membrane protein (G-protein), where it causes two reactions to occur. First the G-protein binds an energetic molecule of GTP and releases a molecule of GDP that was previously bound to it. This molecular exchange in turn causes the G-protein to dissociate into two subunits, G_{α} and $G_{\beta,\gamma}$. This dissociation causes the G_α subunit to present a new active site that binds with the membrane enzyme adenylate cyclase and activates it. Adenylate cyclase catalyzes the synthesis of cyclic AMP (cAMP) from ATP. The GTP on the G_α subunit is then hydrolyzed to GDP, which causes the G_α and $G_{\beta,\gamma}$ subunits to reassociate. Thus the adenylate cyclase becomes inactive and the system is reset to respond to another molecule of hormone. In at least one hormonal response in insects (the eclosion hormone), the enzyme that is activated is not adenylate cyclase but guanylate cyclase, which catalyzes the synthesis of cyclic GMP (cGMP) from GTP. The role of such cyclic nucleotides in insect hormone action has been reviewed by Smith and Combest (1985) and Gilbert et al. (1988).

The cyclic nucleotides (cAMP and cGMP) produced by these reactions are referred to as second messengers (the hormone being the first messenger), and an increase in their intracellular concentration triggers a cascade of biochemical reactions. The first reaction is usually the activation of enzymes (protein kinases and phosphorylases) that, in turn, activate specific other enzymes by phosphorylating them. This second set of activated enzymes often cause the activation of yet additional proteins and enzymes, and this activation cascade eventually initiates new biochemical pathways that alter the cell's physiology.

This cascade of events continues as long as the concentration of the second messenger molecule in the cytoplasm remains elevated. In the case of cAMP, concentration is controlled by the activity of the adenylate cyclase, which regulates cAMP synthesis, and the activity of enzymes called phosphodiesterases, which regulate cAMP breakdown. In experimental endocrinology it is often possible to mimic the action of a hormone on a particular target tissue by exposing that tissue to elevated concentrations of exogenous cAMP, by simple injection of cAMP or a more stable analog such as dibutyryl cAMP. It is also often possible to mimic the effects of a hormone by causing an elevation of the endogenous cAMP level by inhibiting the activity of the phosphodiesterases with drugs such as theophylline, theobromine, or caffeine.

Some peptide hormones (as well as some biogenic amines such as serotonin) do not cause the synthesis of cyclic nucleotides but act by increasing the intracellular calcium ion concentration, either by opening membrane-bound calcium ion channels that allow extracellular calcium to enter the cell, or by causing the release of stored calcium from the endoplasmic reticulum. The release of Ca^{++} from the endoplasmic reticulum is mediated by a special second messenger system (figs. 2.4, 2.5). The hormone-receptor complex, again acting through a G-protein, activates a phospholipase that causes the breakdown of a complex molecule in the cell membrane, phosphatidylinositol 4,5-biphosphate (PIP_2). One of the products of this breakdown is triphosphoinositol (IP_3), a soluble molecule that acts as a second messenger and causes the release of Ca^{++} from the endoplasmic

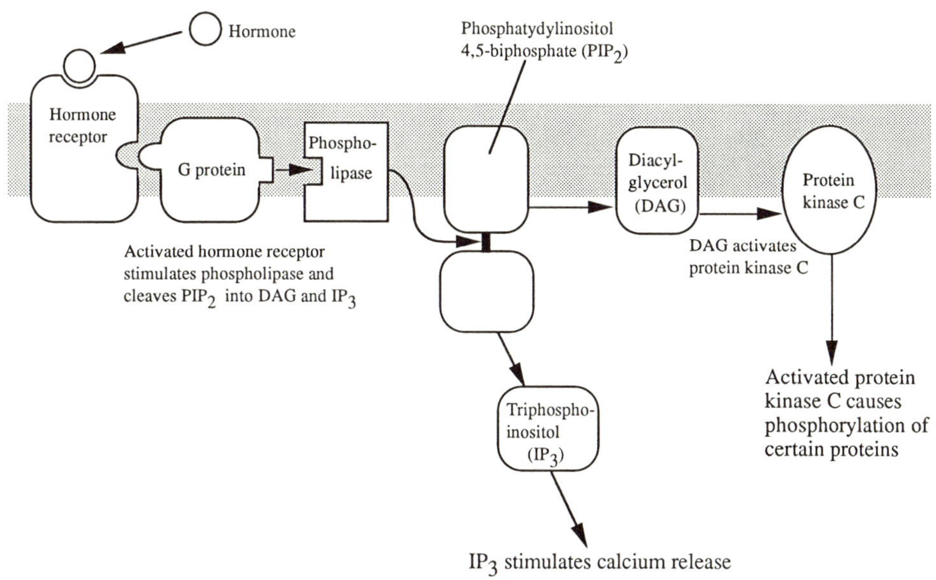

Figure 2.4. Diagrammatic summary of a hormonal response pathway using triphosphoinositol (IP_3) as second messengers. Triphosphoinositol is produced by the hydrolysis of phosphatidylinositol 4,5-biphosphate (PIP_2), a common plasma membrane component. Hydrolysis of PIP_2 into its subcomponents IP_3 and DAG is stimulated when a hormone receptor is activated, and requires interaction with a G protein and a phospholipase. IP_3 and DAG can each stimulate a cellular response via different pathways.

reticulum. The other breakdown product of PIP_2, diacylglycerol, activates a membrane-bound enzyme, protein kinase C, which, like other kinases, can activate other enzymes in the cytoplasm by phosphorylating them. The increase in the intracellular Ca^{++} concentration activates certain calcium-binding proteins in the cytoplasm, and these, in turn, can initiate enzyme activation cascades, much like the ones induced by cyclic AMP (fig. 2.5). Just as in cyclic nucleotide second messenger systems, there are a variety of pharmacological agents that can mimic the calcium and IP_3-mediated responses, such as calcium ionophores (for instance A23187), which allow entry of extracellular calcium ions, and certain phorpbol esters that mimic IP_3. Some endocrine-response systems require the action of both cyclic nucleotides and calcium ions. Good summary accounts of the cell biology of hormone action are given by Darnell et al. (1990), Barritt (1992), and Albers et al. (1989).

In insects, the mode of action of the eclosion hormone and the prothoracicotropic hormone (PTTH) have been best studied so far. The eclosion hormone (which provokes a complex series of behaviors; chapter 9) ap-

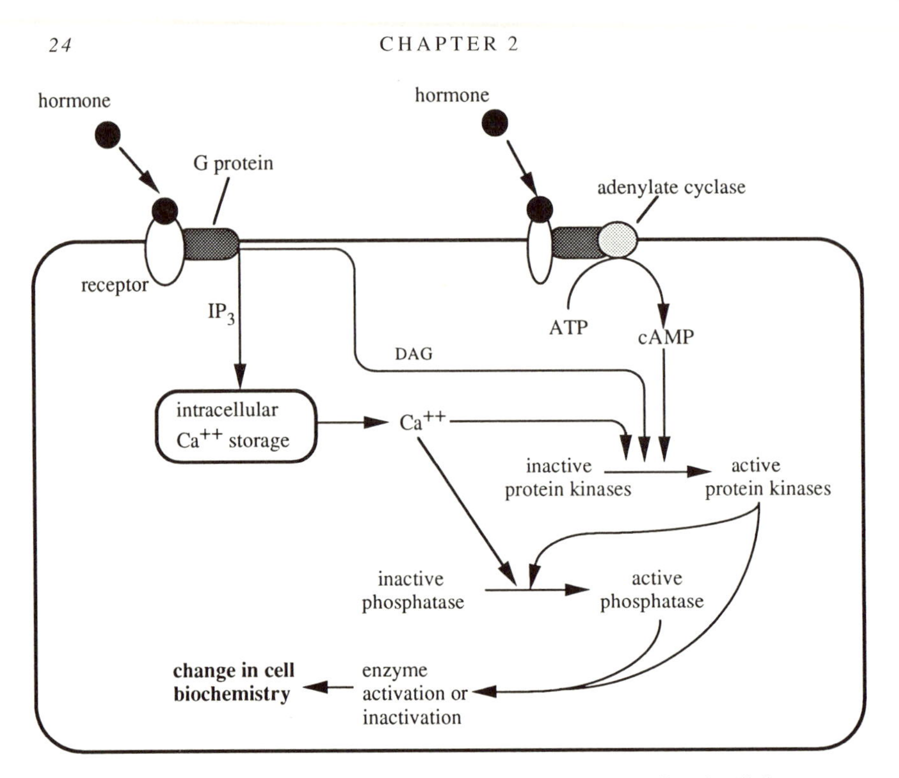

Figure 2.5. Mechanisms of action of second-messenger-mediated cellular response to peptide hormones. The hormone-receptor complex typically binds to a G protein which can, in turn, interact with other cell surface molecules to generate intracellular messengers. Two possible pathways are shown. IP_3 can stimulate the release of intracellular calcium, which can play a role in activating certain enzymes (some cells are also able to get extracellular calcium to enter the cell in response to hormones, probably via calcium channels in the plasma membrane). Alternatively, an activated hormone receptor can interact with a G protein complex and activate adenylate cyclase. This stimulates the intracellular synthesis of cAMP, which acts as a second messenger to alter the cell's biochemistry. Some hormones, for instance the eclosion hormone, cause the activation of a membrane-associated or cytosolic guanylate cyclase instead, and use cyclic GMP (cGMP) as the second messenger. The intracellular messengers generally turn on new biochemical pathways in the cell by activating either a protein kinase or a phosphatase, or both. Numerous variations on these hormone-stimulated cellular mechanisms occur in different species and different cell types (e.g., Barritt, 1992).

pears to act via a cGMP second messenger system (Truman et al., 1979), while signal transduction for the prothoracicotropic hormone (which stimulates ecdysteroid synthesis in the prothoracic glands) involves both cAMP and calcium ions as second messengers (Meller et al., 1988). In the latter case it may be that PTTH actually stimulated the opening of membrane channels in the cells of the prothoracic glands that allow extracellular calcium to enter the cell. The exact roles of calcium ions and cAMP in stimulating ecdysteroid synthesis appear to differ in different developmental stages, and these control mechanisms are still under active investigation (Smith and Pasquarello, 1989).

An important difference in the way in which nonpolar and polar hormones induce a specific cellular response is that the former stimulate the *activation or inactivation of genes*, resulting in de novo synthesis of enzymes or regulatory proteins or in the disappearance of previously active proteins, while the latter act via the *activation of preexisting enzymes* and proteins. In each case, the tissue-specificity of the response to the hormone occurs because tissues either differ in the preexisting enzymes that are activated, or differ in the kind of nuclear receptors they have and the genes that this receptor activates. These biochemical differences among target tissues account for the fact that each tissue responds in a unique and characteristic way to a given hormone. Thus the same hormone can have very different effects on different target tissues, and can also differ in its effect on different species, depending entirely on the evolutionary adaptations of the response mechanism in the target cells. Among the insect hormones, the juvenile hormones have probably undergone the greatest evolutionary divergence of function; they have been found to be used as triggers for such widely divergent response mechanisms as metamorphosis, caste determination, and ovarian maturation, as we will see in the chapters that follow. In addition, the juvenile hormones are unusual in that they are known to act via both major molecular mechanisms described above. In some systems the juvenile hormones act by binding to a nuclear receptor in the manner of a steroid hormone, while in others they act via the activation of a phosphatidylinositol-mediated second messenger system (Palli et al., 1991).

Regulatory and Developmental Hormones

In addition to their division into groups according to their action at the cellular level, hormones can be divided into two categories according to the general characteristics of their physiological effect. Hormones that are involved in the control of metabolic and homeostatic processes are referred to as regulatory hormones. The titers of these hormones fluctuate more or less continuously as they adjust the animal's physiology to deal with current conditions. Regulatory hormones are often component parts of physiological feedback circuits. They also often occur as antagonistic pairs, with opposite effects on a given physiological process so that both positive (stimulatory) and negative (inhibitory) control can be exercised. Hormones that control the onset of irreversible physiological switches are referred to as developmental hormones. Developmental hormones usually act as triggers that initiate more or less elaborate alterations in the animal's physiology and biochemistry. While they generally do not modulate or fine-tune a physiological response, developmental hormones may be secreted periodically to initiate episodic developmental events, such as the molting cycle and the reproductive cycle.

The distinction between regulatory and developmental hormones is a functional one. Both types of hormones act via the mechanisms outlined above, and a few hormones can have both functions in different species or at different times in development. Juvenile hormone, for instance, acts as a developmental switching hormone during larval life, but has a regulatory function in vitellogenin (yolk protein) synthesis in some adult insects.

Hormone-sensitive Periods

In the course of the past fifteen years it has become clear that the target cells of a hormone are not receptive to stimulation by the hormone at all times. In several cases the target cells are only responsive to stimulation within a narrow temporal window just around the time the hormone is normally secreted. In *Manduca sexta*, responsiveness to bursicon and the eclosion hormone at the end of adult development begins only a few hours before the actual release of these hormones (Reynolds et al., 1979). The prothoracic glands of final instar larvae of *Manduca* are unresponsive to stimulation by the prothoracicotropic hormone during the first half of the instar. Responsiveness of the prothoracic glands in this case is inhibited by the high titer of juvenile hormone during the early portion of the instar and develops only when the juvenile hormone disappears (Watson and Bollenbacher, 1988). The corpora allata of adult females of the roach *Diploptera punctata* vary in their sensitivity to allatostatin (an inhibitory neurohormone from the brain) in the course of the reproductive cycle (Pratt et al., 1990). The pheromone gland of *Heliothis zea* also varies its sensitivity to the pheromone biosynthesis-activating neurohormone in a cyclical fashion, so that the gland is only sensitive to stimulation at night, and nearly completely insensitive during the daylight hours (Christensen et al., 1991).

Perhaps the most complex temporal (and spatial) pattern of hormone sensitivity is found in the responsiveness of various tissues to juvenile hormones (Nijhout and Wheeler, 1982). Juvenile hormones control, among others, the metamorphosis of insects. But they do so, not in a concentration-dependent manner as was long believed, but by causing switches in the developmental program of various organs and tissues at discrete times during the molting cycle (see chapter 5). Various polyphenisms, such as caste determination in social insects, are likewise regulated by developmental switches that occur only during relatively brief and well-defined juvenile hormone-sensitive periods (see chapter 8).

Hormone-sensitive periods presumably come about through the modulation of hormone receptors in the target cells, or through modulation of the intracellular response mechanism. Riddiford and Truman (1993) have described a complex temporal pattern of expression of ecdysteroid receptor molecules, ecdysteroid-induced transcription factors, and juvenile hormone-binding proteins in various tissues of *Manduca* during larval growth and metamorphosis. Some of these molecules are expressed only during one or two days of the three-to-four week life cycle, others appear and disappear several times in succession over a period of days or weeks, and the patterns differ in different tissues. It is reasonable to assume that this diversity in the temporal and spatial patterns of receptor and transcription factor expression is associated with the regulation of an equally diverse response pattern to the hormones.

Our view of of hormonal regulatory mechanisms is, therefore, gradually changing. It used to be thought that hormones were active signals that instructed passive target cells to undertake certain preprogrammed actions. Under this view, the system that controls the hormones is ultimately in control of the developmental and physiological processes that the hormones regulate. It is now clear that the target cells are far from passive players in the process of endocrine regulation. The target cells themselves control in large measure whether and how they will respond to a hormone. Furthermore, their hormone responsiveness changes dynamically in a pattern that is often closely correlated with periods of hormone secretion. The mechanisms that control the responsiveness of target cells and the mechanisms that coordinate this responsiveness with changes in the hormone titers are as yet poorly understood.

Methods in Insect Endocrinology

In order to prove that a hormone is involved in the regulation of a physiological event, it is necessary to demonstrate that removal of a source tissue or endocrine gland for the hormone abolishes the effects, *and* that replacement of the source, or of a soluble extract of the source tissue, stimulates or reestablishes the desired physiological effects. Because most of an insect's endocrine centers are in the head, and because of the insect's tolerance to the procedure, it is customary to approach the first characterization of a possible endocrine effect by placing a blood-tight ligature around its neck or thorax. This divides the body into two isolated compartments, and one is able to see if the effect being studied is now restricted to one of the compartments. If that proves to be the case, the active compartment is presumed to contain the source for the putative hormone. If a previously known endocrine gland or neuroendocrine center exists in that compartment, the next step is to remove it surgically to see if that too abolishes the physiological effect. Since the brain is the major neuroendocrine center in the head, investigators generally test the effect of removing the brain on the physiological response they are studying. Such an operation is not as damaging as it might seem. An insect's regulatory physiology is remarkably decentralized and its respiration is passive, relying almost entirely on diffusion. Thus most insects survive quite well without a brain, though they can generally neither eat nor locomote normally. Some insects, such as the silk moth *Bombyx mori*, can metamorphose, mate, and lay eggs, all in the absence of a brain. It is this hardiness to operations which would be lethal in a vertebrate that makes insects such ideal experimental organisms. If removal of the presumptive source of the hormone (a gland or a ganglion of the central nervous system) abolishes the desired physiological effect, then it is necessary to demonstrate that reimplantation of that organ rescues the animal and reestablishes the normal physiology. In cases where the brain is the imputed endocrine source, it is standard to reimplant the brain into the abdomen. If such a "loose-brain" animal gives a normal physiological response, this implies that the brain is an endocrine source and that nervous connections are not necessary for normal regulation. Loose-brain animals and certain types of gland reimplantations, however, often give ambiguous responses, either because nervous input is necessary for stimulation or inhibition of hormone production, or because the neurohemal organs (e.g., corpora cardiaca) are not carried along so that normal release of the hormone cannot occur.

The next step in the investigation of a possible endocrine control mechanism is usually the injection of a homogenate or a crude extract of the putative endocrine center into an animal whose endocrine centers have been removed surgically, to determine whether this provokes the desired physiological reaction. Injections of crude gland or nervous system extracts may not always "rescue" a particular physiological or developmental response, if the hormone is present at too low a concentration (the corpora allata, for instance, synthesize but do not store juvenile hormone, and homogenates of these glands generally have no detectable juvenile hormone effect). In such cases it is often possible to do a rescue experiment by parabiosing the experimental animal whose gland has been removed, with a normal unoperated animal. Because insects have an open circulatory system, parabiosis is a relatively easy procedure to do on them. Parabiosis effectively joins the circulatory systems of the two animals so that the intact "donor" can provide hormones in approximately the correct concentration and for the required time period to produce a physiological effect in the experimental animal.

Neuroendocrine hormones are usually small polypeptides that are relatively resistant to degradation by brief exposures to high temperatures while most proteins and enzymes in a cell are not, so it is usually expedient to determine early on whether the hormone activity of a gland or brain extract is inactivated by brief boiling. While resistance to heating by no means proves that the active factor is a neurohormone, it is generally considered to be an encouraging sign that warrants follow-up investigations. It turns out that crude extracts of nervous systems are often quite toxic. Injections of crude extracts can therefore produce nonspecific pharmacological effects and may even kill the experimental animal. Heat treatment of such extracts prior to injection has the added advantage of denaturing most enzymes and proteins and often renders the extracts nontoxic.

If preliminary tests support the hypothesis that a hormone may be present, it becomes necessary to develop an assay method for the putative hormone by which its effects can be unambiguously characterized and quantified. Such an assay can then be used to study the normal pattern of secretion of the hormone and the physiological processes that stimulate its secretion. A sensitive and specific assay is also absolutely essential to keep track of the hormone during its purification and chemical identification. Excellent accounts of the procedures used in the assay, isolation, and identification of a number of insect hormones are given in Miller (1980).

Hormone Assays

In most cases the first assay method when dealing with an unknown hormone is a bioassay. In a bioassay the hormone is injected directly into an animal to observe its physiological effect, or it is dissolved in a medium in which an appropriately responsive tissue can be cultured. An ideal bioassay is one that produces a graded response to the hormone rather than an all-or-none response. The response of the assay preparation is then translated into an arbitrary numerical scoring scale by which the relative activity of the hormone in the test extract can be quantified. It is customary to refer to hormone concentrations as "titers" because they are usually determined indirectly by progressive dilution (titration) of the hormone sample being tested until its activity in the bioassay is no longer detectable. A suitable bioassay should be able to detect a hormone at approximately the concentration at which it normally occurs in the animal. If excessively high concentrations of a hormone are required to provoke a detectable response it is safer to assume that the response is pharmacological (see below) and may be nonspecific.

When investigating the possible role of a hormone in stimulating (or inhibiting) a particular physiological or developmental event, one needs to be mindful of the possibility of nonspecific responses. When a tissue is exposed to an excessively high concentration of a hormone it may respond in an abnormal or pathological fashion. Such a "pharmacological effect" of a hormone can usually be distinguished from a normal or "physiological effect" by means of a dose-response curve. Physiological responses can generally be measured to be proportional to the dose of hormone, within a certain range of concentrations. If it is necessary to inject an unusually large dose of hormone to get a particular response (large, that is, relative to the amount of a similar hormone that would be needed to provoke a response in another system), and if that response is not graded with dosage of the hormone, then it is reasonable to suspect one is dealing with a pharmacological effect.

Insect hormones normally circulate at extremely low concentrations. The typical range of concentrations for juvenile hormone is 0.1 to 50 nanograms (10^{-9} grams) per milliliter of hemolymph. Ecdysteroids typically occur at concentrations of 10 to 3000 nanograms per milliliter. It is difficult to obtain more than a few microliters of hemolymph from most insects without doing them undue damage, so that if an assay is to be useful for determining the concentration of these hormones in individual insects, it must be able to detect quantities as small as a few picograms (10^{-12} grams) of the hormones. In practice, individual hormone assays are done only on insects from which a sufficient quantity of hemolymph can be obtained (0.1 grams body weight or larger). Assays on small insects are done by either pooling hemolymph from several individuals or by extracting the hormone from whole-body homogenates.

Assay methods for a variety of neurohormones are described in Miller (1980). Assays for juvenile hormone (JH) take advantage of the fact that this hormone readily penetrates the unbroken cuticle. Topical application of JH, usually dissolved in oil, wax, or acetone, can prevent metamorphosis of the integument at the site of application and this is the basis for most bioassays for this hormone. The response to topical application can be improved by wounding the cuticle at the test site. This is the basis for the well-known wax-wound test, usually performed on pupae of the moth *Galleria mellonella*. In this assay the cuticle of *Galleria* is cut and the wound sealed with a small droplet of wax in which the JH sample to be tested is dissolved. The *Galleria* wax-wound test can detect approximately 5 picograms of JH. Another sensitive bioassay for JH is the *Manduca* black larva assay (Truman et al., 1973), which can detect 0.01 nanograms of the hormone. Biological assays for ecdysteroids are usually done by injection into ligated isolated abdomens of a small insect to minimize dilution (for a review of biological assay methods, see Cymborowski, 1989). If the injected material induces a molt in the isolated abdomen this is taken as evidence of ecdysteroid "activity." In addition, ecdysteroids can be assayed with considerable sensitivity in vitro on cultured pieces of epidermis that can be induced to undergo at least a partial molt, or in cultured cell lines, such as the *Drosophila* K_c line, which exhibit morphological and biochemical changes in response to ecdysteroids (Cherbas et al., 1980). Biological assays for ecdysteroids can at best detect quantities in the range of 1–10 nanograms, while radioimmunoassays can detect 1–10 picograms of hormone in a sample.

Once a hormone has been purified, identified, and synthesized, it is possible to use the pure hormone as a "standard" in the assay method and calibrate the titer or the numerical scoring scale in terms of the actual concentration of hormone required to provoke each grade of effect. It is then also possible to determine whether the assay method is specific to the hormone in question or whether it also responds to chemical analogs, precursors, or breakdown products of the hormone. If the assay proves to be reasonably specific, it can then be used as an experimental tool to investigate the pattern of hormone secretion, the presence of the hormone in other species, and so forth. If the chemical structure of the hormone is known, then purely chemical assay methods, using chromatography and mass spectroscopy, can be developed (Dahm et al., 1976; Horn and Bergamasco, 1985). Such chemical methods are often more sensitive and specific than bioassays or radioimmunoassays and are available, for instance, for the identification of the juvenile hormones and ecdysteroids, but because of the specialized equipment and techniques required they are not in widespread use.

As was alluded to above, the most sensitive and convenient method of identification and quantitative assay of a hormone is the radioimmunoassay. If the hormone is sufficiently large to be immunogenic, or if it can be linked to an appropriate carrier protein, it may be possible to make specific antibodies to it (Warren and Gilbert, 1988). If a highly specific antibody can be made, then it is generally possible to develop a very sensitive and specific radioimmunoassay. This method measures the competitive binding to the antibody of a sample of unknown identity and concentration and a sample of radioactively labeled hormone of known concentration. With an antibody of high specificity and high affinity, and a sufficiently radioactive hormone standard, this method can detect picogram quantities of a hormone. Radioimmunoassays are routinely used in the quantitative analysis of ecdysteroids and several of the peptide neurohormones. Protocols for ecdysteroid radioimmunoassays can be found in Bollenbacher et al. (1983), Warren et al. (1984), Warren and Gilbert (1988), and Reum and Koolman (1989). The ecdysteroid radioimmunoassay can be used to detect ecdysteroid secretion by fragments of prothoracic glands in vitro, and this technique has been developed as an indirect but sensitive assay for PTTH activity (Bollenbacher et al., 1979, 1980; Warren et al., 1988). There are now also several excellent radioimmunoassays for juvenile hormones (Granger and Goodman, 1988; Huang et al., 1993). However, a continuing problem with radioimmunoassays for juvenile hormone is that the antibodies often cross-react with several neutral lipids so that careful sample preparation and extraction of all cross-reacting lipids is essential. This brings up one of the main drawbacks of the radioimmunoassay, namely, the potential cross-reactivity of the antibodies to nonspecific ligands in the sample (Granger and Goodman, 1988). Such nonspecific ligands can be precursors or breakdown products of the active hormone, or even unrelated compounds that happen to have similar chemical side groups as the portion of the target hormone that the antibody recognizes. In juvenile hormone radioimmunoassays, for instance, nonspecific binding to neutral lipids is a significant source of false positives, and this varies from species to species (N. A. Granger, pers. comm.); different kinds and degrees of sample preparation may therefore be necessary, depending on the source of the hormone. It is absolutely essential that all the cross-reactivities of the antibody be known and that compounds that might give a false positive be removed from the sample or otherwise accounted for. Investigators differ in the rigor with which they attempt to account for the possible cross-reactivities of their antibodies, but it is clear that hormone titer data from studies that did not take nonspecific binding of the antibody into account may be suspect.

Today, assays for peptide hormones also take advantage of powerful recombinant DNA and antibody technologies. If the amino acid sequence of a hormone is known, it is possible to synthesize nucleic acid probes that can recognize the gene that codes for the hormone, or that can recognize the hormone's mRNA. Such probes, together with highly specific antibodies, can then be used to clone the gene and to study the temporal and spatial pattern of hormone synthesis and localization, and its interaction with target tissues at the cellular and molecular levels.

CHAPTER 3

METABOLISM AND HOMEOSTASIS

METABOLISM, the biochemical processes associated with the utilization of nutrients, and homeostasis, the maintenance of a constant internal environment, are important aspects of animal physiology. An active and accurate regulation of metabolic processes is particularly important for small-bodied and highly active animals which, because of their small volume, have little storage and buffering capacity with which to smooth out the almost constant fluctuations in supply and demand. Flight places particularly high demands on the metabolic machinery, as insect flight muscle is the most metabolically active tissue known in the animal world (Weis-Fogh, 1952, 1964). The periodic molts during larval growth place large episodic demands on carbohydrate metabolism to supply building blocks for the chitin (a complex nitrogen-containing polysaccharide) that makes up the bulk of the new cuticle. At other times in the life cycle the production of large volumes of eggs requires a massive mobilization of lipids and proteins. Most insects undergo periods of migration and diapause at some time during their life cycle, and these events likewise place extreme demands on nutrient acquisition, storage, and utilization.

Most metabolic processes within cells are regulated by the laws of mass action, by processes that control the entry of chemicals into cells, and by the induction or repression of specific enzymes (usually by their substrates or metabolites). In addition to these controls, which are intrinsic to each cell, there are a variety of extracellular signals, such as hormones and neurotransmitters, that are produced and controlled at centralized sites and affect metabolism in distant tissues. So far, it appears that in insects hormones are primarily involved in stimulating large-scale mobilization of resources (such as carbohydrates and lipids). There is little evidence that they participate in the control of detailed aspects of metabolism.

When we think of homeostasis we generally refer to processes that maintain certain biochemical or physiological variables within a relatively narrow optimal range of values. To control such a variable efficiently and precisely requires, as a rule, positive as well as negative regulation so that active stimulation occurs when the value of the variable falls below a certain set point, whereas active inhibition occurs when the value of the variable rises above a set point. Many physiological processes are under such dual control. For instance, thermoregulation generally requires the capacity to both heat and cool, while (in vertebrates at least) the heart rate is controlled by separate nerves that speed up and slow down the rate, respectively. Homeostatic mechanisms that involve control via the endocrine system likewise generally require two hormones: a stimulating (or tropic) hormone, and an inhibitory hormone. In insects, water balance is under such dual hormonal control, as we will see below. General reviews of metabolism and homeostasis in insects can be found in Kerkut and Gilbert (1985, vols. 4 and 10). Below we will deal only with those aspects that are known to be regulated by hormones.

Fat Body and Hemolymph

The insect fat body performs many of the functions that in vertebrates are performed by the liver (Dean et al., 1985). The fat body is, as its name implies, the principal organ for the storage of lipids, but like the liver it also stores carbohydrates (largely in the form of glycogen). Like the liver, the fat body is also the site where these stored products are metabolized, used to build new biological molecules, and exported for use as the energy sources and biochemical building blocks by all other cells in the body. While clusters of fat-body cells can have a definite morphology, the fat body as a whole is a relatively loosely organized tissue with a large surface area in contact with the hemolymph.

The composition of the hemolymph is regulated by several physiological mechanisms. The inorganic ions and water content are regulated largely by the ionic mechanisms located in the Malpighian tubules, midgut, and hindgut. The organic molecules (such as sugars, lipids, amino acids, and proteins) are regulated almost entirely by the synthetic and metabolic activities of the fat body. The fat body is the source of all the circulating carbohydrates that are used for energy-requiring reactions and for building the cuticle. It is also the source of nearly all the amino acids in the hemolymph that are used by other cells to manufacture proteins. The fat body is also the source of the circulating lipids used in biosynthesis and in energy metabolism, and of the vitellogenins (large lipoproteins) that make up most of the volume of the eggs.

Carbohydrate Metabolism

Carbohydrates provide the principal source of energy-rich compounds that power metabolism and that form the material basis of chitin in the cuticle. Glycogen, a long linear polymer of glucose, is the principal storage form of carbohydrates in insects. It is found in most tissues of the body but is most abundant in the fat body. Trehalose, a disaccharide of glucose, is the

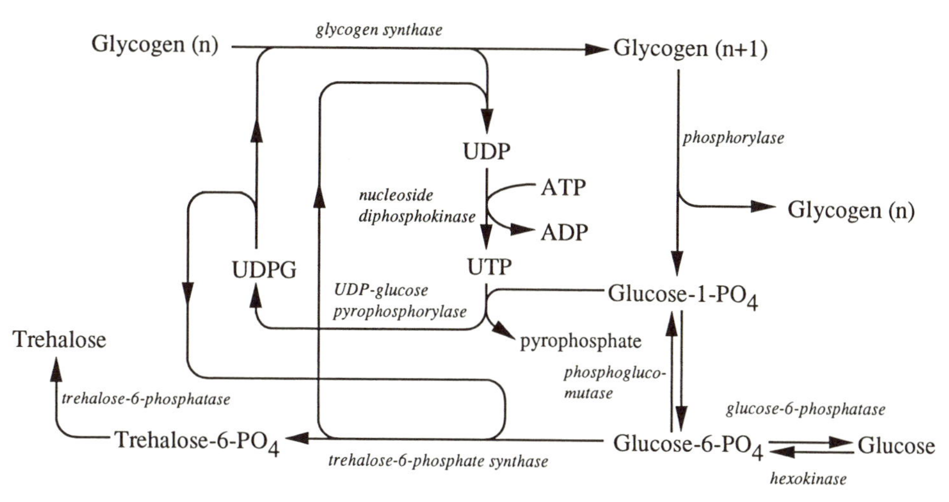

Figure 3.1. The biosynthetic pathways for glycogen and trehalose in the insect fat body. (After Steele, 1980, 1985.)

main form of carbohydrate that circulates in the hemolymph. The biosynthetic pathway by which glycogen and trehalose are interconverted in the fat body is shown in figure 3.1 (Steele, 1980, 1985).

Steele (1961) discovered that homogenates of the corpora cardiaca of the cockroach, *Periplaneta americana*, contain a factor that causes a dramatic elevation of trehalose in the hemolymph when injected. This factor was first referred to as the hyperglycemic hormone, but is now called hypertrehalosemic hormone. Hypertrehalosemic hormone stimulates the breakdown of glycogen via activation of the enzyme, phosphorylase (fig. 3.1), which is the rate-limiting enzyme in the breakdown pathway of glycogen (Mordue and Goldsworthy, 1969; Goldsworthy, 1970; Ziegler, 1979; Gäde, 1981; Steele, 1982, 1985). The hormone acts by stimulating synthesis of the second messenger cAMP in the fat-body cells, and its activity is enhanced by simultaneous injection of theophylline (Gäde and Holwerda, 1976; Hanaoka and Takahashi, 1977). Extracts of the corpora cardiaca also stimulate trehalose production in isolated fat body cultured in saline (Wiens and Gilbert, 1967; McClure and Steele, 1981).

Not all species of insects seem to use trehalosemic hormone in the same way. Extracts of the corpora cardiaca from *Locusta migratoria*, for instance, have no effect on its own hemolymph trehalose level, but have a strong hypertrehalosemic effect when injected into *Periplaneta americana* (Chalaye, 1969). *Periplaneta* is apparently extraordinarily responsive to hypertrehalosemic factors produced by other insects because corpus cardiacum extracts from a fairly broad range of species, including nonorthopteroids, induce a rise of trehalose levels in its hemolymph (Steele, 1985). This suggests that hypertrehalosemic hormone is not species-specific, but it does not explain why many species whose corpora cardiaca stimulate hypertrehalosemia in *Periplaneta* do not themselves respond to the hormone.

The lack of a hypertrehalosemic effect of corpora cardiaca extracts in *Locusta* has been ascribed to the presence of a hypotrehalosemic hormone (a hormone that reduces the trehalose level) in this species, because when locusts are neck ligated (so that head-derived factors no longer have a physiological effect), subsequent injection of an extract of their corpora cardiaca *does* have a marked *hyper*trehalosemic effect (Loughton and Orchard, 1981). There is also evidence for negative regulation of hemolymph trehalose levels in the fly *Calliphora erythrocephala*. When adult *Calliphora* are decapitated, the trehalose level in their hemolymph rises. The same result is obtained when only the medial neurosecretory cells of the brain are removed surgically. The trehalose levels in such operated animals are brought down again upon injection of an extract of the medial neurosecretory cells (Normann, 1975). This indicates the presence of a neurosecretory hormone that depresses the concentration of hemolymph trehalose: a hypotrehalosemic hormone. No such hormone has yet been isolated, however, and the existence of a hypotrehalosemic hormone has been called into question by Veenstra (1989a), who suggests that simple constant catabolism of trehalose (and sequestration by the fat body) is sufficient to account for its decline at times when it is not being stimulated by hypertrehalosemic hormone. The rise in trehalose in decapitated *Calliphora* could then be due ro derepression of hypertrehalosemic hormone production.

The hypertrehalosemic hormone is a neurosecretory hormone and is produced by the intrinsic neurosecretory cells of the corpora cardiaca. Interestingly, the hormone is structurally similar to the adipokinetic hormone (see below) and, together with that hormone, is a member of an extensive family of structurally related neuropeptides (the adipokinetic/red-pigment-concentrating, or AKH/RPCH, family of hormones) that have a variety of physiological functions in insects and other animals (Box 1). The RPCH, for instance, is a hormone in crustacea that affects coloration by controlling the movement of pigment granules in (red) chromatophores.

In addition to the trehalosemic hormones, it is possible that the neurohormone bombyxin, from the brain of *Bombyx mori*, may have a role in the regulation of carbohydrate metabolism. Bombyxin has a great deal of similarity, both in its amino acid sequence as well as its tertiary structure, to vertebrate insulin and the family of peptides known as insulinlike

BOX 1

The Adipokinetic Hormone (AKH)/ Red-Pigment-Concentrating Hormone (RPCH) Family

Structure and Nomenclature: Adipokinetic hormones (AKH) and hypertrehalosemic hormones (HTH) are structurally related, and the amino acid sequences of several of them have been determined; all are small polypeptides of 8 or 10 amino acids. Interestingly, the structure of these hormones is very similar to that of the crustacean red-pigment-concentrating-hormone (RPCH), which regulates the expansion of pigment in red chromatophores. More than a dozen neuropeptides (of unknown function) have been identified with structures closely related to those of AKH and RPCH. They form a widespread and functionally diverse family of neuropeptides (called the AKH/RPCH family) that also probably include a variety of peptides that regulate the heartbeat (cardioaccelerating factors [CAH], periplanetins, corazonin). The structures of some selected hormones in the AKH/RPCH family are given below:

Species	*Hormone*	*Structure*
Schistocerca gregaria	AKHI	pGlu--Leu-Asn-Phe-Thr-Pro-Asn-Trp-Gly-ThrNH2
Locusta migratoria	AKHI	pGlu--Leu-Asn-Phe-Thr-Pro-Asn-Trp-Gly-ThrNH2
Locusta migratoria	AKHII	pGlu-Leu-Asn-Phe-Ser-Ala-Gly-TrpNH2
Schistocerca gregaria	AKHII	pGlu-Leu-Asn-Phe-Ser-Thr-Gly-TrpNH2
Manduca sexta	AKH	pGlu-Leu-Thr-Phe-Thr-Ser-Ser-Trp-GlyNH2
Heliothis zea	AKH	pGlu-Leu-Thr-Phe-Thr-Ser-Ser-Trp-GlyNH2
Gryllus bimaculatus	AKH	pGlu-Val-Asn-Phe-Ser-Thr-Gly-TrpNH2
Carausius morosus	HTH	pGlu-Leu-Thr-Phe-Thr-Pro-Asn-Trp-Gly-ThrNH2
Heliothis zea	HTH	pGlu-Leu-Thr-Phe-Ser-Ser-Gly-Thr-Gly-AsnNH2
Blaberus discoidalis	HTH	pGlu-Val-Asn-PheSer-Pro-Gly-Trp-Gly-ThrNH2

(continued on next page)

BOX 1 (cont.)

Species	*Hormone*	*Structure*
Periplaneta americana	CAHI (periplanetin CCI)	pGlu-Val-Asn-Phe-Ser-Pro-Asn-TrpNH2
Periplaneta americana	CAHII (periplanetin CCII)	pGlu-Leu-Thr-Phe-Thr-Pro-Asn-TrpNH2
Periplaneta americana	Corazonin	pGlu-Thr-Phe-Gln-Tyr-Ser-Arg-Gly-Trp-Thr-AsnNH2
Pandalus (Crustacea)	RPCH	pGlu-Leu-Asn-Phe-Ser-Pro-Gly-TrpNH2

The AKHs do not appear to have species-specific actions. *Locusta* (Orthoptera) AKH, for instance is very active in *Manduca* (Lepidoptera). Furthermore, the same hormone can have very different functions in different life stages; for instance, the AKH of *Manduca* controls lipid metabolism in the adult and carbohydrate metabolism in the larval stage (where it thus acts as HTH). Given the overall similarity in structure of the members of this family of peptide hormones, it is not clear what feature of their structure accounts for the differences in their biological functions. Much of the diverse regulatory functions may be due to differences in the receptor and transduction mechanisms of the target cells. There is a slight sequence similarity between AKH and the amino-terminal portion of glucagon.

SOURCE: In all species tested, AKH can be extracted from the corpora cardiaca. In the Orthoptera the AKH and the hypertrehalosemic hormone are both products of the intrinsic neurosecretory cells of the corpora cardiaca.

REFERENCES: Fox and Reynolds, 1990; Gäde, 1990; Goldsworthy et al., 1986; Orchard, 1987; Veenstra, 1989b,c; Ziegler et al., 1985; Ziegler et al., 1990.

growth factors (IGF) (Iwami, 1990). In vertebrates, insulin and IGFs are involved in the regulation of carbohydrate metabolism and cell growth and proliferation, and exogenous insulin has been shown to stimulate cell growth and proliferation in insects as well (Davis and Shearn, 1977; Mosna, 1981). Bombyxin also stimulates cell division in insects (Mercola and Stiles, 1988) and may play a role in insect growth and metabolism that parallels the functions of IGFs in vertebrates. Additional information on the structure and biology of Bombyxin is presented in chapter 5.

Lipid Metabolism

In animals, energy is most efficiently stored in the form of lipids. Lipids yield more than twice as many calories per gram as do carbohydrates. Moreover, because they are largely hydrophobic they carry little water of hydration; thus lipids can be stored more compactly than carbohydrates and most other biological molecules. Lipids also have a wider importance in metabolism beyond that of energy storage because, when the fatty acids of lipids are broken down, their acetyl groups (via acetyl Coenzyme A) can serve as the basis for the biosynthesis of nearly all other major kinds of bio-organic molecules.

The control of lipid metabolism in insects has been studied primarily in migratory locusts which, because of their long and intense migratory flights, put extreme demands on their energy storage and retrieval mechanisms. In locusts, and probably all insects, lipids are released from the fat body into the hemolymph as diacylglycerides (that is, glycerol with two fatty acids; lipids are stored in the fat body as triglycerides). In the hemolymph, lipids do not occur freely but are bound to carrier proteins called *lipophorins* that greatly enhance the carrying capacity of the hemolymph for these otherwise poorly soluble molecules.

When the migratory locust *Schistocerca gregaria* begins migratory flight, it has a high metabolic rate, burning primarily carbohydrates and some lipids. After about 30 minutes the carbohydrates in the hemolymph are exhausted and further flight is fueled entirely by lipids (Weis-Fogh, 1952). Stored lipids can sustain flight for several hours before they too are exhausted, at which point the animal can no longer fly and must feed to replenish its reserves (this need to replenish energy is the root cause of the destructive power of a swarm of locusts). Soon after the onset of flight the level of lipids in the hemolymph begins to rise; it reaches a peak at about 30 minutes and then slowly declines to a slightly lower steady state for the remainder of flight (Spencer and Candy, 1974; Cheeseman and Goldsworthy, 1979). The use of lipids for flight metabolism is not restricted to locusts. In the moth *Manduca sexta*, lipid mobilization begins almost immediately after the onset of flight and appears to be the principal fuel used for both short- and long-distance flights (Ziegler and Schulz, 1986).

Mayer and Candy (1969) and Beenakkers (1969) were the first to show that injection of extracts of the corpora cardiaca of *Locusta* and *Schistocerca* caused an increase in the hemolymph lipid concentration roughly similar to that which is seen during flight, and which persists for several

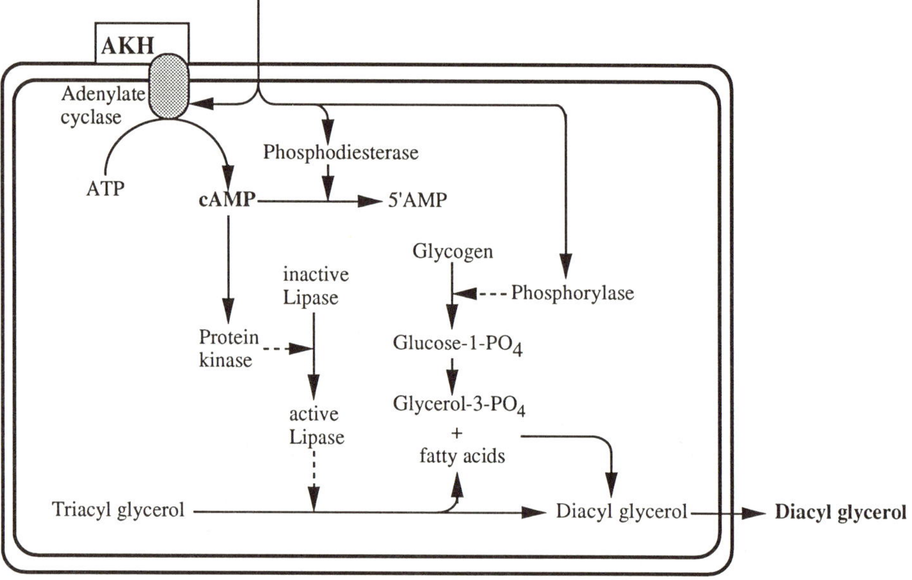

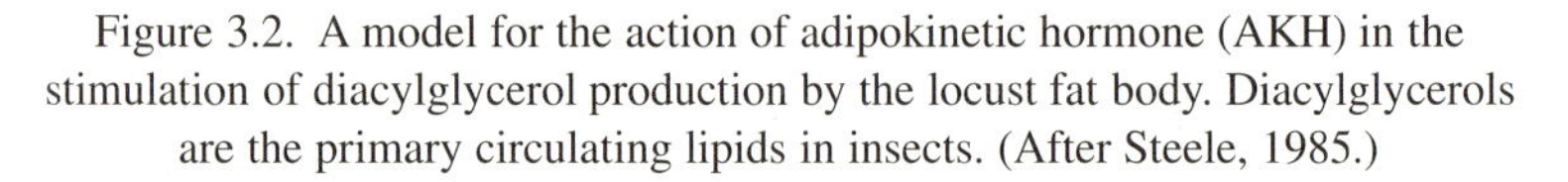

Figure 3.2. A model for the action of adipokinetic hormone (AKH) in the stimulation of diacylglycerol production by the locust fat body. Diacylglycerols are the primary circulating lipids in insects. (After Steele, 1985.)

hours after the injection (Goldsworthy et al., 1972b). The active factor in the corpora cardiaca extract has been identified as adipokinetic hormone (AKH). The adipokinetic hormone acts on the fat body via a second messenger system (chapter 1) and causes release of stored lipids into the hemolymph. Part of this reaction involves the breakdown of stored triglycerides into diacylglycerides (fig. 3.2).

In *Locusta*, AKH is produced by the intrinsic neurosecretory cells of the corpora cardiaca. Secretion of AKH by the corpora cardiaca is believed to be controlled by conventional neurons from the brain, since severing the nervous connection between brain and corpora cardiaca (the NCCI and NCCII; see fig. 1.6) prevents the flight-induced elevation of hemolymph fat levels (Goldsworthy et al., 1972a; Jutsum and Goldsworthy, 1977), while electrical stimulation of these nerves in isolated corpora cardiaca causes the release of AKH (Orchard and Loughton, 1981). There is evidence for a feedback regulation of AKH production. A rise in the concentration of hemolymph lipids inhibits AKH release by the corpora cardiaca. Interestingly, a decline in the hemolymph trehalose concentration also triggers AKH secretion, while injection of trehalose inhibits the rise of hemolymph lipids during flight. This inhibitory effect of injected trehalose can be negated by simultaneous injection of corpora cardiaca extracts (Cheeseman and Goldsworthy, 1979; Van Der Horst et al., 1979). The secretion of AKH thus appears to be sensitive to the current metabolic demands of the insect. The control mechanism that regulates AKH secretion appears to be geared to favor trehalose utilization whenever it is available, so that lipids are released into the hemolymph only as the trehalose concentration declines. Evidence for the existence of hypolipemic factors is equivocal (Steele, 1985), and the negative regulation of hemolymph lipid levels may be due simply to catabolism and a constant sequestration by the fat body.

Physiological stress and injury can cause a rise in the hemolymph concentration of both lipids and trehalose. This stress-induced mobilization of lipids and carbohydrates is caused by release of the neurotransmitter, octopamine, into the hemolymph from neuromuscular junctions or directly from sites in the central nervous system (Gole and Downer, 1979; Orchard et al., 1981; Steele, 1985). The hyperlipemic effect of octopamine appears to act via the same second messenger system that is stimulated by AKH. The response to octopamine is more rapid than the response to AKH, however, and may well be an adaptation for the swift supply of energy resources during periods of acute need. When stress occurs as a predictable event in an insect's life cycle, however, special physiological mechanisms can evolve to deal with it. In the moth *Manduca sexta*, for instance, the normal stress of periodic starvation during molting causes a rise in hemolymph trehalose that is regulated by the secretion of a trehalosemic hormone (Siegert and Ziegler, 1983).

Water Balance and Diuresis

Small animals such as insects have a relatively high surface-to-volume ratio which makes them particularly susceptible to changes in their external environment. They also have a relatively small absolute body volume, which, as we noted above, gives them little buffering capacity with which to resist those changes. Because of their terrestrial habits, insects are particularly susceptible to dehydration, and water conservation is at a premium. Herbivorous and blood-feeding insects, by contrast, feed on materials that are water-rich and thus experience a periodic excess water load that must be efficiently eliminated. The regulation of water is therefore a physiological function of particular importance in insects. As in other animals, the water content of the body in insects is regulated indirectly, by controlling the osmotic pressure and ionic composition of the hemolymph.

The Malpighian tubules regulate the osmotic and ionic characteristics of the hemolymph by pumping ions from the hemolymph into their lumen. In the majority of insects K^+ ions are the primary ions transported (Bradley, 1985). The Malpighian tubules of bloodsucking insects transport Na^+ as well as K^+, and in the tse-tse, *Glossina morsitans*, they appear to transport Na^+ exclusively (Maddrell, 1969; Gee, 1976; Phillips, 1981; Bradley, 1985). When ions are transported, water follows by osmosis, though the pathway of water and ions in this osmotically driven fluid transport mechanism of the Malpighian tubules is very complex (Berridge, 1982; Berridge and Oschman, 1969; Phillips, 1981). Water absorbed into the Malpighian tubules enters the hindgut. The most posterior portion of the hindgut, the rectum, is specialized for the reabsorption of water. The rectum contains active transport pumps for Cl^-, Na^+, and K^+ that are particularly concentrated in specialized areas called the *rectal pads* (Phillips et al., 1982). In a number of species the distal ends of the Malpighian tubules are closely applied to the rectum adjoining the rectal pads to form a complex called the cryptonephridial system, which is extraordinarily efficient in extracting water from the rectal contents (Ramsay, 1964; Grimstone et al., 1968). The cryptonephridial system is most highly developed in species that live in dry habitats or that feed on relatively dry foods (Chapman, 1985).

The transport activities of the Malpighian tubules and hindgut must be regulated independently to maintain a relatively constant internal osmotic and ionic environment in the face of a widely varying input of water and ions. This is accomplished in most insects by means of a diuretic hormone, which increases the ion pumping (and thus the water transport) rate of the Malpighian tubules, and an antidiuretic hormone, which primarily affects the pumping activity of the hindgut.

Diuretic hormone (DH) was discovered by Maddrell (1963, 1964) working with the bloodsucking bug *Rhodnius prolixus*. *Rhodnius* feeds exclusively on blood and takes very large blood meals of up to twelve times its own body weight. Immediately after a blood meal, *Rhodnius* begins to produce a copious urine and excretes virtually the entire plasma volume of the ingested blood (about half the ingested volume) over a period of about an hour. Maddrell found that when isolated Malpighian tubules were incubated with a drop of hemolymph from a newly fed animal, the tubules began to produce urine rapidly (fig. 3.3). He found that removal of the thoracico-abdominal ganglionic mass prevented diuresis after a blood meal, and that extracts of the posterior portion of this ganglionic mass stimulated a vigorous diuresis by isolated Malpighian tubules.

In *Rhodnius*, DH is released from the abdominal ganglion when the abdomen is stretched by a blood meal. Stretch receptors in the abdominal nerves are activated when the abdomen is distended by a large blood meal. Severing the abdominal nerves between stretch receptor and ganglion abolishes secretion of DH, but severing the connection between the abdominal ganglion and the thorax or brain does not (Maddrell, 1964). Thus abdomi-

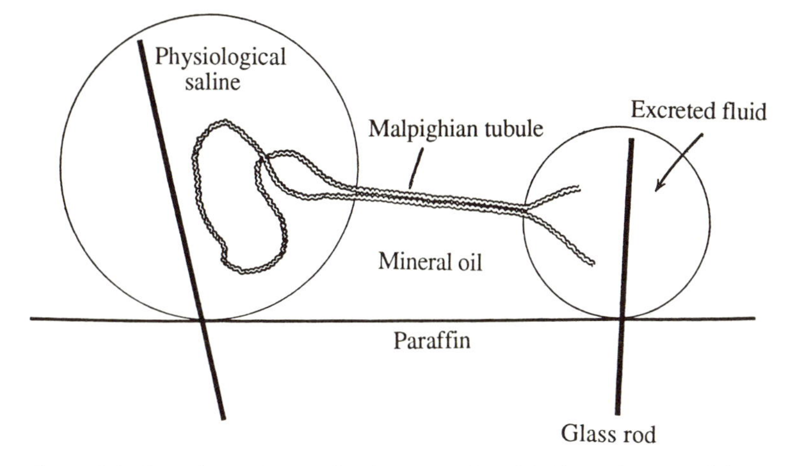

Figure 3.3. Experimental setup for the study of Malpighian tubule physiology. A length of tubule is isolated in a droplet of physiological saline held by a thin glass rod under mineral oil. The free ends of the tubule are pulled into the mineral oil by a second rod. Stimulants and test substances can then be introduced by a needle into the saline droplet. The secreted fluid accumulates as a droplet around the free ends of the tubule. The volume of secreted fluid can be measured and samples can be withdrawn for analysis. (Modified from Maddrell, 1969. Reprinted with permission of *J. Exp. Biol.* and Company of Biologists Ltd.)

nal stretch receptors appear to stimulate the neurosecretory system in the abdominal ganglion directly and the brain is unnecessary in this response.

Maddrell and Nordmann (1979) have described a mechanism by which the DH of the bloodsucking bug *Rhodnius prolixus* can regulate both the volume of the hemolymph and the rate of diuresis, while eliminating the enormous water load that accompanies a blood meal. Maddrell and Nordmann note that three transport epithelia are involved in diuresis. First the midgut absorbs an iso-osmotic solution of NaCl from the blood meal into the hemolymph. Then the distal portions of the Malpighian tubules secrete an iso-osmotic solution of NaCl and KCl from the hemolymph into their lumen. Finally the distal portions of the Malpighian tubules recover much of the KCl from this urine and return it to the hemolymph. The urine that is eliminated thus consists of a hypo-osmotic solution of mostly NaCl. The DH controls the rate of all three transport processes, and the rate of all three increases dramatically immediately after a blood meal as DH secretion begins. In order to maintain a relatively constant hemolymph volume, the rate of fluid transport from the gut into the hemolymph must obviously be closely matched to the rate of fluid transport from the hemolymph into the Malpighian tubules. As it is, under maximal stimulation by DH the midgut can transport fluid more rapidly than the Malpighian tubules can

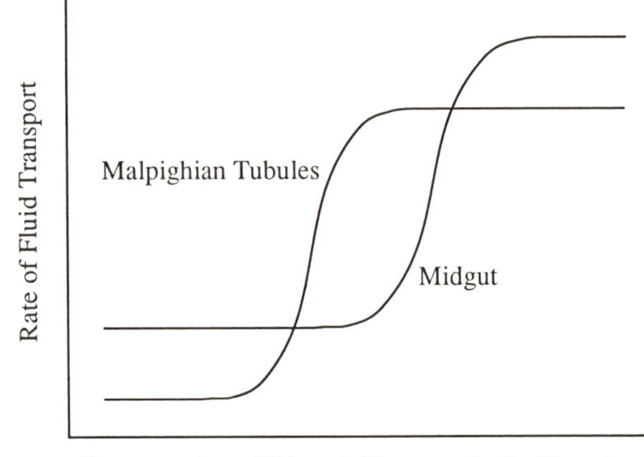

Figure 3.4. Hypothetical regulatory mechanism for water balance in *Rhodnius*. Differences in the dose-response curves for diuretic hormone-stimulated fluid transport by the midgut and Malpighian tubules ensure that the Malpighian tubules secrete only as much fluid as is made available by resorption from the midgut, as discussed in the text. (Redrawn from Maddrell and Nordmann, 1979.)

pump it out. Maddrell and Nordmann (1979) proposed that a balance between the two transport rates is maintained by the differential sensitivity of the midgut and Malpighian tubules to the concentration of DH (fig. 3.4). The Malpighian tubules are more sensitive than the midgut to low concentrations of DH. Thus as the hemolymph volume increases by transport of water from the gut contents, the DH becomes diluted and the midgut begins to pump less actively. Meanwhile the Malpighian tubules remain highly active and now remove water from the hemolymph at a higher rate then it enters from the gut. Then, as the hemolymph volume decreases, the concentration of DH rises and the transport rate from the gut increases. As long as there is a constant amount of DH in the hemolymph, this mechanism will maintain a constant hemolymph volume during excretion.

Diuretic hormones have also been shown to play a role in the rapid diuresis after a blood meal in mosquitoes (Nijhout and Carrow, 1978; Williams and Beyenbach, 1983), as well as in the day-to-day regulation of water balance in nearly two dozen other species of insects (Phillips, 1983; Spring, 1990; Wheeler and Coast, 1990). Diuretic hormone plays an important role in many herbivorous insects such as *Locusta* and *Schistocerca* whose food tends to have a high water content that needs to be eliminated efficiently. The DH of *Locusta* has been identified as a peptide of 14 amino acids (Box 2) that binds to antibodies against vertebrate arginine-vasopressin, suggesting a certain sequence similarity with that hormone (Proux et al., 1987; Schooley et al., 1987). Osmoreceptors in the foregut and in the

BOX 2

Diuretic and Antidiuretic Hormones (DH and ADH)

Structure and Nomenclature: The diuretic hormones of insects are neurosecretory peptides. They are quite heterogeneous and do not appear to belong to a common family. In the Orthoptera several DHs have immunological cross-reactivity with the vertebrate hormone arginine-vasopressin. One of the DHs from *Locusta* has been identified and is an antiparallel dimer of two identical 9-amino acid peptides held together by two disulfide bonds:

```
                    Cys-Leu-Ile-Thr-Asn-Cys-Pro-Arg-Gly-NH2
                     |                   |
NH2-Gly-Arg-Pro-Cys-Asn-Thr-Ile-Leu-Cys
```

A diuretic hormone from *Manduca sexta* has been successfully purified and shown to be a polypeptide of 41 amino acids without disulfide bonds:

H-Arg-Met-Pro-Ser-Leu-Ser-Ile-Asp-Leu-Pro-Met-Ser-Val-Leu-Arg-
Gln-Lys-Leu-Ser-Leu-Glu-Lys-Glu-Arg-Lys-Val-Lys-Ala-Leu-
Arg-Ala-Ala-Ala-Asn-Arg-Asn-Phe-Leu-Asn-Asp-IleNH_2

This DH has a certain amount of sequence similarity to several vertebrate neurohormones (corticotropin-releasing hormone, urotensin I), and to the toxin sauvagine (which also is a potent diuretic in vertebrates) from the skin of a South American tree frog. The *Manduca* DH also stimulates diuresis in *Pieris rapae*. The action of DH is mimicked in vitro by the neurotransmitter 5-hydroxytryptamine (serotonin) in all insects where it has been studied.

There are at least two different kinds of ADHs in the Orthoptera. One of these is probably identical with the chloride transport stimulating hormone (CTSH). The other has immunological cross-reactivity to a group of insect neuropeptides called neuroparsins, whose normal biological function is still unclear.

Sources: The source of DH is quite diverse in different groups of insects. Several parts of the central nervous system usually contain DH activity. In *Periplaneta* DH activity is found in the terminal abdominal ganglion, while in *Glossina*, *Anopheles*, and *Rhodnius* the mesothoracic ganglionic mass is the principal source. The neurohemal sites for the *Rhodnius* DH occur along the proximal portions of the abdominal nerves. In *Carausius*, *Locusta*, *Schistocerca*, *Pieris*, *Papilio*, and several other insects the brain appears to be the chief source of DH.

The two types of ADH are both found in the corpora cardiaca. The CTSH-like hormone appears to be a product of the cerebral neurosecretory cells. The neuroparsinlike ADH may be a product of the intrinsic neurosecretory cells of the corpora cardiaca.

References: Kataoka et al., 1989; Maddrell, 1966; Maddrell and Phillips, 1975; Mordue and Morgan, 1985; Phillips, 1983; Phillips et al., 1986; Proux et al., 1987; Spring, 1990; C. H. Wheeler and Coast, 1990.

central nervous stem are believed to be involved in the regulation of DH secretion in *Periplaneta*. In other species the stretch caused by bloating due to excessive water gain is believed to be the principal stimulus for DH secretion (Phillips, 1983), although how such a crude sensory mechanism could be used for accurate water regulation is not known.

Diuresis has been studied in several other groups of insects. Many Lepidoptera, for instance, produce a significant volume of urine immediately after emergence of the adult. This diuresis is stimulated by a hormone that can be extracted from the brain and corpora cardiaca (Dores et al, 1979; Kataoka et al., 1989). A diuretic hormone of *Manduca sexta* has been isolated that has a high activity in *Pieris rapae* (but not in *Manduca*); it has been identified as a 41-residue peptide with a certain amount of sequence similarity to several vertebrate neurohormones.

Water regulation in several insects appears to be under dual control. The elimination of water is stimulated by DH and inhibited by an antidiuretic hormone (ADH). The ADH does not control secretion by the Malpighian tubules, however, but stimulates the recovery of water in the rectum (Herault et al., 1985; Phillips, 1983; Phillips et al., 1986). ADH is a neurosecretory hormone and has been found most concentrated in the corpora cardiaca in *Periplaneta* and *Locusta*. The ADH of *Periplaneta* has been isolated and characterized as a peptide with a molecular weight of about 8 kD (Goldbard et al., 1970).

There are several kinds of ADHs in insects. The glandular and neurohemal portions of the corpora cardiaca appear to differ in the principal type of ADH activity they contain (Herault et al., 1985). The neurohemal portion of the corpora cardiaca in *Locusta* and *Schistocerca* contains an antidiuretic factor identical to the chloride transport stimulating hormone (CTSH), which may be a product of the medial neurosecretory cells of the brain. The intrinsic neurosecretory cells of the glandular portion of the corpora cardiaca also contain an antidiuretic factor that may be identical to a subclass of small polypeptides called neuroparsins (Fournier and Girardie, 1988; Spring, 1990).

Muscle-Stimulating and Cardioactive Peptides

A variety of neuropeptides that stimulate strong muscle contraction have been identified in insects. Some of these, the so-called cardioactive peptides discussed below, act specifically on the heart and appear to regulate the rate of heartbeat. Others have strong effects on a variety of test muscles, though their role in the normal physiology of insects is still poorly understood. The leukopyrokinins are among the best studied myotropic peptides (Holman et al., 1986, 1987; Schoofs et al., 1990). The leukopyrokinins have a significant sequence similarity to the neuropeptides that are involved in the stimulation of pheromone production (see chapter 6), and to the diapause hormone of *Bombyx mori* (see chapter 7).

Proctolin

Proctolin is a neuropeptide in insects that occupies a special place in insect endocrinology because it seems to act as both a neurotransmitter and a neurohormone, and because, in spite of its dramatic and readily demonstrable effect on insect muscle, its exact role in an insect's normal physiology remains somewhat of a mystery. Proctolin was first discovered as a factor that could be extracted from the foregut and hindgut of the cockroach *Periplaneta americana*, and induced a slow, strong contraction in the longitudinal muscles of the cockroach hindgut (Brown and Starratt, 1975; Starratt and Brown, 1975).

Proctolin is a pentapeptide: Arg-Tyr-Leu-Pro-Thr. It appears to be a neurotransmitter in several kinds of insect visceral and skeletal muscle. It is also found in Crustacea, where it again appears to be both a neuromuscular neurotransmitter and a circulating neurohormone. Much of the evidence for its role in neuromuscular physiology comes from experiments that first detect proctolin chemically in a particular muscle and then demonstrate that exogenous proctolin causes that muscle to contract. Release of proctolin from motorneurones at neuromuscular junctions has been specifically demonstrated in the coxal depressor muscle by Adams and O'Shea (1983). Immunocytochemical studies of proctolin have shown that it is also present in several ganglia of the central nervous system of cockroaches and locusts (O'Shea and Adams, 1986).

Cardioactive Peptides

Changes in the rate of the heartbeat are known to accompany various physiological events in insects. Acceleration of the heartbeat, and an acceleration of the accessory pulsatile organs at the base of the wings, accompanies inflation of the wings after adult emergence in Lepidoptera (Wasserthal, 1975; Moreau and Lavenseau, 1975; Truman, 1985). Tublitz and Truman (1981, 1985) have partially purified two peptides with molecular weights of 500 D and 1000 D that control the rate of the heartbeat. These cardioaccelerating peptides (CAP) are neurosecretory products of the abdominal ganglia and are released from perivisceral organs (Tublitz and Truman, 1985). In adult *Manduca sexta*, the CAPs are active in accelerating the heart rate during wing inflation immediately after adult emergence, and during flight (Tublitz, 1989). During the larval stage one of these molecules, CAP_2, has a completely different function, namely in controlling muscular activity of the hindgut. At the end of larval life and in preparation for pupation, insects void their entire gut contents during a fairly brief period of time. This voiding behavior is accompanied by a great increase in intestinal motility, particularly of the hindgut. Tublitz et al. (1992) have shown that this increased hindgut motility is stimulated by CAP_2, but that this factor probably does not act via the hemolymph because there is no accompanying acceleration of the heartbeat. It may be that during the larval stage CAP_2 is released from nerve endings directly onto or near the hindgut. The CAPs appear to exert their effect via an inositol triphosphate pathway and the release of intracellular calcium (Tublitz et al., 1991). Some of the peptides of the AKH/RPCH family have been shown to have a cardioaccelerator effect in several insects and correspond in molecular size to the cardioaccelerating factors from *Manduca* (Orchard, 1987).

In *Periplaneta americana* and *Carausius morosus*, a hormone from the corpora cardiaca, called neurohormone D, has been implicated in the control of the heart rate. The heart of *Periplaneta* is also heavily innervated, however, and the relative roles of this innervation and of neurohormone D in controlling the heart rate under natural conditions remain unclear (Mordue and Morgan, 1985). Neurohormone D also has a stimulatory activity on the Malpighian tubules, and affects color change in *Carausius*. A variety of other neurogenic compounds are known to accelerate the heartbeat. Among these are acetylcholine, 5-hydroxytryptamine, octopamine, and proctolin (Mordue and Morgan, 1985). In view of this diversity of stimulating factors, all of which also affect other physiological systems in the insects, it appears that cardioacceleration may be a secondary effect, perhaps serving to enhance the rate transport of hormones, substrates, or metabolites required for these various physiological functions.

The beating rate of the accessory heart at the base of the antennae of *Periplaneta americana* is stimulated by proctolin and inhibited by octopamine. While proctolin does not appear to be present in these accessory hearts, octopamine has been found at high concentrations and may constitute the native inhibitor (Pass et al., 1988; Hertel and Penzlin, 1992). In addition, Hertel and Penzlin (1992) have shown that the neuropeptide allatostatin A2 (see Box 7) can antagonize the effects of proctolin as long as it is presented prior to the exposure to proctolin.

In recent years several investigators have used antibodies to vertebrtate peptide hormones to detect and isolate cross-reacting substances in insects. Such studies have led to the isolation of a number of insect neuropeptides, several of which have muscle-stimulating or cardioaccelerating effects in test insects. Among these are the members of the leukosulfakinin family (Veenstra, 1989b), which are peptides with ten amino acids that cross-react with antibodies to vertebrate gastrin and cholecystokinin, and have muscle-stimulating effects on insects. Among the cardioaccelerators are the periplanetins (of eight amino acids) and corazonin (of ten amino acids), all of which belong to the AKH/RPCH family of neuropeptides (Scarborough et al., 1984; Veenstra, 1989c, and see Box 1).

Neurogenic Amines as Endocrine Secretions

Several neurogenic amines that in some systems act as neurotransmitters are also released as neuro "hormones" by certain insect nerve cells. Among these are octopamine and serotonin. Octopamine mimics the action of adipokinetic hormone and the pheromone-releasing hormone, PBAN. Serotonin (5-hydroxytryptamine) can mimic the action of diuretic hormone in *in vitro* preparations, stimulates salivary-gland secretion, and causes plasticization of the cuticle in some insects. The exact sites of origin of circulating neurogenic amines and their natural roles in physiological regulation are poorly understood at present.

CHAPTER 4

THE DEVELOPMENTAL PHYSIOLOGY OF GROWTH, MOLTING, AND METAMORPHOSIS

THE DEVELOPMENTAL endocrinology of insects deals primarily with the regulation of growth, molting, and metamorphosis. While hormones control all three processes, the secretion of those hormones can, in turn, often be controlled by the size of the larva or by its stage in the molting or metamorphic cycle. In other words, growth, molting, metamorphosis, and the hormones that control these processes are all parts of an integrated regulatory feedback system. To understand the endocrinology it is necessary to understand the developmental physiology, and vice versa. Although endocrinology is the primary subject matter of this book, a full appreciation of endocrine regulation requires a fairly detailed understanding of the processes of growth, molting, and metamorphosis. Presenting a fully integrated view of the regulation of growth, molting, and metamorphosis in a single chapter would be unwieldy and would require many digressions to deal with special cases. Such an account would certainly lack coherence. Instead, I will present in this chapter the general physiology of growth, molting, and metamorphosis, indicating where these developmental processes interact with the endocrine system. Some sections in this chapter, such as the treatment of allometry, may at first seem like an excessive digression, but we are finding that allometric constants can be affected by the endocrine system and, therefore, that the overall form of insects may be at least under partial endocrine control. The next chapter (chapter 5) will deal explicitly with the endocrine regulation of molting, metamorphosis, and some aspects of growth. Later chapters on diapause (chapter 7) and polyphenisms (chapter 8) will revisit some of these topics in a different context and also introduce other aspects of the control of growth.

The bodies of arthropods and sea urchins and the brains of vertebrates are each encased in essentially nonliving external skeletons. In order for the tissues within to grow, the volume and surface area of the external skeleton must increase. Several different solutions to the problem of growth of an external skeleton have evolved in these taxa. The plates that make up the tests of sea urchins and the skulls of vertebrates are not perfectly fused. Cells in the sutures between plates allow the plates to grow in surface area by accretion at their edges. In addition, sea urchin tests and vertebrate skulls also have a surrounding epidermis which, together with an internal epithelial layer, can mold the shape of the growing skeletal plates by external deposition and internal absorption. In this way the volume and shape of skulls and tests are continually altered to keep up with the growth of their respective contents. Insects have solved the problem of growth within a nonliving exoskeleton in two different ways. The primitive method by which their integument grows, shared with all other arthropods, is by molting. This involves the periodic casting away of the old confining cuticle and the manufacture of a new larger one. In addition, the soft-bodied larvae of some of the holometabolous insects have evolved a mechanism of intercalary growth, quite different from that of vertebrate skulls and sea urchin tests, whereby the unsclerotized portions of their cuticle can actually increase in surface area without molting.

All insects must molt, however, because the size and shape of hardened, sclerotized cuticle is fixed and unchangeable. As a consequence, growth and any change in the shape or texture of the cuticle can occur only if and when a new cuticle is manufactured. The molting cycle of insects is therefore fundamental to any change in size or form.

The Molting Cycle

The terms *molt* and *molting* have an unfortunately imprecise meaning. They can refer to the entire process whereby a new cuticle is made within the old one, or they can refer merely to the final step in that process, the casting off of the old cuticle (exuvium). This last process does have a technically correct name: *ecdysis* (or *eclosion* in the case of adult insects). For the purposes of our discussion we will use the term *molting cycle* to indicate the sum total of the processes associated with molting, beginning with the first detectable endocrine and physiological events that herald the onset of a new stage, and ending with growth of cuticle during the intermolt period. This term acknowledges the fact that molting is a continuous process; the insect integument constantly changes during larval growth and similar events recur in a cyclical fashion.

The insect molting cycle consists of an elaborate sequence of events designed to build a new, larger cuticle within the confines of the old one. During this process, much of the old cuticle is digested for reuse. Various steps in the molting cycle are designed to protect the new cuticle from digestion by the enzymes that attack the old cuticle, while keeping the former soft and pliable so it can expand rapidly when the old cuticle is finally shed.

The face of the epidermis that is in contact with the cuticle bears a dense array of microvilli. During the intermolt period the epidermis is tightly attached to the cuticle at specialized junctions, called *plaques*, that occur at the tips of these microvilli (Locke, 1974, 1985, 1990; Locke and Huie, 1979). The first step in the molting cycle is the separation of the cuticle from the epidermis (fig. 4.1B). This process is called *apolysis* and is corre-

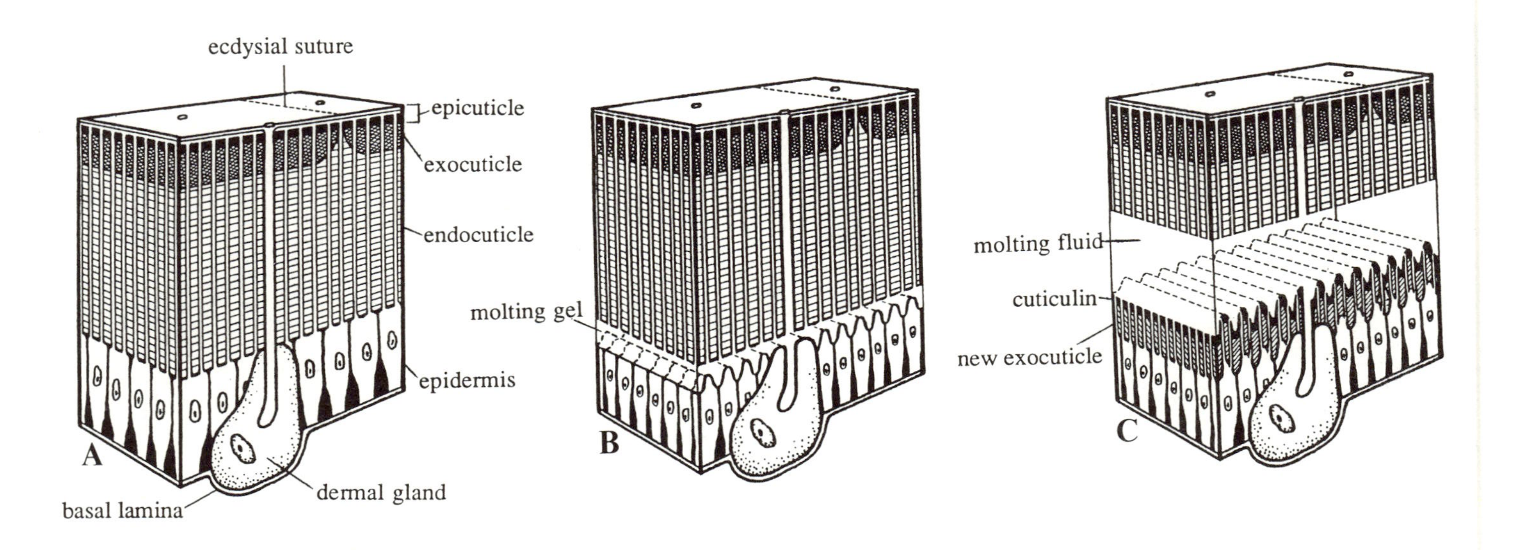
ecdysial suture
epicuticle
exocuticle
endocuticle
epidermis
dermal gland
basal lamina
A
molting gel
B
molting fluid
cuticulin
new exocuticle
C

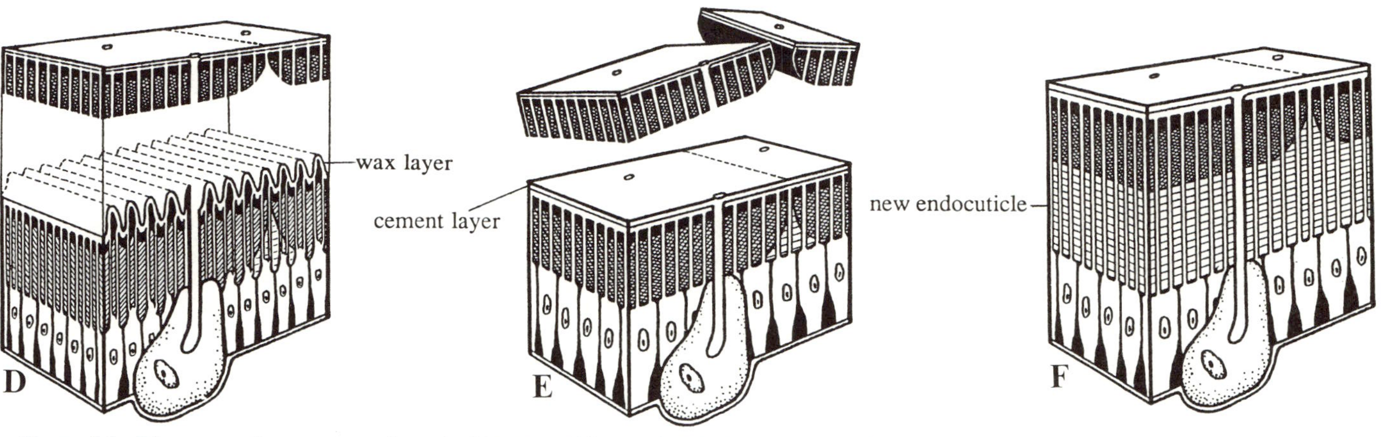

Figure 4.1. Diagrammatic summary of a typical insect molting cycle. (A) Intermolt cuticle showing its various layers. (B) The molting cycle starts with apolysis, the secretion of a molting gel between the epidermis and old cuticle, and a round of cell division. The increased cell number throws the surface of the epidermal cells into fine folds. (C) The epidermal cells secrete a new cuticulin layer and begin to secrete a new exocuticle; the molting gel is activated, becomes fluid and begins to digest the old endocuticle. (D) Digestion of the endocuticle is complete; the wax layer is secreted through the pore canals (shown here as two vertical channels per cell; in reality they are much narrower on this scale and more numerous). (E) The old exocuticle breaks at an ecdysial suture; the cement layer of the epicuticle is secreted by dermal glands. (F) New endocuticle secretion continues through the next intermolt. (Illustration by Paul Kendra.)

lated with, and presumably due to, the disappearance of the attachment plaques. The space between the epidermal cells and the cuticle becomes filled with a gelatinous solution, the *molting gel*, which contains, in an inactive form, the enzymes that will eventually digest part of the old cuticle. At about the time that the molting gel is secreted, the epidermal cells undergo a period of mitosis and cell division. The epidermal cell population becomes denser, the cells more columnar, and their apical surface is thrown into a series of fine folds (fig. 4.1B,C). When cell divisions are completed, the epidermal cells secrete the cuticulin layer of the new epicuticle. Cuticulin is made up of lipoproteins that quickly become cross-linked and sclerotized with other proteins. This tanning process renders the cuticulin layer refractory to dissolution by acids, solvents, and enzymes, although it remains quite permeable to water and many small molecules.

After the cuticulin layer is laid down, the enzymes in the molting gel become activated, probably through an activation factor secreted by the epidermal cells. Through the action of these enzymes the molting gel becomes liquefied and is now called the molting fluid. The molting fluid contains proteases and chitinases that now begin to digest the old endocuticle. Insects differ in the exact timing of molting gel activation and the degree to which the old cuticle is digested; in some species major digestion does not occur until late in the cycle, presumably to afford continued protection to the animal within. In some insects the innermost lamellae of the old endocuticle are resistant to digestion and remain as a thin so-called *ecdysial membrane* between the old and new cuticles. It is believed that the epidermal cells are protected from the action of the digestive enzymes in the molting fluid by the new cuticulin layer they have secreted. As the old endocuticle is digested, the epidermal cells begin to secrete the exocuticle of the new cuticle (fig. 4.1C). The digestive products of the old endocuticle, primarily N-acetylglucosamine (from chitin) and small peptides and amino acids, are continually resorbed through the nascent cuticle and epidermal cells, and can therefore be reused in building the new exocuticle. Eventually the entire endocuticle is digested and resorbed. The enzymes of the molting fluid cannot digest the sclerotized old exocuticle, however, and the old exocuticle is the only portion of the cuticle that is eventually shed at ecdysis (fig. 4.1E). As the exocuticle makes up only 10% to 20% of the total cuticle in most insects, this means that most of the cuticular proteins and carbohydrates are recovered by digestion and reabsorption.

The epidermal cells stay in communication with the internal portions of the growing new cuticle via a system of pore canals. Depending on the species, there may be 50–200 pore canals per epidermal cell. Pore canals are extracellular; they are not lined by a plasma membrane, but they contain one or more fine filaments that run from the cytoplasm of the epidermal cell to the epicuticle. Locke (1974) has suggested that the filaments may be instrumental in keeping the pore canal open during cuticle formation by inhibiting polymerization of cuticular macromolecules in their vicinity. The filaments may also be the means by which the epicuticle and the nascent cuticle are anchored to the epidermal cells.

A few hours before ecdysis, the molting fluid is resorbed and air fills the space between the new and old cuticles. Most of the fluid is absorbed directly through the new cuticle and epidermal cells, but it has been shown that in *Bombyx mori* neck ligation prevents full resorption of the molting fluid, which suggests that some of it may be resorbed by drinking. Soon after resorption of the molting fluid and prior to ecdysis, a complex mixture of hydrocarbons and resins is secreted via the pore canals. These spread out over the surface of the cuticulin and form the wax layer of the new epicuticle (Locke, 1974; Hepburn, 1985). This thin layer almost completely abolishes the permeability of the new integument to water and accounts for the insect's resistance to water loss through the cuticle.

Ecdysis from the old exocuticle involves an elaborate sequence of stereotyped behaviors. Most insects begin ecdysis by swallowing air or water. This increases the internal pressure and eventually causes the old exocuticle to split, usually along the dorsal midline of the head and thorax. Splitting of the cuticle occurs along preformed lines of weakness, the *ecdysial sutures*. These are areas in the cuticle where the endocuticle penetrates deeply into the exocuticle (fig. 4.1A) so that when the endocuticle has been completely digested only a very thin exocuticle remains. A sequence of peristaltic movements and abdominal rotations gradually slips the old cuticle backward towards the tip of the abdomen. When the animal is free of its old cuticle it usually assumes a characteristic posture that allows its appendages to extend freely, and remains quiescent until the new exocuticle becomes sclerotized and hardened.

At the time of ecdysis the new exocuticle is soft, pliable, and very extensible. In many species of insects extensibility of the new cuticle is further enhanced by the action of two hormones, eclosion hormone and bursicon, that are secreted just prior to ecdysis (see chapter 5). These hormones cause the release of plasticizing factors into the cuticle, probably via the pore canals. The plasticizing factors are believed to weaken some of the noncovalent bonds that hold cuticular proteins and chitin molecules together (Reynolds, 1985). The extent to which the new cuticle can expand during ecdysis is determined entirely by the fine folds laid down in the outer cuticulin layer at the beginning of the molt cycle (fig. 4.1B,C). During the pre- and postecdysial expansion the epicuticle stretches to a nearly flat surface. Cuticles that must remain extensible during the intermolt period, such as those of the larval abdomen of the bloodsucking bug *Rhodnius prolixus*, retain a finely puckered cuticulin layer, which allows for some additional expansion when the animal takes a blood meal. The cuticulin layer of soft-bodied holometabolous larvae, such as caterpillars, is also finely corrugated at the beginning of an instar and gradually smooths out as the animal grows and the cuticle expands. If such provision for surface amplification of the epicuticle is not made, any further expansion in volume during the intermolt would be restricted to the amount that the abdomen can extend by telescoping.

Immediately after expansion of the cuticle dermal glands, whose canals exit via pores on the surface of the cuticle, secrete a thin fluid containing a complex mixture of carbohydrates, lipids, and proteins. This fluid spreads quickly to cover the entire body surface, and then dries to form the cement layer of the epicuticle. In most insects the dermal glands that secrete the cement layer are small and evenly spread across the body surface, but in the caterpillars of Lepidoptera the cement layer is produced by a small number of enormous unicellular glands called Verson's glands. There is one pair of these glands per segment; they can be nearly 2 mm in diameter and probably constitute the largest cells in the Insecta. While the cuticle is expanding and the cement layer is being deposited, the hormone bursicon is acting on the epidermal cells, causing them to secrete a variety of phenolic compounds. These phenolics permeate the still soft cuticle via the pore canals. Within the cuticle they undergo oxidation and then, with the aid of various phenolases, they form cross-links between cuticular proteins that sclerotize and harden the new cuticle (Lipke et al., 1983; Sugumaran, 1988; Hopkins and Kramer, 1992).

After ecdysis and hardening of the new exocuticle, the epidermal cells begin secretion of the new endocuticle. In many insects endocuticle deposition continues throughout the entire intermolt period, so that the cuticle grows progressively thicker. Endocuticle deposition stops when apolysis for the next molting cycle begins. The chitin molecules of the endocuticle are laid down with a preferred orientation. It has been shown that a parallel orientation of chitin fibers greatly enhances the strength of the cuticle over a random fiber orientation. Parallel orientation in a single direction, however, strengthens the cuticle in only that one direction. Insects have solved this problem by periodically altering the orientation of the chitin fibers that are laid down by the epidermis, so that the cuticle develops a layered plywoodlike structure. In many cuticles, the orientation of the chitin fibers in each successive lamella is rotated by a constant angle (fig. 4.2). This gives rise to the so-called *helicoidal* structure of cuticle, which is revealed as rows of concentric arcs of fibers in oblique sections of cuticle (fig. 4.2; Neville, 1967, 1984). Many insects deposit helicoidal cuticle only during the night, while during the day they make a cuticle in which the chitin fibers of successive lamellae are all parallel (fig. 4.2A). This gives the endocuticle a layered structure with alternate helicoidal and parallel chitin fibers. These layers can be easily demonstrated in sections of cuticle (from

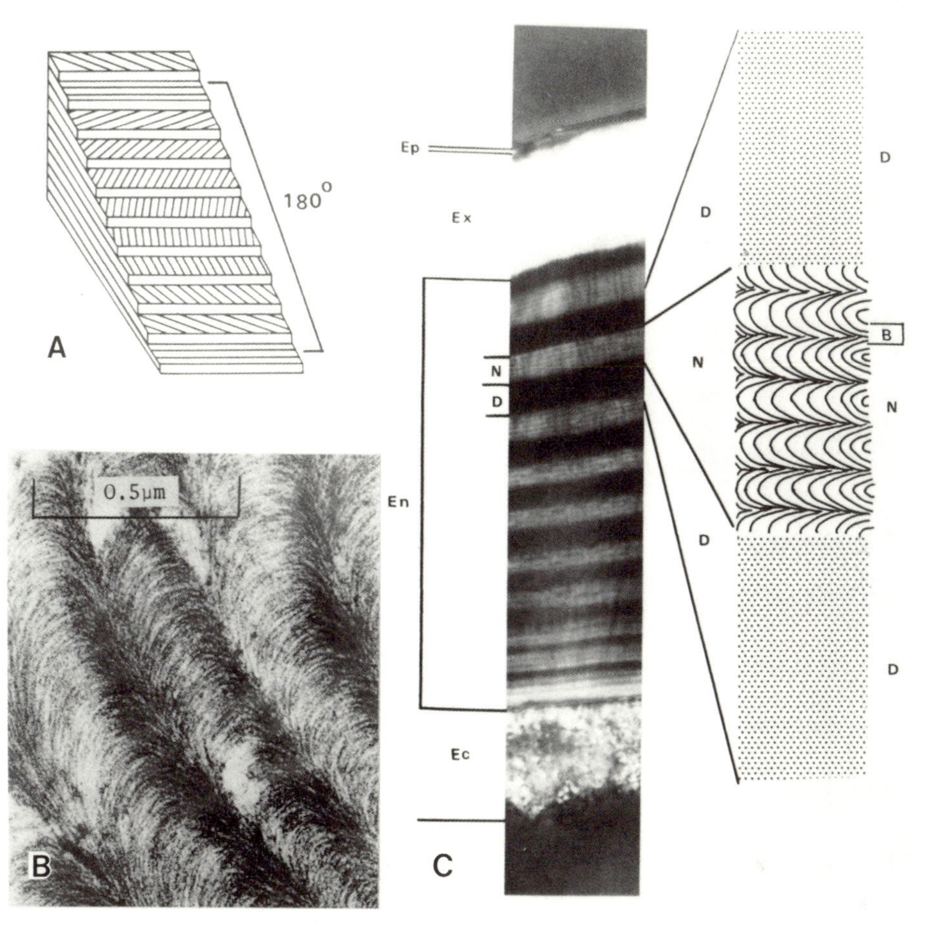

Figure 4.2. Microsopic structure of the insect cuticle. (A) Helicoidal structure of the endocuticle comes about by sequential deposition of layers with parallel arrays chitin crystallites whose preferred orientation changes progressively. (B) Electron micrograph of an oblique section through a layered structure, as in (A), produces arclike figures of fibers. (C) Section through an entire cuticle, showing the fine structure of the various regions; the left panel gives the view with a light microscope under crossed polarizers, the right panel is a diagrammatic representation of the view produced by an electron microscope. Ec, epidermal cell; En, endocuticle; Ep, epicuticle; Ex, exocuticle; D, layers deposited during the day; N, layers deposited during the night; B, bright "lamella" produced at each 180° turn of the chitin crystallite orientation. (From Neville, 1984. Reprinted with permission of Springer-Verlag.)

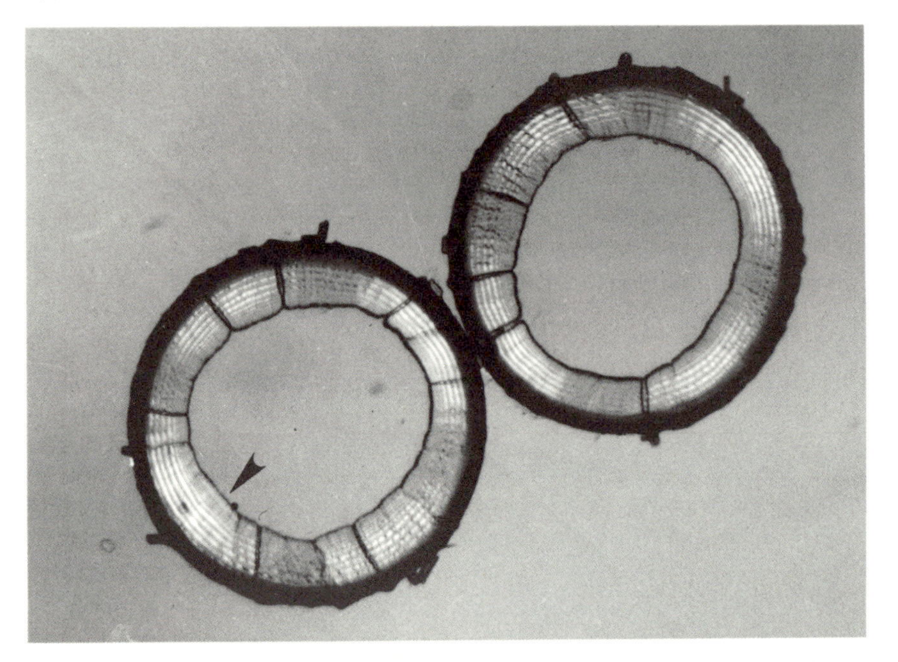

Figure 4.3. Cross section of legs of *Oncopeltus fasciatus*, viewed under crossed polarizers, showing birefringence of alternating daily and nightly layers of cuticle. Ten paired daily rings can be counted at point indicated by arrow. (From Dingle et al., 1969. Reprinted with kind permission of *J. of Insect Physiol.* and Pergamon Press Ltd.)

which the protein has been removed by soaking in hot NaOH) by taking advantage of the fact that the parallel arrays of chitin fibers are birefringent. By placing a section of endocuticle between crossed polarizers, an alternation of bright and dark layers is observed due to the fact that the different layers rotate the polarized light in different ways (fig. 4.3). Each pair of layers represents a day's growth and can thus be used to determine the age of an insect since its last molt (Neville, 1983).

All parts of the body that are lined with cuticle participate in the molting cycle. This includes the linings of the foregut, the hindgut, and the tracheal system. Tracheal molting presents a special problem because the tracheal system is composed of very thin, branching blind tubes that are attached to the cuticle of the outer body wall only at the spiracles. The tracheal cuticle goes through much the same sequence of events during a molt as the cuticle of the body wall. The taenidial ridges are established very early in the molt cycle as spiral ridges in the cuticulin (Locke, 1958). The inside of these ridges becomes filled with a mass of exocuticle, while the exocuticle around the remainder of the tracheal wall remains very thin. A relatively thin endocuticle is deposited during the intermolt, and is digested again in the early stages of the molting cycle. At ecdysis the tracheal exocuticle with its taenidial ridges is pulled out through each spiracle as a continuous strand.

Many parts of the insect integument have a fairly complex and finely detailed surface texture. The most extreme and best known example is the elaborate and finely sculptured structure of butterfly scales. These fine structures, some so closely spaced and regularly arranged that they act as interference reflectors for visible light (and account for the metallic structural colors of many butterflies), arise as fine folds in the cuticulin, just like the taenidia of the tracheal system. It is believed that stresses produced by regular arrays of microtubules under the cell membrane immediately below the cuticulin throw the cell surface and the cuticulin layer into very regular and finely spaced folds. These folds are subsequently stabilized by sclerotization and by being filled with exocuticle (Overton, 1966; Ghiradella, 1974, 1985).

"Hidden Phases" in the Molting Cycle

The only portion of the molting cycle that is visible to an outside observer is the ecdysis, when the old exocuticle is shed and a new larger or different-looking animal emerges. The period between successive ecdyses is commonly called the intermolt period, stage, or instar (the latter term has become preferable among insect physiologists). Technically, however, it is difficult to pinpoint when a new instar begins, and the choice of starting point is in large measure arbitrary. For instance, a fully formed pupa within the old larval skin a few hours before ecdysis is really a pupa, not a larva. But when exactly did it become a pupa? In our discussion of the molting cycle we began with apolysis. Apolysis is, however, not the first physiological step in a molting cycle, since it is preceded by the secretion of the molting hormone, which, in turn, is preceded by the secretion of the prothoracicotropic hormone, whose secretion, in turn, may be provoked by abdominal stretch, which depends on nutrition, and so on. The cyclical nature of growth and molting clearly precludes the definition of an objective beginning and endpoint. The resolution to this dilemma is to assume (arbitrarily) that the new instar begins when apolysis takes place, because this is the time that the animal becomes "freed" from the cuticle of the previous stage (Hinton, 1958).

While the new instar is still enclosed within the cuticle of the previous one it is referred to as a *pharate instar*. Thus by this definition the pupal stage begins upon apolysis of the cuticle of the last instar larva, and until the pupa ecdyses from the old larval skin it is called a *pharate pupa* (the term *prepupa* is also commonly used to designate this stage in insect metamorphosis). When a pupa undergoes apolysis it becomes a pharate adult. The period between pupal apolysis and adult eclosion is called *adult devel-*

opment. It is during this period of adult development that the recognition of the pharate (or hidden) stage becomes particularly important from a developmental viewpoint, because of the massive reorganization of internal and external anatomy that takes place throughout this period even though from the outside the animal simply looks like a "pupa."

Growth of Soft Cuticle

Not all parts of the cuticle become sclerotized after ecdysis. Intersegmental membranes generally remain unsclerotized, as does most of the body wall of so-called soft-bodied insect larvae such as the caterpillars of Lepidoptera, the maggots of Diptera, and the grubs of Coleoptera and Hymenoptera. Intersegmental membranes are not only flexible, but also stretchable. Those in the abdomen of female grasshoppers can stretch to twelve or more times their resting length during oviposition when the abdomen is used to dig a deep oviposition chamber (Vincent, 1975a,b, 1981). In *Locusta*, stretch of the intersegmental membranes is facilitated by the action of plasticizing factors that somehow loosen the weak bonds between adjacent chitin and protein fibers in the cuticle and allows them to slip past each other.

The unsclerotized abdominal cuticle of many larvae is also stretchable. Larvae of *Rhodnius*, for instance, take enormous blood meals, and their abdominal wall stretches by a linear factor of nearly three to accommodate that volume (Reynolds, 1975). During a blood meal the abdominal cuticle becomes plasticized through the release of a neurotransmitter from nerve endings throughout the abdominal wall. The neurotransmitter 5-hydroxytryptamine (serotonin) causes plasticization of the cuticle when it is injected into an unfed *Rhodnius* larva, which suggest that this may be the natural plasticizing factor (Reynolds, 1974, 1985).

The abdominal cuticle of soft-bodied larvae of the Holometabola (insects with complete metamorphosis) also stretches as the animal grows, but, unlike the intersegmental membranes and soft cuticle of Hemimetabola, these cuticles also grow along with the animal during the intermolt period. Between molts these larvae can grow as much as ten-fold in mass and two- to threefold in any linear dimension without smoothing out the folds and creases in their cuticle, thus suggesting that there must be significant growth of the cuticle between molts. It is generally impossible to stretch the cuticle of such larvae by mechanical means without tearing it (Locke, 1974), and it has been shown that the increase in surface area during the intermolt period is caused by actual material growth of the cuticle (Wolfgang and Riddiford, 1981; Locke, 1985).

Two mechanisms of cuticular growth have been described: intussception, and plastic deformation accompanied by appositional growth. Growth by intussception comes about through the diffuse incorporation of new protein into the existing cuticle. The new material is believed to enter the cuticle by way of the pore canals. At least a portion of the cuticular growth in *Drosophila*, *Sarcophaga*, and *Calpodes* occurs by intussception.

The cuticle of larvae of *Manduca* grows by plastic deformation and apposition. Plastic deformation is somewhat like that of other soft cuticles that stretch, but in *Manduca* this is facilitated by a special arrangement of the fibrillar material into columns perpendicular to the thickness of the cuticle (Wolfgang and Riddiford, 1981). During the first half of the instar each epidermal cell secretes one of these vertical cuticular columns, in addition to the normal horizontal lamellar endocuticle. Later in the instar the epidermal cells secrete only the normal lamellar cuticle. As the larva grows, expansion of the cuticle occurs mostly in the cuticular columns. The vertical orientation of chitin and protein fibrils in these columns allows the columns to expand by reorientation of these fibrils into a more horizontal direction, parallel to the surface. During expansion the horizontal lamellae become thinner, which suggests that they too stretch to some degree. While the outer layers of the cuticle expand, more cuticle continues to be deposited basally so that the thickness of the cuticle continues to increase even as its total area increases. This growth in thickness is important because it enables the cuticle to withstand the increasingly large hydrostatic pressures developed within the growing caterpillar. It is likely that a special provision needs to be made to plasticize the cuticle in the vertical columns in order to allow them to expand without rupturing (Reynolds, 1975), but the mechanism by which this might occur has not yet been discovered. The limit of potential expansion in this cuticle, as in all others, is set by the degree of folding of the inextensible epicuticle. In early fifth instar larvae of *Manduca* the epicuticle of the soft portions of the body is much more deeply corrugated than that of sclerotized portions of the cuticle (Wolfgang and Riddiford, 1981), suggesting a preadaptation for significant expansion and growth.

Triggers for Molting

In general, species in the more primitive taxonomic orders undergo a considerably larger number of larval molts than those in the more advanced orders (table 4.1). The sharpest division in this regard is between the insects with incomplete metamorphosis (the Hemimetabola; see below) and those with complete metamorphosis (the Holometabola). At this taxonomic level there is no relation between the final size of the adult insect and the number of larval molts required to grow to that size. The insects with the largest number of larval molts (the Ephemeroptera and Plecoptera) are small to medium sized, while the largest insects, such as the African goliath

TABLE 4.1
Number of Larval Molts Typical of Various Orders of Insects

Order	*Common Name*	*Number of Larval Molts*
Collembola	Spring-tails	4–5 (up to 50 as adults)
Thysanura	Silver-fish	9–13 (dozens as adults)
Ephemeroptera	Mayflies	27–44
Plecoptera	Stoneflies	21–32
Odonata	Dragonflies	9–14
Blattodea	Roaches	5–11
Isoptera	Termites	4–10
Mantodea	Mantids	4–8
Dermaptera	Earwigs	3–5
Orthoptera	Grasshoppers, etc.	4–9
Embioptera	Embiids	3
Psocoptera	Booklice	5
Anoplura	Lice	2
Hemiptera	True bugs	2–6
Neuroptera	Lacewigs, etc.	2
Coleoptera	Beetles	2
Mecoptera	Scorpionflies	3
Siphonaptera	Fleas	2
Diptera	Flies	2–3
Trichoptera	Caddisflies	5–6
Lepidoptera	Butterflies and moths	2–10 (usually 4 or 5)
Hymenoptera	Bees, wasps, ants, etc.	2–4

SOURCE: Data from Williams (1980).

beetles and South American rhinoceros beetles, are also the ones with the smallest number of larval molts. The ability of Holometabola to grow to large sizes in relatively few larval molts is associated with the fact that their larvae are soft bodied and have cuticles that can grow during the intermolt. Within a given taxonomic group such as an order or family, the number of larval molts is relatively constant. Variation in the number of molts at this level is determined largely by nutrition and the growth rate of the individual.

How is the timing of a molt controlled? Under optimal conditions of growth and nutrition, molting cycles occur with predictable regularity, but when conditions become less than optimal the temporal pattern of molting becomes irregular in most insects. In general, insects molt only when they grow (though there are exceptions to this rule), and this implies that some physiological process associated with growth is likely to be involved in triggering the onset of a molting cycle.

Each molting cycle begins with the secretion of the prothoracicotropic hormone (PTTH), a neurohormone from the brain. PTTH, in turn, stimulates the secretion of the molting hormone, ecdysone (see chapter 5 for a discussion of the hormonal control of molting and metamorphosis). The secretion of ecdysone is necessary and sufficient to initiate apolysis, mitoses in the epidermis, and the secretion of a new cuticle. The control over the timing and frequency of molting must, therefore, reside in the physiological mechanism that controls the secretion of PTTH by the brain.

The physiological control over PTTH secretion is best understood in the Hemiptera. Wigglesworth (1934) recognized that larvae of the bloodsucking reduviid bug *Rhodnius prolixus* always initiated a molting cycle upon taking a full meal of blood. Apolysis takes place two to three days after a blood meal, and ecdysis to the next stage follows fifteen or twenty-eight days after the blood meal in the fourth and fifth larval instars, respectively. A small meal does not provoke molting, nor does a series of small meals given in succession. Thus the nutrients acquired by feeding do not seem to be important for molting. Only meals that exceed a critical size (about 100 mg for fifth instar larvae) trigger the onset of a molt. Wigglesworth showed that severing the ventral nerve cord between head and thorax prevented the response to a blood meal and concluded that simple stretch of the abdomen due to the blood meal provided the required stimulus to the brain to initiate the secretion of PTTH. A meal of saline also induces apolysis and mitoses in the epidermis (Beckel and Friend, 1964), demonstrating clearly that the nutritive value of the meal is not important to initiate a molt (although nutrients are required to complete the prolonged transformations of a molt cycle). The neurophysiology of the abdominal stretch receptors that provoke molting in *Rhodnius* and a related bloodsucking reduviid, *Dipetalogaster maximus*, have been studied by Nijhout (1984) and Chiang and Davey (1988). They showed that the normal stretch receptors in the abdominal musculature do not produce the required long-term stimulation. In *Dipetalogaster* it is the stretching of the main trunk of the abdominal nerves that produces a continued and persistent train of action potentials. The abdominal nerves usually stretch to nearly one-and-a-half times their resting length after a blood meal. In *Rhodnius* abdominal distension is perceived slightly differently. Here a small portion of each abdominal nerve is sensitive to applied pressure. This receptor area is believed to produce action potentials when it is pressed against the body wall by a filled midgut. These two types of stretch receptors are unique in that they produce action potentials continuously for hours and probably days without adaptation, as long as they remain stretched or pressed.

Stretch-induced molting is not restricted to Hemiptera such as *Rhodnius* that take single large meals. Larvae of the milkweed bug *Oncopeltus fasciatus*, a hemipteran with quite conventional feeding habits, begin a molting cycle only after they reach a sharply defined critical weight in each instar (Nijhout, 1979, 1981). In last instar larvae, ecdysis to the adult takes place 6–7 days after the critical weight is reached, irrespective of the prior or subsequent feeding pattern. Unlike *Rhodnius*, the critical size is not achieved through a single large meal but by a gradual growth in body mass. It is possible to make a larva that is still well below its critical weight molt simply by injecting it with small volume of saline. The saline injection expands the abdomen to produce the necessary degree of stretch, and the animal subsequently molts to a perfectly normal though miniature adult. The critical weight is therefore a measure of the size at which a critical degree of abdominal stretch is achieved.

The critical weight is not identical for all larvae but depends on their body size at the beginning of the instar. Body size of a given instar is usually defined as the dimension of a sclerotized and inextensible part of the body, such as the width of the head, or the length of a leg segment. Thus an *Oncopeltus* larva with a small body size has a lower critical weight than a larva with a larger body size (fig. 4.4), as would be expected if, during growth, stretch receptors are triggered when the abdomen is expanded to a precise critical multiple of its size at the beginning of the instar. The stretch receptors of *Oncopeltus* have not yet been identified.

The control over the onset of molting of insects outside the Hemiptera is not nearly as clear or as simple as the two cases just described. In the dermestid beetle *Trogoderma glabrum*, for instance, stretch is certainly not involved in the control of molting. When larvae of *Trogoderma* are starved they continue to molt, though at a much reduced frequency, and such individuals actually become smaller with each molt (Beck, 1971a,b, 1972). Last instar caterpillars of the moth *Manduca sexta* initiate their molt to the pupal stage when they pass a well-defined critical weight (the endocrine events that follow attainment of the critical weight in *Manduca* are considerably more complicated than those of *Rhodnius*, and will be discussed later in this chapter). As in the case of *Oncopeltus*, this critical weight depends on the size of the larva at the outset of the instar. For animals growing under ideal conditions the critical weight is almost exactly 5 grams. But if animals experience food deprivation in earlier instars their size at the outset of the final instar can vary widely and their critical weight varies in direct proportion to that size (fig. 4.5).

The extraordinary regularity in the relation between body size and critical weight suggests the operation of a relatively simple size-monitoring mechanism. But in *Manduca*, artificial stretch of the body wall is completely ineffective in provoking a premature pupal molt, nor can a decrease in body stretch by bleeding and starvation immediately after the critical weight is reached prevent the normal onset and continuation of a molt (Nijhout, 1981). It appears therefore that caterpillars of *Manduca* do not use abdominal stretch as a cue for body size. If larvae of *Manduca* are

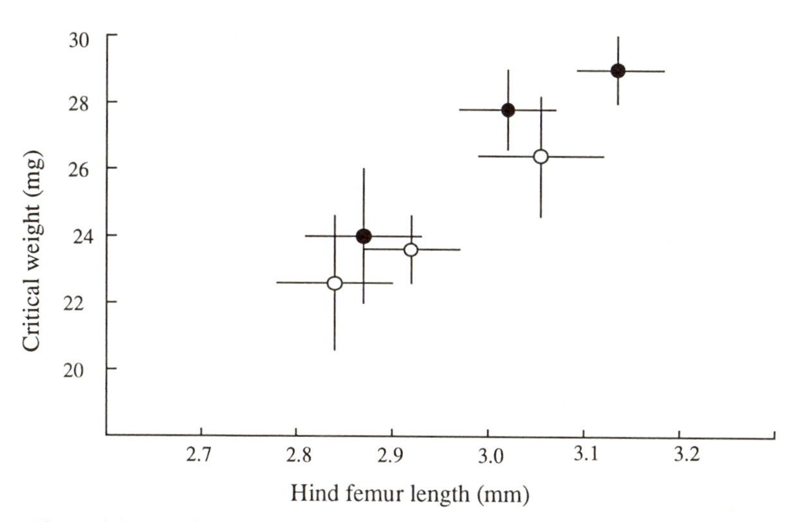

Figure 4.4. Relationship between the critical weight for molting and body size (indexed by length of the hind femur) of six cohorts of final instar larvae of *Oncopeltus fasciatus*. *Black*, males. *White*, females. (Redrawn from Nijhout, 1979).

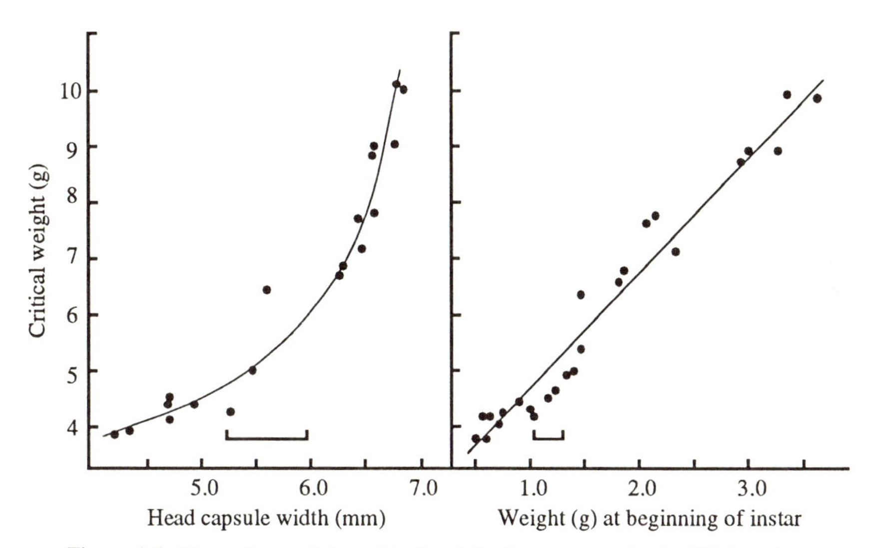

Figure 4.5. Dependence of the critical weight for metamorphosis (CA inactivation) in final instar larvae of *Manduca sexta*, on the size of the individual. The size of an individual can be indexed by the width of head capsule (*left*), or by its weight at the outset of the instar (*right*). Horizontal bars are the ranges of head capsule sizes and initial weight found under optimal growing conditions in the laboratory. Under such conditions the critical weight is about 5 grams. (Redrawn from Nijhout, 1981.)

starved before they reach their critical weight, they will eventually pupate after a delay of one or two weeks. The passage of time can apparently override any specific stimulus to molt in *Manduca* as in *Trogoderma*, and this appears to be the case for many other insects as well. For instance, when penultimate instar larvae of the moth, *Galleria mellonella*, are fed a protein-free diet, they fail to grow but undergo a number of irregularly timed stationary larval molts (Allegret, 1964).

Growth

Sclerotized and unsclerotized parts of the insect cuticle grow in different ways during a molt cycle. Unsclerotized parts of the cuticle can stretch and grow during the entire intermolt period, but sclerotized portions of the cuticle can increase in size only during the brief time around ecdysis. Sclerotized cuticle therefore grows in a stepwise or episodic fashion. Since the larval stages of most insects have sclerotized cuticles on most of their body, the insect as a whole grows in what, from the outside, appears to be a stepwise fashion.

The maximum dimensions to which the cuticle can expand during ecdysis is determined by the degree of pleating of the sclerotized cuticulin layer of the epicuticle. This epicuticular pleating is a function of the degree of cell-surface amplification of the epidermal cells during the early stages of a molt and is not connected in any obvious way to cell division or nutrition. The maximum expansion allowed by the epicuticle is probably seldom realized. It is clear from electron-microscopical studies that nearly all cuticles, whether they are soft or sclerotized, retain a certain degree of surface sculpturing in their epicuticle. Soft cuticles can grow by stretching this fine pleating, but in cuticles that become sclerotized this is not possible and postecdysial expansion is the only determinant of the size of new stage.

The actual degree of expansion of the cuticle at ecdysis varies widely among the insects and is a species-specific character. Within a species, however, the cuticle of any given part of the body expands by a nearly constant factor over its size during the previous stage. This constancy of the growth ratio between instars is known as *Dyar's rule*, and the growth ratio is referred to as Dyar's coefficient, or simply the coefficient of growth. The consequence of having incremental growth with a constant growth coefficient at each increment is that the overall growth of insects is exponential. When the logarithm of size is plotted against instar number, a straight line is obtained. This is the case no matter what measure of size is used. Different structures grow at different relative rates, that is, their size increases by different factors during each molt, and therefore their slopes on a semilogarithmic plot against instar number will be different. Figure 4.6 shows plots for several structures in caterpillars of *Manduca*. Dyar's rule holds not only for the linear measures of sclerotized structures

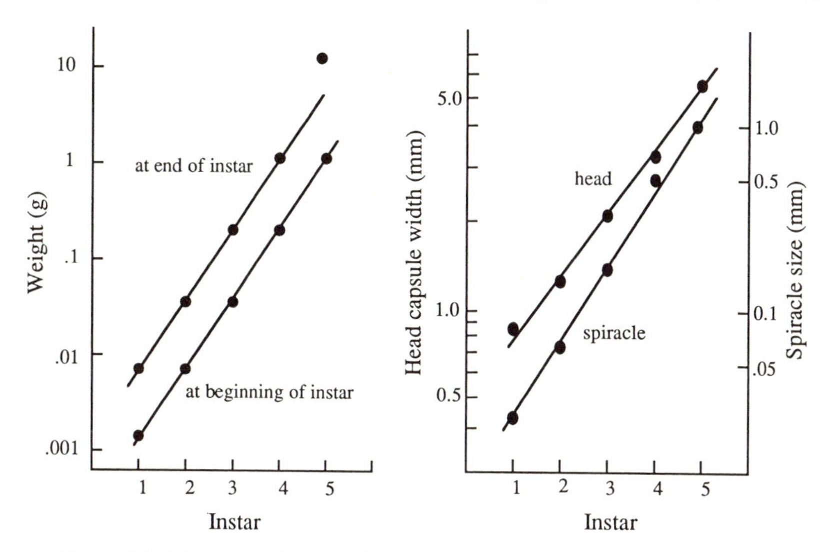

Figure 4.6. Exponential growth in larvae of *Manduca sexta* under laboratory conditions. The weight and dimensions of body parts increase by a constant factor during each instar.

but for many other features of size and growth as well. For instance, the left side of figure 4.6 shows a plot of the weight at the beginning of each instar against instar number in *Manduca*; it is clear that these measures of growth also increase by a constant factor in each instar. Not all species have a constant growth coefficient during their entire larval life. Figure 4.7, for instance, illustrates the case of a stonefly, *Neoperla clymene*, in which the growth coefficient of larger animals is less than that of smaller animals. In

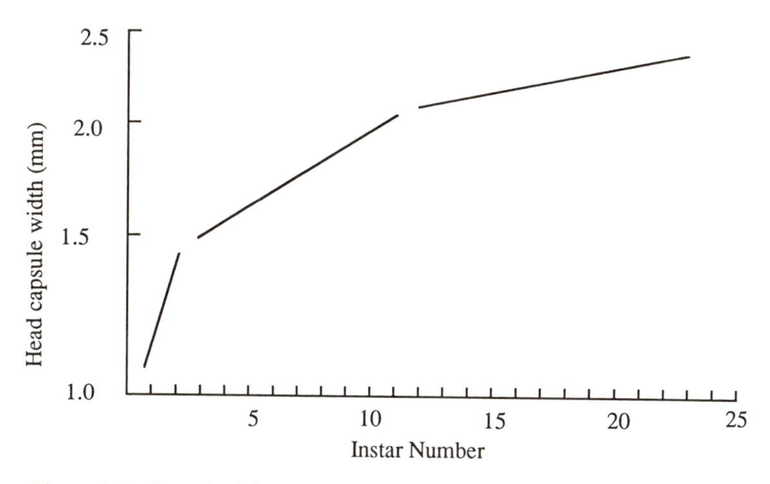

Figure 4.7. Growth of the stonefly, *Neoperla clymene*, illustrating stepwise changes in the growth coefficient. (Redrawn from Vaught and Stewart, 1984.)

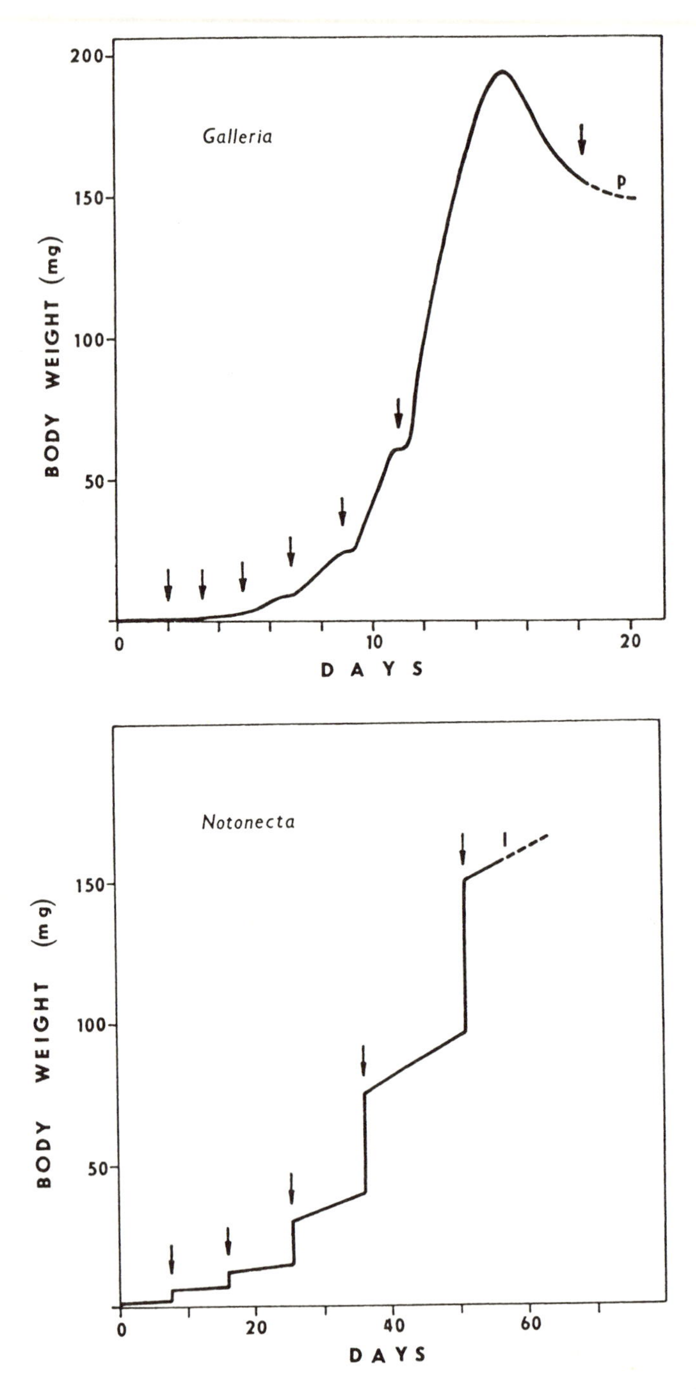

Figure 4.8. Growth in body weight during larval development in *Galleria mellonella* (Lepidoptera) and *Notonecta glauca* (Hemiptera). Arrows indicate ecdyses. *Galleria* gains weight constantly during the instars and loses a small amount each ecdysis. *Notonecta* gains weight more abruptly, probably by water intake, during each ecdysis. (From Sehnal, 1985. Reprinted with kind permission of Pergamon Press Ltd.)

this case the growth coefficient diminishes in two abrupt steps, between the second and third, and between the eleventh and twelfth instars, and remains constant during the intervening instars. Interestingly, there appears to be a systematic difference in the growth coefficients between holometabolous and hemimetabolous insects, such that the former grow with nearly twice the size increment during each molt as do the latter (Cole, 1980). This difference in growth increments is probably due to the fact that in holometabolous insects the cuticle can grow, and may explain why holometabolous insects tend to have few larval instars (table 4.1).

When an insect expands its cuticle upon ecdysis it does so by swallowing air or water, and when the newly expanded cuticle hardens, this expansion volume is available for the growth of internal tissues and organs. Tissue growth and the accumulation of reserves in the fat body occur throughout the intermolt period. As a consequence, the growth in biomass, or dry weight, is nearly continuous, punctuated only by brief periods of stasis when the animal stops feeding while it molts, and a slight loss in weight at each ecdysis due to loss of the exuvium. Plotted against instar number, weight increases exponentially throughout larval life, as might be expected from the fact that weight at the time of molting increases exponentially. Under optimal conditions of growth, weight also increases exponentially with time (fig. 4.8). Under less than ideal growing conditions, exponential growth in real time no longer holds, even though the growth coefficient at each ecdysis stays constant (or nearly so), so that growth is still exponential when measured by instar number. Only when a larva is severely undernourished, so that it cannot expand its cuticle to the full extent possible after ecdysis, does the exponential relation between instar number and body size break down. The ability to break Dyar's rule experimentally and alter the dimensions of a larva at the time of molting has proven a useful tool for investigating the physiological control of metamorphosis, as we will see later.

Metamorphosis

The molting cycles during which an immature insect transforms into a sexually mature form are called metamorphosis. The morphological changes that accompany metamorphosis may be slight, involving only the addition of external genitalia, or they may be so pervasive that the morphological differences between larval and adult forms are greater than those between major taxonomic groups among the insects.

The insects are divided into three major evolutionary lineages that are characterized by the degree of morphological change that accompanies their metamorphic molt (fig. 4.9).

The Ametabola. These are insects without significant metamorphosis, in which the body form of larvae and adults is identical except for the devel-

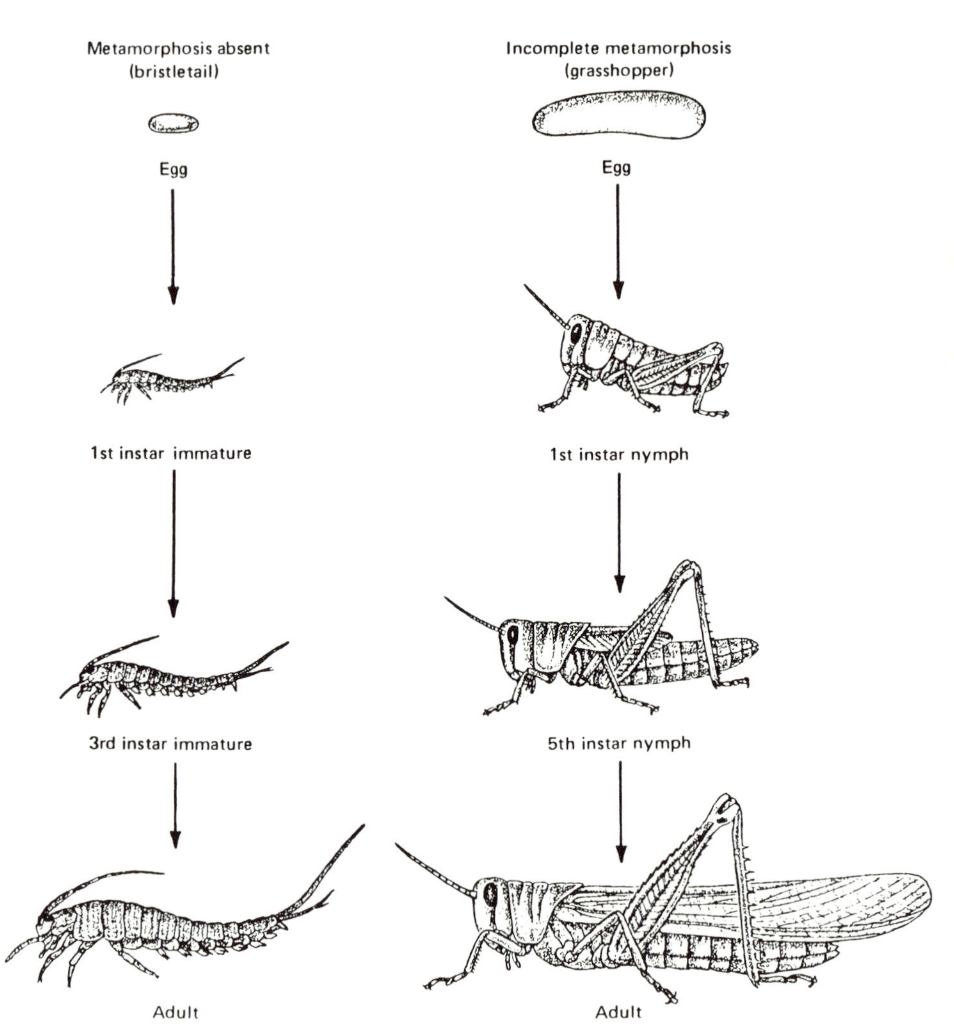

opment of external genitalia and the internal organs of reproduction. All Ametabola are also Apterygota, and vice versa.

The Hemimetabola. These are insects with a so-called incomplete metamorphosis in which, in addition to the previous features, adults differ from larvae in having functional wings. Depending on the taxonomic group, there can be considerable morphometric differences between larvae and adults, as well as great differences in the pigmentation, bristle pattern, and structure of the cuticle. The Hemimetabola are also known as the Exopterygota.

The Holometabola. These are insects with so-called complete metamorphosis. Here, development of adult segmental appendages, wings, and eyes

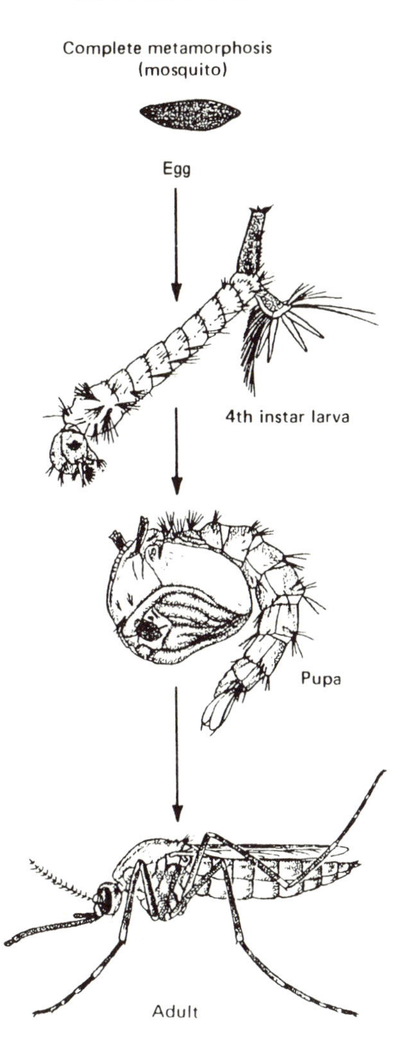

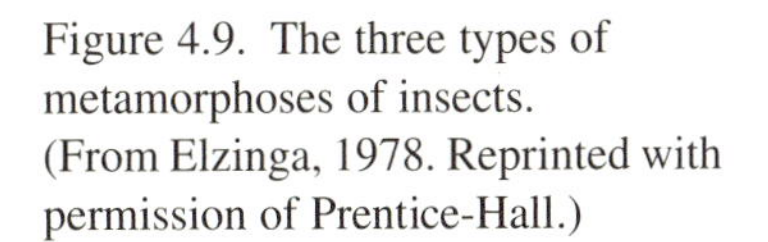
Figure 4.9. The three types of metamorphoses of insects. (From Elzinga, 1978. Reprinted with permission of Prentice-Hall.)

is suppressed during the larval stage. The anlagen for these adult structures are maintained as small undifferentiated pockets of cells, the imaginal disks, underneath the larval integument. At metamorphosis, the imaginal disks undergo rapid growth and differentiation, and become externalized to form the adult appendages. Metamorphosis in the Holometabola requires two molting cycles, the first one to externalize the imaginal disks and a second one to allow for the differentiation of the adult morphology. Considerable restructuring of internal organs is also common in holometabolous metamorphosis. The Holometabola are also known as the Endopterygota, in reference to the internal development of their wing primordia.

Progressive Larval Differentiation (Heteromorphosis)

Changes in morphology and in the structure, texture, or color of the cuticle occur at almost every larval molt. Often these changes are so small as to be hardly noticeable unless one is especially attuned to them; but in some instances they are sufficiently dramatic to have attracted widespread notice. Young caterpillars of many Papilionidae, for instance, are black and white and mimic bird droppings, but in the course of two or three molts they transform into conventional green caterpillars. The ornamental tubercles of many caterpillars of the Saturniidae often change dramatically in number, shape, and coloration from instar to instar. The larvae of Thysanura acquire a body covering of scales after their third molt, and all Hemimetabola develop ever larger external wing pads during their later larval stages. The most extreme cases of heteromorphosis are found in some of the insects that have endoparasitic larval stages. The Meloidae (blister beetles) have a slender, long-legged, and active first instar larva (triungulin) that searches for an appropriate host and transforms into a typical beetle grub in the course of two molts. The larvae of Mantispidae undergo a similar extreme heteromorphosis, from an early active larval form to an inactive and parasitic grublike form. Some authors reserve the term "heteromorphosis" only for this last type of development, which is also called "hypermetamorphosis"; here we use heteromorphosis in its broader sense to refer to all changes in larval form through development.

Hemimetabolous Metamorphosis

With the exception of the development of the wings and external genitalia, the metamorphosis of the Hemimetabola is a simple extension of larval heteromorphosis. The morphological changes that occur during metamorphosis are often no more extensive than the more extreme forms of larval heteromorphosis, and sometimes much less so. The most severe changes in hemimetabolous metamorphosis involve changes in the proportions of various parts of the body, and the elaboration of various flanges and projections on the body wall. The flattened and heavily armored digging legs of cicada larvae are transformed into the more conventional-looking slender and cylindrical legs of the adult. In the Reduviidae the abdomen becomes wider and flatter and develops flattened lateral expansion flanges that allow the abdomen to swell during feeding and egg development. In some Ephemeroptera, Odonata, and Plecoptera the body shape changes from a flattened rounded shape to an elongated cylindrical one. Membracidae develop spectacular projections on their pronotum that, in some tropical genera, can become as large as the rest of the body. Fulgoridae develop a variety of projections, horns, and bulges on their heads. Secondary sexual structures, such as the timpani of cicadas and the various stridulatory organs of Orthoptera and Hemiptera, also develop during the metamorphic molt. Together with the appearance of wings and external genitalia (usually claspers in males and ovipositors in females), these changes in morphology can dramatically alter the overall appearance of an insect in the course of a single molting cycle.

The internal anatomy also changes during metamorphosis, but these changes have been little studied. Internal organs such as the digestive system and Malpighian tubules change little during metamorphosis, as adults usually retain the same feeding habits as the larvae. The most extensive metamorphic changes in internal organs occur in the elaboration of the thoracic musculature required for flight, the rearrangement of parts of the nervous system to deal with this new musculature, and the development of gonads, their ducts, and accessory structures.

Holometabolous Metamorphosis

The origin of holometabolous metamorphosis is believed to lie in the progressive evolutionary divergence of larval and adult specialization in the course of insect phylogeny. In the ametabolous and hemimetabolous insects, larvae and adults have similar if not identical feeding habits and, with the exception of Ephemeroptera, Odonata, and Plecoptera, are usually adapted to live in very similar habitats. Metamorphic changes in the hemimetabola are mostly designed to add the structures required for dispersal, courtship, and reproduction to the general larval body plan. Something very different happens in insects with holometabolous metamorphosis.

During the early stages of embryonic or larval development in holometabolous insects, small pockets of cells invaginate from the epidermis at various locations along the body. These pouches of cells are the tissues from which the compound eyes, the appendages of head (antennae and mouthparts), thorax (walking legs) and abdomen (external genitalia), and the wings of the adult insect will be made. They are called the *imaginal disks*. The imaginal disks remain undifferentiated and grow slowly during larval life. In spite of their continuity with the general epidermis, the imaginal disk cells do not secrete a cuticle, nor are the mitoses of their cells synchronized with those of the epidermis during the molting cycle. Instead, the imaginal disks grow slowly but continuously throughout larval life. They undergo a sudden, nearly explosive period of growth during the latter half of the last larval instar, and secrete their first cuticle during the first metamorphic molting cycle that transforms the larva into a pupa.

The morphology of the imaginal disks is quite diverse. The leg imaginal disks of Lepidoptera and Coleoptera are little more than placodes or pads of enlarged epidermal cells during most of larval life. The disk for the head

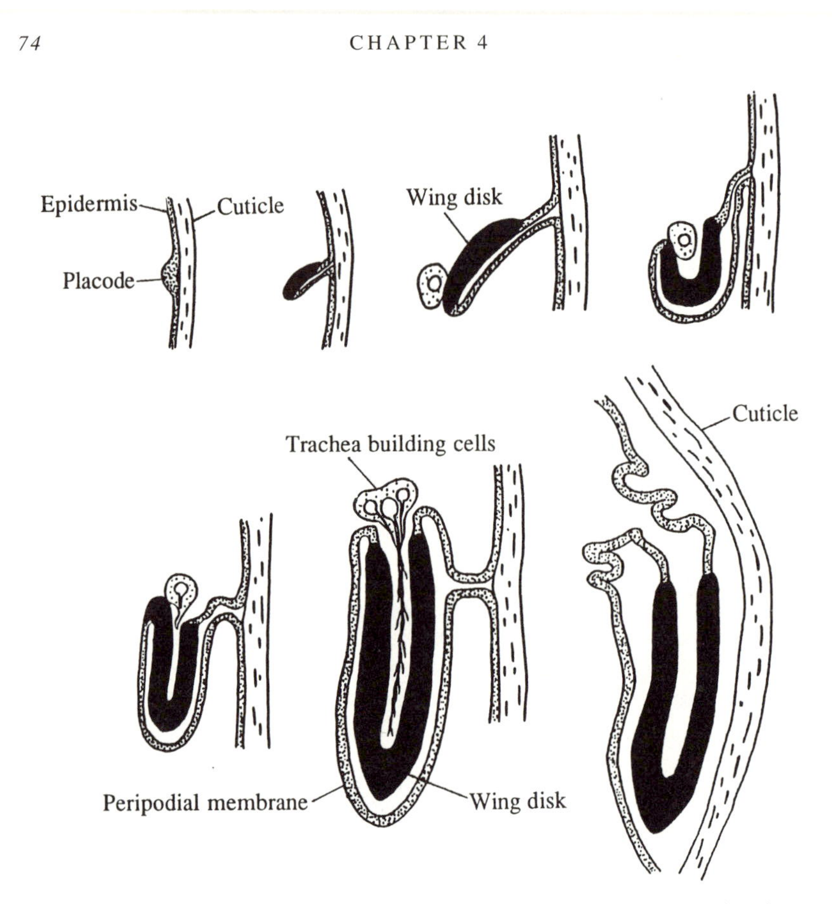

Figure 4.10. Diagrammatic view (in cross section) of the development of a wing imaginal disk of Lepidoptera. Development begins as a placode of epidermal cells that invaginates to form a pouch. The pouch invaginates again to form a double-layered disk (*black*) and a surrounding peripodial membrane. Prior to pupation the disk evaginates and comes to lie between the epidermis and the old cuticle. (Modified from Nijhout, 1991.)

appendages and external genitalia in most orders form into fairly compact balls or short cylinders that lie well beneath the integument, and their continuity with the epidermis is not always easy to detect. The wing imaginal disks are by far the best studied, from both a morphological and developmental perspective, because they are generally the largest of the imaginal disks and because the wings have traditionally been of particular interest for insect systematics and evolution. Cross-sectional views of a developing and an evaginated lepidopteran wing imaginal disk are shown in figure 4.10. The disk begins as a placode of thickened epidermis. This placode invaginates, and the invaginated pouch folds upon itself to produce a four-layered structure. The inner two layers become closely pressed together

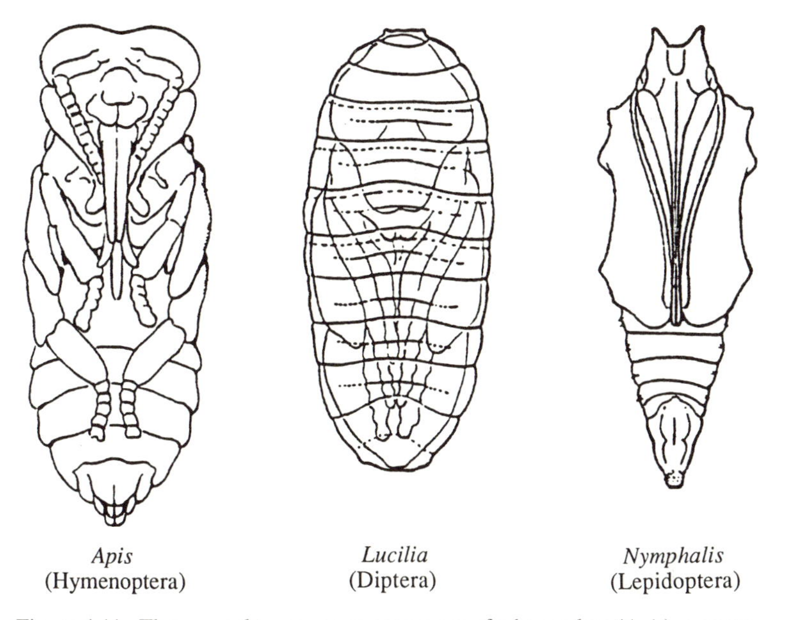

Figure 4.11. Three pupal types: an exarate pupa of a honey bee (*Apis*), a coarctate pupa, which is an exarate pupa inside a puparium, of a blow fly (*Lucilia*), and an obtect pupa of a butterfly (*Nymphalis*). (From Weber, 1954. Reprinted with permission of Gustav Fischer Verlag.)

and are the imaginal disk proper, while the outer two layers form a thin sheath called the *peripodial membrane*. During the last larval instar a system of tracheae penetrates between the two cell layers of the imaginal disk along a system of lacunae (channels between the two cell layers) that form the initial stages of the wing venation (Kuntze, 1935; Nijhout, 1985).

Development of the other imaginal disks follows the general pattern of the wing disks, differing from it mainly in shape and in the pattern and degree of folding. All disks are surrounded by peripodial membranes and are believed to remain connected to the epidermis via the pore formed by the original invagination. During the second half of the last larval instar, the disks begin to grow at a greatly accelerated rate, and in some species initial differentiation begins. Shortly after apolysis to the pupal stage, each disk evaginates through the original pore and comes to lie beneath the cuticle (fig. 4.10). The disks continue to grow vigorously after this evagination and quickly take on the shape of the pupal appendages. When ecdysis takes place, the insect now has the fully formed wings and appendages characteristic of the pupa.

The pupae of Holometabola have the general body form of the adult insect. The pupal appendages are, however, shorter, more rounded and much less detailed in their morphology than those of the adult (fig. 4.11).

In the pupae of most orders, the appendages lie free from the body wall, and these pupae are called *exarate*. In the pupae of Lepidoptera, however, the appendages are glued to each other and to the body wall to form a pupa with a smooth, almost streamlined outline, called an *obtect* pupa. Soon after the pupal ecdysis (unless diapause intervenes), a new molting cycle begins during which the epidermal cells of body wall and appendages secrete a cuticle with the characteristics of the adult stage. Cell divisions during this molting cycle establish the final dimensions of the appendages and the various morphological details of the adult form.

The only major deviation from this general scheme of insect metamorphosis occurs in the higher Diptera (suborder Cyclorrhapha). The morphology of their imaginal disks is more derived and their imaginal disks are also proportionally much larger than in other Holometabola and much more compact in form. The leg imaginal disks, for instance, are not cylindrical but flattened and built up as a series of concentric layers of cells. Upon expansion during pupation these concentric layers expand in a telescope-like fashion to form the elongated leg (Gehring and Nothiger, 1973; Oberlander, 1985). The wing imaginal disks, which remain connected to the epidermis via a long tubular stalk, form not only the adult wing, but also nearly the entire lateral and dorsal portions of the adult thorax (figure 4.12 shows a comparison of the wing imaginal disks of a lepidopteran and cyclorrhaphan dipteran). The imaginal disks of the compound eye and antenna are fused into a single large structure which upon pupation also forms most of the anterior and dorsal surface of the adult head. The ventral portions of head and thorax are made up of tissues from the imaginal disk for the mouthparts and legs, respectively.

In larval Cyclorrhapha the epidermal cells of head, thorax, and abdomen undergo programmed cell death at the time of pupation, quite in contrast to the situation in other Holometabola where the general body epidermis is carried through from the larval to the adult stage. The entire head and thorax of adult Cyclorrhapha are made up of tissues derived from expansion of the imaginal disks. The abdominal epidermis degenerates after secreting a pupal cuticle, and the adult abdomen develops from novel imaginal disklike structures found only in the Cyclorrhapha, the *histoblasts*. Histoblasts are small clusters of cells that bud off from the abdominal epidermis during embryonic development and remain undifferentiated and attached underneath the epidermis throughout larval life (fig. 4.13). In *Dacus* there is one pair of histoblasts per abdominal segment, while in *Drosophila* there are three clusters (called nests) of histoblasts per segment (Anderson, 1972; Oberlander, 1985). Because the larval epidermal cells degenerate and the imaginal disk and histoblasts require time and support to build the new adult body wall, a special provision is made to support and protect the body during pupation. Pupation in the Cyclorrhapha takes

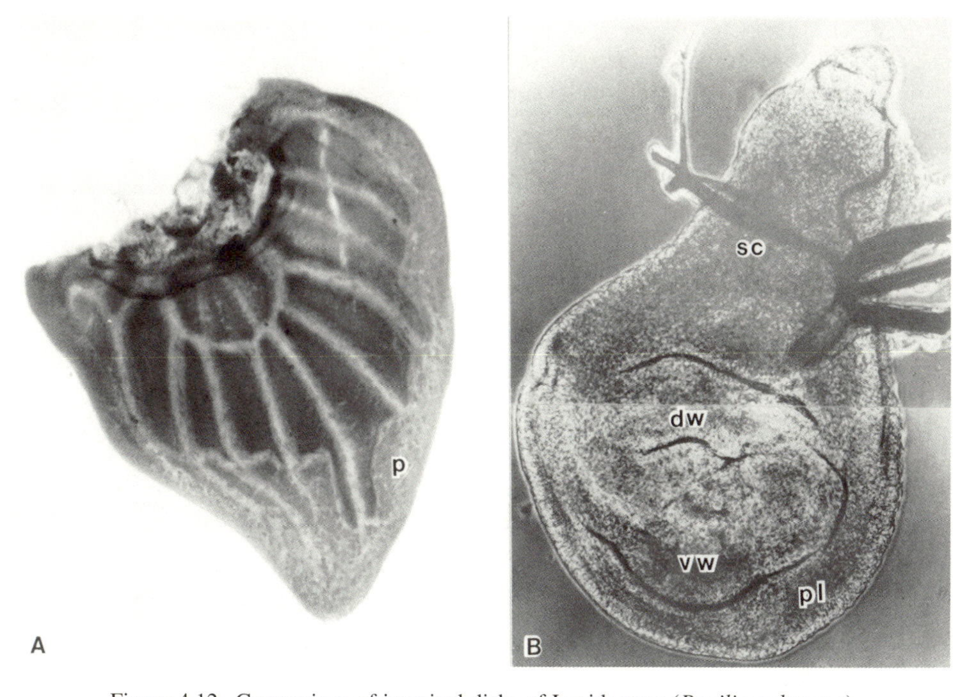

Figure 4.12. Comparison of imaginal disks of Lepidoptera (*Papilio polyxenes*) and cyclorrhaphan Diptera (*Drosophila melanogaster*). In *Papilio*, the imaginal disk contains only the wing. A rim of peripheral cells (p) undergoes programmed cell death after pupation, leaving the adult wing shape. In *Drosophila* the disk contains the future wing as well as large portions of the thoracic sclerites (see fig. 4.13); vw, ventral portion of wing; dw, dorsal portion of wing; pl, thoracic pleural sclerites; sc, thoracic scutal sclerites. (A, after Nijhout, 1991; B, photo from Oberlander, 1985. Reprinted with kind permission of Pergamon Press Ltd.)

place within the cuticle of the last larval instar. Prior to apolysis the larva contracts into a barrel shape and the soft and pliable cuticle becomes tanned and sclerotized. This hardened larval skin is called the *puparium*, and the entire process of pupation and adult development takes place within.

The internal organs of Holometabola also undergo substantial changes at metamorphosis (Whitten, 1968). The larval midgut of Diptera, Lepidoptera, and many Coleoptera breaks down and is rebuilt from small areas of generative cells at the junctions with foregut and hindgut and from nests of imaginal cells scattered throughout the midgut epithelium. The salivary glands, prothoracic glands, and most of the larval muscles break down during or shortly after pupation (Locke, 1985; Lockshin, 1985). The fat body usually changes its morphology considerably as the reserves it con-

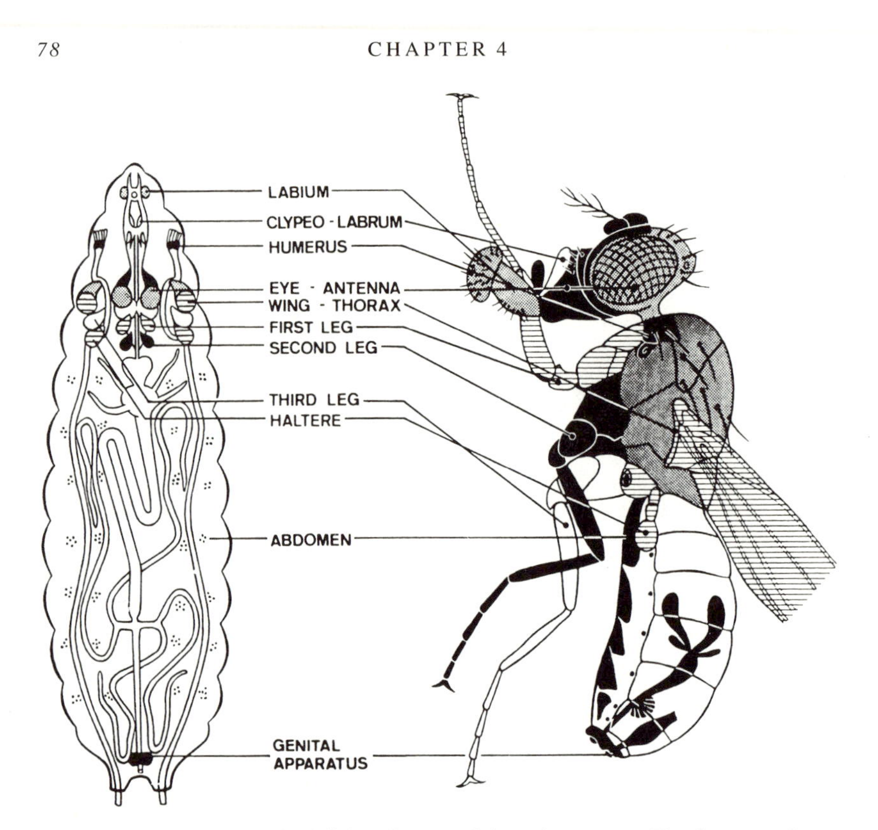

Figure 4.13. The imaginal disks of *Drosophila melanogaster*. The figure on the left shows the arrangement of imaginal disks and histoblasts (dots) in a larva. The figure on the right shows the corresponding adult structures that develop from the disks. (From Nöthiger, 1972.)

tains are used to build the adult structures. The massive flight musculature of the thorax develops, as do the muscles for the new adult mouthparts, appendages, and genitalia. The tracheal system is modified by the loss of branches that service degenerating tissues and the development of new branches to the flight muscles, gonads, and the new appendages. There is also a major reorganization in the gross and fine structure of the central nervous system. In many Lepidoptera and Diptera the ventral nerve cord shortens, and some of the abdominal ganglia may fuse to form large ganglionic masses in thorax and abdomen. Many interneurones in the ganglia degenerate, and new interneurones arise from undifferentiated groups of cells, the neuroblasts (Truman, 1988; Truman and Riddiford, 1989). This reorganization of the cellular structure of the central nervous system is believed to involve the degeneration of neurons involved in larval-specific behaviors and the development of new neural pathways for the control of adult-specific behaviors.

Evolution of Holometabolous Metamorphosis

Like the evolution of insect wings, the evolution of holometabolous metamorphosis has provoked a great deal of speculation and debate for more than one hundred years. These speculations have centered largely on the evolution of the internalization of wings and on the origin and homology of the pupal stage. According to Hinton (1963) it was the internalization of the wings that provided the critical preadaptation for the evolutionary radiation of the Holometabola. Hinton has suggested that internal development of the wings allowed larval stages to assume a burrowing mode of life (in soil, plants, carrion, or living animals), which opened up a wide diversity of new habitats and allowed the great radiation of the Holometabola. While the reasons for the success of this suborder are undoubtedly more diverse and complex than that, Hinton's scenario does emphasize that internal development of wings and appendages probably evolved before the significant divergence of adult and larval morphology.

The evolutionary radiation of the Holometabola is correlated with a progressive divergence of morphology, behavior, and ecology of the larval and adult forms. Divergence of morphology has been most severe in the larval stages; adult Holometabola have a conventional insect anatomy, but their larvae (caterpillars, maggots, grubs, etc.) often have a highly derived and specialized morphology. Such a morphological divergence of two stages in the life cycle can occur only if a workable mechanism exists to manage the transformation between the two forms. Holometabola have done this by physically discarding much of the larval morphology at metamorphosis and by building the adult form from undifferentiated cells held in reserve for that purpose. The pupal stage evolved as a quiescent instar during which these cells could grow into the wings and appendages on the outside of the animal while allowing the musculature required to move those appendages to develop within (Hinton, 1963).

Holometabolous metamorphosis evolved fairly early in insect evolution, during the upper Carboniferous (Carpenter, 1976). The long intervening period of divergence and specialization and the absence of metamorphic information in the fossil record preclude any possibility of reconstructing the path of the evolution of metamorphosis. Phylogenetic analysis methods (Hennig, 1965; Wiley, 1981) should eventually allow us to reconstruct the relationships among modern metamorphic mechanisms, and from these we might be able to deduce the general features of a primitive holometabolous metamorphosis.

An interesting evolutionary problem that arises in holometabolous metamorphosis is one of homology of the larval and adult appendages. Since the adult form is most conservative and therefore probably primitive, the appendages of adult holometabola are *phylogenetically homologous* to those of the Hemimetabola. But since adult appendages arise anew from imaginal disks, they are not *developmentally homologous* to either the larval appendages or those of the Hemimetabola. This paradox of homologies suggests that it will be necessary to reconstruct the evolution of metamorphosis by tracing the relationships among morphologies as well as the relationships among developmental processes.

Allometric Consequences of Holometabolous Metamorphosis

When two parts of an insect's body grow at different rates, or grow with different coefficients, their relative sizes will change. This results in a gradual change of the proportions of those parts, which is most easily detected as a change in shape. The relative growth of different body parts is called *allometry*, and this is also the name for the study of relative growth and its morphological consequences.

Under conditions of continuous growth, allometry among two body parts can be expressed by the relationship $y = ax^b$, in which x and y are the dimensions of two body parts, a is a constant, and b is the *allometric constant* whose dimension is determined by the ratio of the growth constants of the two body parts. This formula was derived by Huxley (1972) and is used extensively in the study of relative growth and morphology in insects and other animals (Gould, 1966; Wilson, 1971; Sweet, 1980; Schmidt-Nielsen, 1984).

When two body parts grow isometrically, there is no change in form with increasing size and the allometric constant, b, is equal to one (the growth constants of both body parts are identical). When insects change their form as they grow, this change can almost always be attributed to changes in the relative sizes of various parts of their body, seldom to the loss of parts or the addition of novel structures. Therefore, most changes in form are allometric and the allometry equation is a satisfactory descriptor of this change in form. The allometric growth equation, $y = ax^b$, can also be written as: $\log y = \log a + b \log x$. This means that when the logarithms of sizes of the two body parts, x and y, are plotted against each other, one obtains a straight line whose slope is the allometric constant, b. Conversely, when the dimensions of two body parts are found to be related as a straight line in double logarithmic plots, this is usually taken as evidence that simple differences in relative growth of parts accounts for the observed changes in morphology during growth.

Double logarithmic plots are also used to study the relative sizes of body parts among adult insects of different sizes. The mandibles of stag beetles (Lucanidae), the horns and forelegs of many scarab beetles, and the heads of ants are much larger relative to the rest of the body in large specimens than in small ones (fig. 4.14A,B). These changes in the relative sizes of

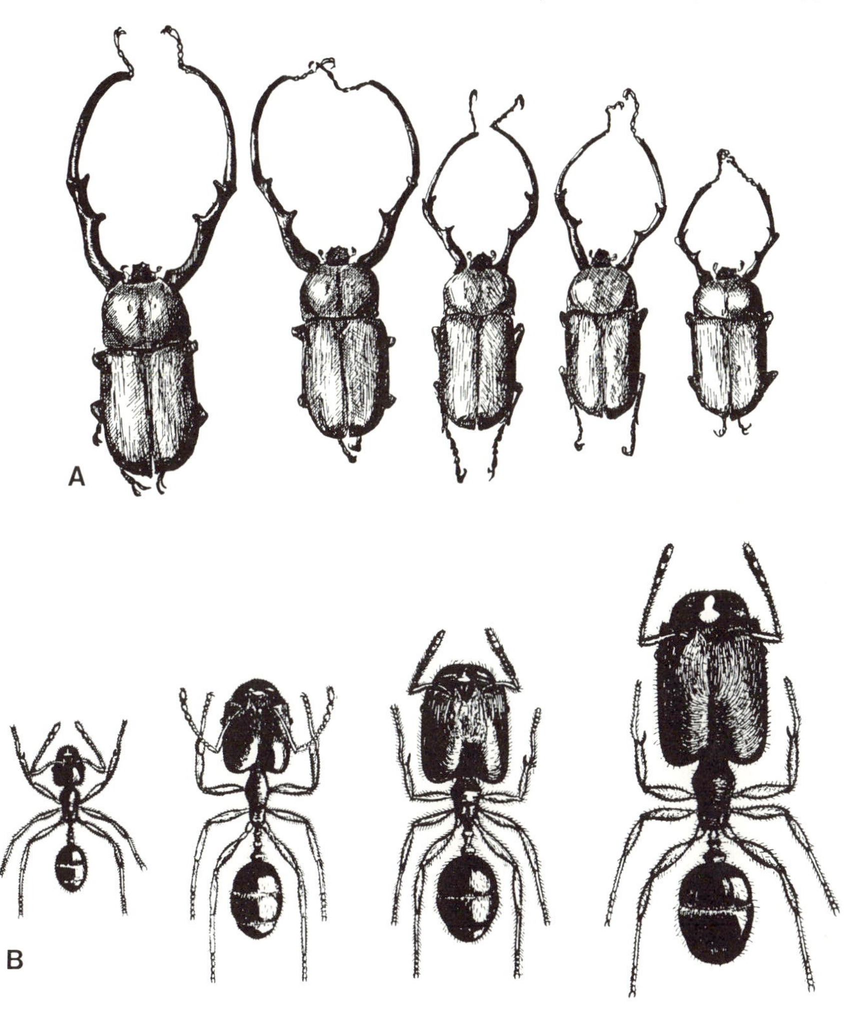

Figure 4.14. (A) Allometry of forelegs in male *Euchirus longimanus* (from Huxley, 1972. Reprinted with permission of Dover Publications.) (B) Allometry of head sizes in the ant, *Pheidole kingi* (from W. M. Wheeler, 1910. Reprinted with permission of Columbia University Press.)

body parts with increasing overall size suggest that the appendages or head must have grown at a higher rate than the rest of the body, so that in animals that have a longer growth period, and which therefore metamorphose at a larger size, they have become disproportionately large. When the sizes of such body parts are plotted as a function of overall body size, one usually gets relationships that look exponential but that seldom form straight lines in a double logarithmic plot (fig. 4.15; see also Wilson, 1971; Huxley,

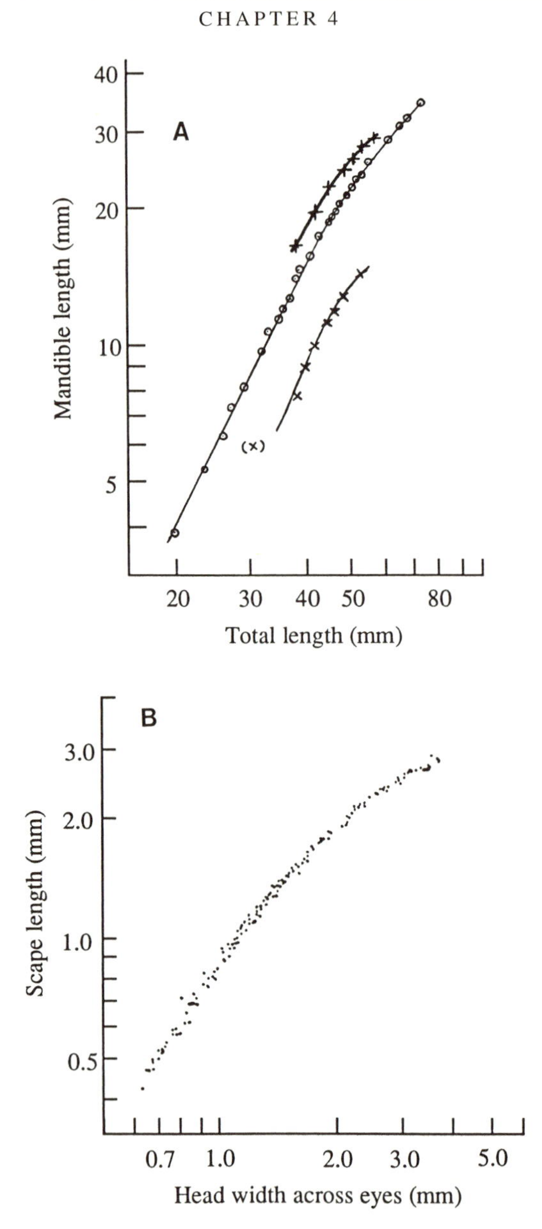

Figure 4.15. (A) Allometric relations of mandible length to body size in three species of stag beetles (Lucanidae), illustrating a diminishing coefficient at large body sizes (from Huxley, 1972. Reprinted with permission of Dover Publications.) (B) Curvilinear allometric relation between the length of the antennal scape and head width in the ant, *Atta texana* (from Wilson, 1953).

1972). The reason for this deviation from the expected linear relationship is that the growth of insects, and in particular the growth of holometabolous insects, violates many of the underlying assumptions of the allometry equation.

The allometry equation assumes continuous and simultaneous growth of the two body parts under consideration. This is seldom the case in insects because the sclerotized parts of their body usually grow discontinuously. For most practical purposes, however, the episodic growth of insects has little or no effect on the allometry relationships of parts, as long as their body parts grow by the same relative amount during each molting cycle. The adult appendages of holometabolous insects do not grow together with the body, however, and the relationship of their size to the overall body size is much more complex than that suggested by the allometry equation.

The imaginal disks that will form the adult appendages grow very slowly during most of larval life, and do not begin their period of rapid growth until late larval life, after the larva is already fully grown. Most of the growth of imaginal disks, in fact, occurs after the larva itself has stopped feeding and growing. This means that the final size of the imaginal disks is a complex function of the final body size because they grow in a closed system; growth of the disks must be at the expense of the rest of the body and, therefore, at the expense of overall body size. Huxley (1972) suggested that the decrease in the slope of the allometry curve of imaginal disk structures at large body sizes (fig. 4.15A) results from an increasing competition for a limited supply of nutrients of a rapidly growing disk so that, as size becomes larger, the disk falls ever farther below its theoretical maximum.

The real situation is likely to be even more complex than that. The rapid growth phase of the imaginal disk immediately prior to pupation is not known to be appreciably longer in large individuals than in small ones. Therefore the disproportionate differences between the appendages of large and small insects must be due to (1) differences in the *duration* of the early period of slow growth of the imaginal disks, so that different disks start their rapid growth phase at different sizes, and (2) to differences in the *rate* at which they grow during their rapid growth phase immediately prior to metamorphosis. A small size advantage at the outset of a period of exponential growth could result in a dramatic difference in final sizes. It is therefore not unreasonable to assume that the early period of growth can also play an important albeit indirect role in the allometry of holometabolous insects. We do not yet know enough about the patterns of relative growth of imaginal disks throughout larval and pupal life, nor about the possible competition among imaginal disks for nutrients during their rapid growth phase, to evaluate the relative significance of the many variables involved in the relative growth of imaginal disk-derived structures. A comparative understanding of the control of growth and development of imaginal disks remains one of the great gaps in our understanding of insect development. For our present purposes, however, it is important to note that the allometric constants by which disks grow can be affected by the developmental hormones (see chapter 8). Consequently, certain changes in shape during development and evolution are intimately tied into the insect's endocrine control mechanism.

Physiological Control of Metamorphosis

The onset of metamorphosis, like that of the molting cycle, is controlled by hormones. The mechanisms of hormonal regulation will be discussed in detail in chapter 5. Here we will be concerned only with the nonendocrine physiological processes by which insects determine the timing of metamorphosis in the course of their life cycle. For all practical purposes these constitute the mechanisms by which insects regulate the secretion of the hormones associated with metamorphosis.

Under optimal conditions of growth, most insects undergo a species-specific number of larval molts and then metamorphose into adults of a species-specific size. When conditions are not optimal, both the number of larval molts and the size after metamorphosis can vary. The manipulation of growth rates to examine the pattern of this variability has revealed much about the physiological control of metamorphosis, although it is clear that a vast amount must still be learned before we can claim fully to understand the control over this important process.

One of the consequences of the exponential growth of insects is that, no matter how many instars an insect goes through, most of the growth occurs during the last larval instar. This also means, of course, that the regulation of final size of the adult is determined in large measure (though not exclusively) by the amount of growth that occurs during the last larval instar. The control of metamorphosis can therefore be divided into two somewhat independent processes: (1) the control over *when* in the final instar to initiate the metamorphic molt (and thus effectively cease growth), and (2) the control over the number of larval instars (i.e., how does an insect know that *this* is final larval instar?). Both processes are normally controlled within fairly narrow limits, even though they can be experimentally altered.

The control of metamorphosis has been best studied in the Lepidoptera. When we discussed the control of molting in *Manduca* we noted that the metamorphic molt is triggered when the larva reaches a well-defined critical weight. At this critical weight the corpora allata cease to secrete juvenile hormone, and this endocrine switch is the first step in a cascade of endocrine and physiological events that result in the metamorphic molt (Nijhout and Williams, 1974b). Under normal conditions in *Manduca* the critical weight is 5 grams. Since it takes about 24 hours for the JH titer to decay completely, and an additional number of hours before the next gate for PTTH release opens (see p. 91), growth actually continues for one to two days after the critical weight is attained. During this period the larva can double its weight, growing to nearly 10 grams. The size of the adult is determined by the weight attained at the time PTTH is released an the molt starts. If the larva grows little after attaining the critical weight, it will metamorphose at a smaller size and form a smaller adult than a larva that grows vigorously during the period after attaining its critical weight.

Thus in *Manduca* the smallest size at which a larva can metamorphose is given by the critical size for molting during the last larval instar. We have seen that this critical size (corresponding to a weight of 5 grams under optimal conditions) is actually a function of the size of the cuticle that was established during the molt *to* the last instar (fig. 4.5). Therefore, the actual size at which metamorphosis begins and the resulting size of the adult are determined by two variables: the critical size, which is influenced by the pattern of growth in the early instars, and the growth that takes place between the time the critical weight is attained and the time the pupal molt actually begins—a period of one to several days.

The mechanism by which a larva determines during which instar the corpora allata will be shut off is not yet fully understood. It is clear that in most and perhaps all insects the number of larval molts that precede metamorphosis is an indeterminate character. If early larval growth is poor the final instar may be reached after a large number of molts, with little growth between each, whereas if larval growth is optimal the growth steps at each molt are larger and the final instar is reached after fewer molts. Whether or not a given instar is to be the final instar is determined at the time of ecdysis to that instar, and this decision is somehow associated with the size at which that molt takes place. When early larval growth of *Manduca* is manipulated experimentally, by feeding animals on nonnutritive diets, or by temporary or periodic starvation, it is possible to cause them to molt at a size much lower than that predicted by Dyar's rule. By varying growth in this way it is possible to make larvae molt over a continuous range of sizes. When the fate of those larvae is examined it is evident that there exists a very sharp size threshold for metamorphosis that is related to the size at which the preceding larval molt occurs (Nijhout, 1975b). Larvae that are smaller than this threshold size will undergo a normal larval molt when next they molt, and those that are larger than this threshold size will shut off their corpora allata and initiate the metamorphic molt during that instar (fig. 4.16). The size of an instar is traditionally given by the width of the

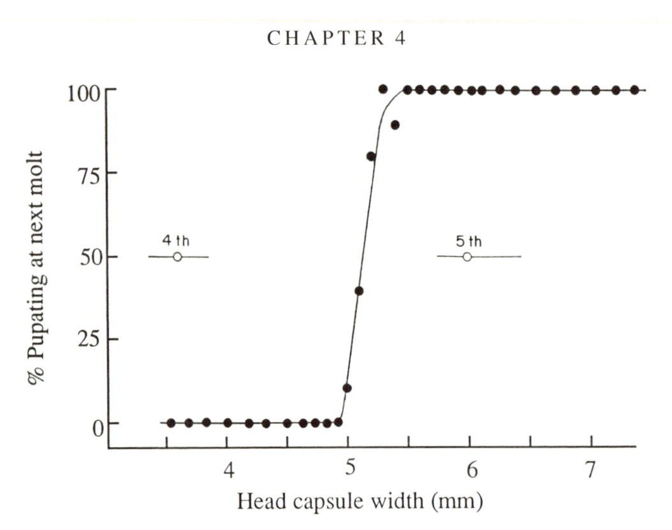

Figure 4.16. Relationship between larval size (indexed by head capsule width) and metamorphosis in *Manduca sexta*. Larval size was manipulated by varying the nutrition of larvae. Larvae with head capsule sizes larger than about 5 mm pupate at the next molt, whereas larvae with smaller head capsules molt to an additional larval instar. Horizontal bars indicate the mean and range of sizes for normal fourth (penultimate) and fifth (final) instar larvae. (From Nijhout, 1975b.)

head capsule, a sclerotized portion of the cuticle that does not grow during the intermolt. By this measure of size, the threshold size for metamorphosis of *Manduca* occurs at a head capsule width of 5.1 mm.

Unlike the critical size for molting, which is a function of body size at the outset of the instar, the threshold size for metamorphosis appears to be an absolute measure. It is completely unaffected by the prior growth history of the animal, as well as the number of molting cycles that have preceded (fig. 4.17). It is not known how the threshold size for metamorphosis is actually measured by the larva. Whatever the physiological process behind the threshold is, it is clearly very accurate and tightly correlated with the size of the animal's cuticle. An interesting consequence of having a fixed critical size for metamorphosis is that starvation early in larval life can result in gigantism at metamorphosis. This paradoxical physiological effect has been noted in a number of insects (Wigglesworth, 1972) and is explained by the fact that the normal size of the penultimate instar is quite far below the threshold for metamorphosis (figs. 4.16, 4.17). Intercalation of an instar due to poor early growth can therefore cause the penultimate instar to be only just below threshold and therefore somewhat larger than a normal penultimate instar. If subsequent growth follows the typical coefficient, size at metamorphosis will be much greater than usual, as shown in figure 4.17.

In larvae of *Galleria* the instar in which metamorphosis occurs is also

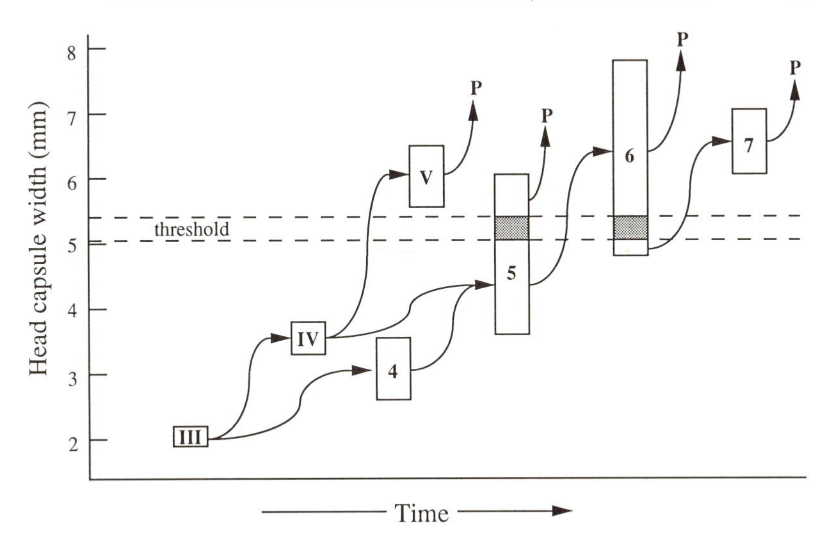

Figure 4.17. Diagrammatic representation of the fate of larvae of *Manduca sexta*, growing under various conditions. Boxes indicate the range of sizes (indexed by head capsule width) in each instar. Roman numerals represent the values in larvae growing under optimal conditions. Such larvae typically pass the threshold size for metamorphosis between the 4th and 5th instar and pupate (P) after the fifth instar. When larvae grow under suboptimal conditions (Arabic numerals), they grow more slowly and exhibit a broader range of sizes, so that only a portion of fifth instar larvae are above threshold. The remaining larvae undergo one or more supernumerary larval molts and pass through a sixth (and seventh) instar before pupating. Animals whose size is barely subthreshold molt to an unusually large final larval instar and eventually form giant adults. (Redrawn from Nijhout, 1975b.)

determined by a threshold mechanism, but in a slightly different way from that in *Manduca* (Allegret, 1964). When larvae of *Galleria* are fed on a protein-free diet before they reach a weight of about 20 mg, they continue to undergo larval molts but do not grow and never metamorphose. Larvae that are fed a protein-free diet after they have passed this critical weight also cease growth, but these larvae all undergo metamorphosis during the current instar, or after one additional larval molt. The possible existence of a critical cuticular size has not yet been investigated in *Galleria*. But the fact that the control over the onset of metamorphosis here appears so far to be somewhat different from that in *Manduca* suggests that different species may have evolved variations on an ancestral mechanism by which they determine which larval instar is to be the last.

Larval heteromorphosis in *Manduca* is also controlled by a threshold mechanism. Early larval instars of *Manduca* have an array of white tubercles on the cuticle of their head capsule, while later instars have head cap-

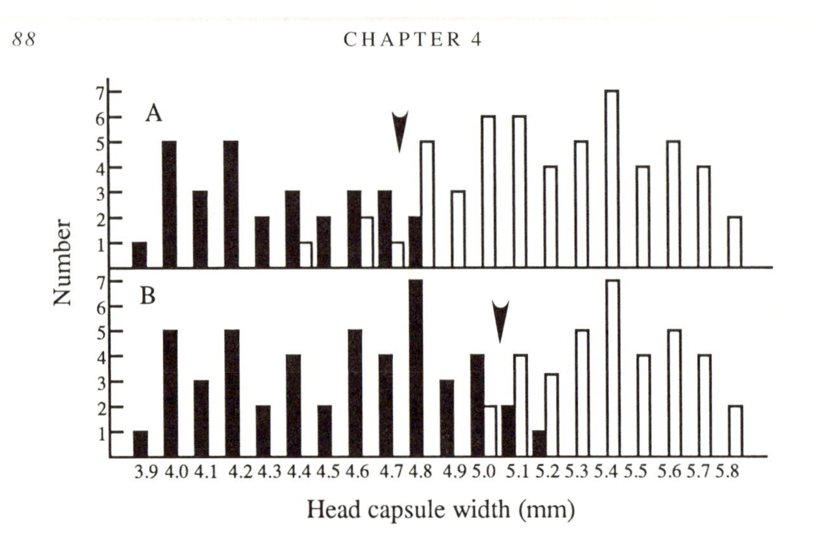

Figure 4.18. Heteromorphosis in larva of *Manduca sexta*. (A) Histogram showing the transition of surface sculpturing of head and thorax in larvae of different sizes: *black*, larvae with white tubercles on their integument typical of the fourth instar; *white*, larvae with a smooth green integument typical of the 5th instar. (B) Histogram showing fate of the same larvae as in (A) at next molt: *black*, larvae that molt to a supernumerary larval instar; *white*, larvae that pupate at next molt. The heteromorphic change of the integument occurs at a different size threshold than metamorphosis (arrows) so that it is possible to have penultimate instar larvae with the cuticle sculpturing of the final instar, which indicates that larval heteromorphosis is not controlled by the same mechanism as metamorphosis. (Redrawn from Nijhout, 1975b.)

sules that are very smooth and green. This heteromorphic transformation occurs at a very sharp size threshold (fig. 4.18). The threshold size for heteromorphosis is not the same as the threshold size for metamorphosis, and the physiological mechanisms that control heteromorphosis (in *Manduca* or any other species) have not yet been discovered.

CHAPTER 5

THE ENDOCRINE CONTROL OF MOLTING AND METAMORPHOSIS

THE ONSET of a molting cycle and the initiation of metamorphosis are indeterminate events in the life of most and perhaps all insects. Both processes are tightly linked to growth and nutrition; insects do not molt by the clock, nor do they count their instars to determine when they will metamorphose, even though under optimal growing conditions the molting cycle occurs with predictable regularity, and the time of metamorphosis can often be predicted to the nearest day. Natural conditions are seldom optimal, and insects have evolved mechanisms to adjust the timing of their molts and the onset of their metamorphosis to their pattern of growth. The physiological variables by which insects assess their growth and their size are thought to be very diverse. Some have undoubtedly evolved as specializations for particular modes of life, while others may have evolved by cueing in on any of a number of convenient physiological correlates of growth or size.

Only in the Hemiptera do we know the ultimate stimulus for molting, namely stretch of the abdomen, as we saw in Chapter 4. The ultimate stimulus for metamorphosis is still not known for any insect. In some Lepidoptera the onset of metamorphosis can be shown to be tightly correlated with the attainment of a particular critical size, but how these insects are able to "measure" their size is not at all clear. A host of physiological and environmental variables has been shown to affect the timing of both molting and metamorphosis. Among these are temperature, humidity, photoperiod, contact with specific substrates (thigmotropism), injury, the presence or absence of specific chemicals or nutrient in the food, pheromones, and juvenile hormone (the effects of these are reviewed in Wigglesworth, 1965; Nijhout, 1981; Nijhout and Wheeler, 1982; Sehnal, 1985; Steel and Davey, 1985). Depending on the species, variation in these parameters can accelerate or delay the onset of molting or metamorphosis in complex ways that are as yet poorly understood. The one thing that these variables have in common, however, is that they directly or indirectly affect the secretion of either the prothoracicotropic hormone (PTTH) or the juvenile hormone (JH). These two hormones are the immediate or proximate regulators of the onset of molting and metamorphosis, respectively, in all insects. The manner in which PTTH and JH regulate the life cycle of insects has been the subject of intense interest for more than half a century, but it is only in the past two decades that details of the mechanism of their action and the cascades of events they set in motion have become evident.

THE HEAD CRITICAL PERIOD AND PTTH SECRETION

Every molting cycle, whether it is a larval, pupal, or an adult molt, begins with the secretion of PTTH by the brain. PTTH stimulates the secretion of the molting hormone, ecdysone, and it is the action of ecdysteroids on the epidermal cells that is responsible for stimulating the biochemical and morphological events of the molting cycle. Once ecdysone secretion is underway, the secretion of PTTH stops and the molting cycle continues without further need for PTTH. Good general assay methods for PTTH have only recently become available, and PTTH titers have not been measured directly in most insects. Most studies on the control of molting in insects begin with an investigation of the time at which the molt becomes independent of the brain. This is most easily done by removing the brain by decapitation or by placing a blood-tight ligature around the neck at progressively later times in a molting cycle. At a certain point in time a few, and then an increasingly larger percentage of such animals will undergo a normal molt. The point in time at which half the animals are able to complete a normal molt in the absence of their brain is called the *head critical period.*

One of the earliest and clearest demonstrations of the head critical period was provided by Wigglesworth (1934). He decapitated larvae of *Rhodnius* at progressively later times after giving them a blood meal and showed that the molt became independent of the head about three days after the blood meal in fourth instar larvae and about seven days after the blood meal in fifth instar larvae (fig. 5.1). It is assumed that PTTH secretion in *Rhodnius*

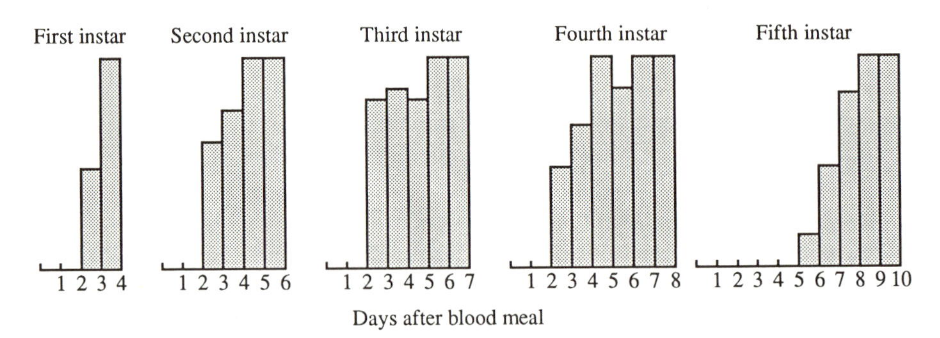

Figure 5.1. Histograms of head critical periods for the five larval instars of *Rhodnius prolixus*. Bar height indicates percentage of larvae which molted when decapitated at various times after a blood meal. (Redrawn from Wigglesworth, 1934.)

begins very soon after the blood meal because ecdysteroid titers in the hemolymph begin to rise shortly after a meal is taken (Steel et al., 1982; Knobloch and Steel, 1987). This means that the onset of a molting cycle in *Rhodnius* requires a three- or a seven-day period of PTTH secretion in fourth and fifth instar larvae, respectively.

Truman (1972) has shown that the timing of PTTH secretion during a larval molt in *Manduca sexta* is gated (controlled) by the photoperiod. In the fourth larval instar, the gate during which PTTH secretion is allowed extends for a 10-hour period from about 1.5 hours before lights-off to about 8.5 hours after lights-off, under a 12L:12D photoperiod. The head critical period must therefore occur sometime during this gate. The exact position of the head critical period within the gate depends on how early in the gate PTTH secretion actually starts, and that is determined by the prior growth history of the larva (Nijhout, 1981).

The duration of PTTH secretion can generally be deduced by assuming that it occupies the interval between the first appearance of ecdysteroids in the hemolymph and the head critical period. Direct measures of PTTH have been obtained by means of a bioassay that measures the stimulated production of ecdysteroids by prothoracic glands in tissue culture (Bollenbacher et al., 1980). By means of such an assay, the period of PTTH secretion during molting in *Manduca* has been shown to be only a few hours in duration (Bollenbacher and Granger, 1985). Thus in *Manduca*, PTTH is present rather briefly and acts like a trigger, while in *Rhodnius* it seems to act in a more sustained fashion in the stimulation of ecdysteroid secretion.

The Prothoracicotropic Hormone

The PTTH is produced by neurosecretory cells in the brain and regulates the secretion of ecdysteroids by the prothoracic glands. PTTH is the first hormone discovered in insects and the first neurosecretory hormone discovered in any animal (Bollenbacher and Granger, 1985). The endocrine role of the insect brain was first recognized by Kopeć (1917, 1922). Kopeć studied the control of molting in the gypsy moth, *Lymantria dispar*. He noted that when a ligature was placed around the middle of the body of early final instar larvae, the anterior portion pupated normally but the posterior portion never did. Kopeć concluded that something in the anterior half of the body was needed for normal metamorphosis. Removal of the brain from a mid-final instar larva likewise inhibited pupation, but removal of the subesophageal ganglion did not. From these experiments Kopeć concluded that the brain was the source of the metamorphosis-stimulating factor, and that the brain did not exert this effect via nerves, since removal of the subesophageal ganglion effectively severed the nervous connections between the brain and the rest of the body.

The endocrine function of the brain was confirmed by Wigglesworth (1934). Wigglesworth parabiosed decapitated nymphs of the bloodsucking bug *Rhodnius prolixus*, which would normally never molt, to intact nymphs that had previously been given a blood meal to induce a molting cycle. He showed that a decapitated (brainless) nymph underwent a molting cycle in parallel with its parabiotic twin. Thus a soluble molt-stimulating factor was transferred to the headless nymph via their shared circulatory system. Wigglesworth (1940) eventually traced the source of this factor to the brain. By implanting various portions of the central nervous system into suitable larvae, Wigglesworth showed that only the anterior central portion of the brain possessed molt-inducing activity. The complete endocrine pathway for molting was elucidated by Williams (1947, 1948a,b) working on the silkmoth, *Hyalophora cecropia.* Williams showed that when "active" brains were implanted into bisected pupae of this moth, the anterior portion underwent a normal molt to the adult stage while the posterior portion remained pupal. He then found that the posterior portions could be induced to molt if they received implants of active brains *and* prothoracic glands. Williams concluded that molting and development required endocrine factors from both the brain and prothoracic glands, and that the factor from the brain, initially called the "brain hormone," stimulated the prothoracic glands to produce the actual molting hormone. Subsequent studies by many other investigators have established that the brain-prothoracic gland axis controls molting in all the insects. When it became clear that the insect brain actually secretes many different kinds of hormones, the "brain hormone" was renamed the prothoracicotropic hormone (PTTH) (Box 3). Its only known function is to stimulate the prothoracic glands to secrete the molting hormone.

The PTTH of *Manduca* and *Bombyx* has been purified and partially characterized. In *Manduca* PTTH exists in two molecular forms: big PTTH with an approximate molecular weight of 28,500 D, and small PTTH with a molecular weight of about 7,000 D (Bollenbacher and Granger, 1985). These two forms of PTTH appear to be molecularly unrelated to each other and also differ slightly in their biological activity. When tested in vitro, small PTTH is more effective in stimulating larval prothoracic glands than pupal glands, but whether the two forms have a differential function in vivo is unknown (Bollenbacher and Granger, 1985). Adult PTTH may be yet a third form of the hormone and is involved in stimulating ovarian ecdysteroid production and vitellogenin synthesis, perhaps analogous to the ovarian ecdysteroidogenic hormone (Bollenbacher and Granger, 1985; Kelly et al., 1986; and see chapter 6).

The PTTH of *Bombyx* is also heterogeneous. A small PTTH of about 5,000 D has been isolated form the brains of adult moths, while a much

BOX 3

PROTHORACICOTROPIC HORMONES
(PTTH)

STRUCTURE AND NOMENCLATURE: The prothoracicotropic hormones are neurosecretory polypeptides. The PTTH of *Manduca sexta* occurs as two classes of molecules: the so-called small-PTTH with a molecular weight of about 7 kD, and the so-called big-PTTH with a molecular weight of about 28 kD. Both small and big-PTTH occur as families of structurally related molecules. Larval prothoracic glands are equally sensitive to small-PTTH and big-PTTH, but pupal glands are much more sensitive to big-PTTH. Big-PTTH undergoes complex processing: the PTTH gene codes for a large polypeptide that is cleaved into three smaller peptides, the largest of which (12 kD) forms dimers that constitute the active hormone. The active PTTH of *Bombyx mori* is also a homodimer, but of slightly larger molecular weight. Partial amino acid sequences have so far revealed no sequence similarities between the *Bombyx* and *Manduca* PTTHs, suggesting they are unrelated molecules (Muehleisen et al., 1993).

SOURCES: Big-PTTH of *Manduca* is produced primarily by a single pair of lateral neurosecretory cells of group III, while small-PTTH is produced by a cluster of four large medial neurosecretory cells of group IIa_2 (figs. 1.5, 5.2). The *Bombyx* PTTH is produced by a pair of lateral neurosecretory cells, possibly the same as those that produce small-PTTH in *Manduca*.

REFERENCES: Agui et al., 1979; Bollenbacher and Gilbert, 1981; Bollenbacher et al., 1984; Iwami et al., 1990; Kawakami et al., 1989, 1990; Mizoguchi et al., 1987, 1990a,b; Nagasawa et al., 1984, 1990; O'Brien et al., 1988; Westbrook and Bollenbacher, 1990.

larger one, of about 22,000 D, has been isolated from pupal brains (Yamazaki and Kobayashi, 1969; Ishizaki and Ichikawa, 1967; Nagasawa et al., 1990). The larger molecule is the active PTTH of *Bombyx*. The smaller molecule is probably not a true PTTH and is now called *bombyxin*. Bombyxin has been shown to consist of a family of polypeptides with extensive sequence homology to vertebrate insulins and insulinlike growth factors (Iwami et al., 1989). Bombyxins are produced by a group of medial neurosecretory cells in the brain. Interestingly, bombyxin has a strong prothoracicotropic effect in the saturniid moth *Samia cynthia*, but has only a very weak effect on the prothoracic glands of *Bombyx* and *Manduca*. The normal physiological role of bombyxin is still unclear. Mizoguchi (1990) has suggested that there may be two independent evolutionary developments of PTTH function, one of the smaller bombyxin-like molecules from the medial neurosecretory cells, the other of larger big-PTTH-like molecules from the lateral neurosecretory cells, and that different species (or different stages in one species) use one or the other molecule as their principal control of the prothoracic glands. The fact that bombyxin stimulates *Bombyx* prothoracic glands only at a concentration some 800 to 1600 times that required for normal stimulation in *Samia* suggests that this weak effect in *Bombyx* is pharmacological, and that its prothoracicotropic action may be only a laboratory phenomenon (Kiriishi et al., 1992). Based on its structural similarity to insulinlike growth factors, bombyxin may prove to have an important role in the control of cell growth and differentiation.

The neurosecretory cells that produce PTTH in *Manduca* have been identified using an antibody to PTTH (fig. 5.2). PTTH in this species is primarily produced by a single pair of neurosecretory cells in the lateral portion of the brain, the cells of group III (see figs. 1.5 and 5.2; O'Brien et al., 1988; Westbrook et al., 1993). Surgical ablation experiments in other insects such as *Rhodnius*, *Locusta*, *Bombyx*, and *Hyalophora* indicate that PTTH in these species is produced by one or more of the medial neurosecretory cells of the brain, while in *Aeschna* and *Megoura* small groups of neurosecretory cells that lie at a short distance from the medial neurosecretory cells have been implicated in the production of PTTH. Using antibodies to big-PTTH, Westbrook et al. (1993) have shown that in *Manduca*, PTTH can actually be found in several different types of neurosecretory cells, depending on the developmental stage. In the larval and adult stages PTTH is most concentrated in the group III cells but is also detectable in the ventromedial neurosecretory cells of group V (which are also the source of the eclosion hormone; see chapter 9). In the embryo and larval stages PTTH also occurs in neurosecretory cells of the frontal ganglion and subesophageal ganglion. The widespread presence of PTTH (or a sufficiently closely related molecule to be recognized by anti-PTTH antibodies) in different neurosecretory cells and at different developmental stages parallels the findings on the distribution of neuropeptides in vertebrates, where spatial and temporal differences in neuropeptide production may be involved in modulating local developmental and physiological processes. Westbrook et al. (1993) suggest that in insects, too, the ability to release single regulatory peptides in different microenvironments and at different

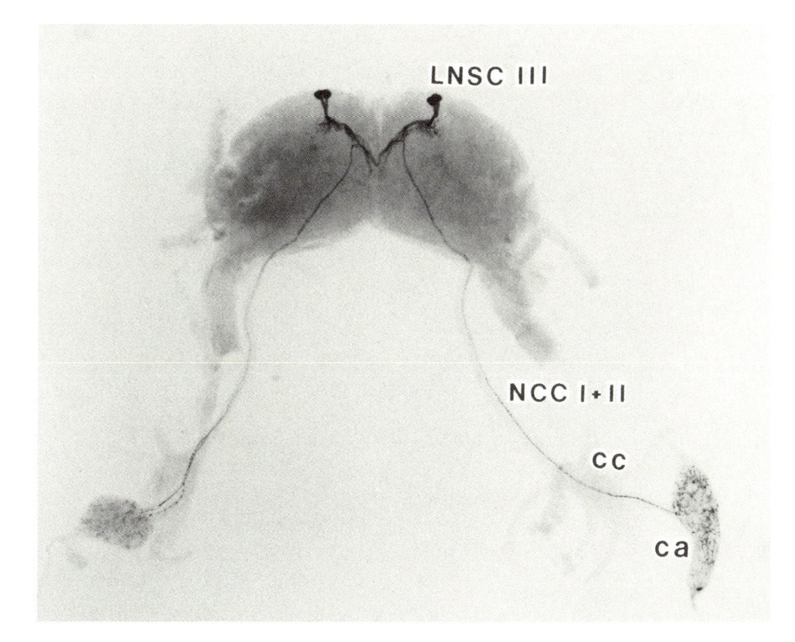

Figure 5.2. Brain-retrocerebral complex of a fifth instar larva of *Manduca sexta*. Immunohistochemical staining with an antibody to PTTH. The lateral neurosecretory cells of group III (LNSC III) are the primary sources of this hormone, which is transported via the nervi corporis cardiaci I and II (NCC I+II) and released from the corpus allatum (ca). The corpus cardiacum (cc) is not a neurohemal organ for PTTH. (Photo courtesy of A. L. Westbrook and W. E. Bollenbacher.)

developmental stages may be an efficient way to control a wide array of physiological and behavioral processes.

The neurohemal organ for PTTH in most insects is the corpus cardiacum (CC). The Lepidoptera appear to be an exception. In *Manduca*, for instance, the neurosecretory axons from the PTTH cells pass through the CC and terminate in the corpus allatum, which serves as the neurohemal site for the hormone (fig. 5.2; Nijhout, 1975a; Agui et al., 1980).

The Role of Ecdysteroids in the Molting Cycle

Ecdysteroid titers in the hemolymph begin to rise almost immediately after the initiation of PTTH secretion. The period of ecdysteroid secretion is usually fairly prolonged, lasting one to five days depending on the species and on the point in the life cycle that the molt occurs. As a rule, the period

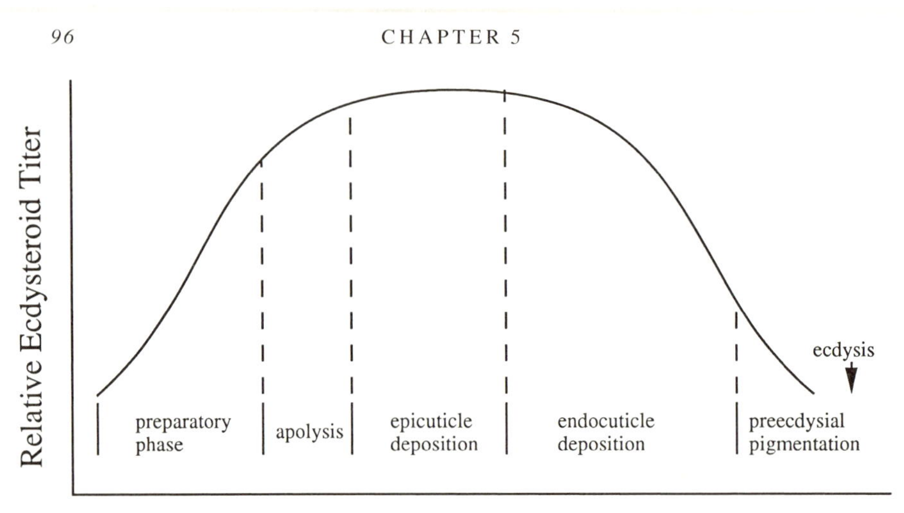

Figure 5.3. Correlation of cellular events during a molting cycle with the rise and fall of the ecdysteroid titer. (Redrawn from Riddiford, 1985.)

of ecdysteroid secretion occupies nearly the entire pharate period of a molting cycle (see chapter 4), beginning shortly before mitosis and apolysis and ending shortly prior to ecdysis. Thus in the Holometabola, ecdysteroid secretion during the pupa-to-adult molt is much more prolonged than during a larva-to-larva molt, in accord with the fact that the entire molting process of adult development during the "pupal" stage (actually the pharate adult) takes much longer than does a typical larval molt.

There is a correlation between the phases of an ecdysteroid peak and the various physiological and morphological events in the molting cycle (Riddiford, 1985, 1989). While there are numerous variations on the general theme, the following generalizations can be made at this time (fig. 5.3). During the early portion of the rising phase of ecdysteroid titer there is a complex pattern of DNA and RNA synthesis in the epidermal cells, accompanied by various cytoplasmic changes, called the "preparatory phase." Mitosis usually takes place during the latter portion of this phase. Apolysis occurs during the latter half of the rising phase of the ecdysteroid peak, and epicuticle deposition occurs at about the time that the maximum of the peak is reached. Exocuticle deposition occupies nearly the entire period of the falling phase of the ecdysteroid peak.

There is, however, some variability in the relationship between the ecdysteroid titer and the molting cycle, even within an individual. The cellular events of the molting cycle are not perfectly synchronous in all parts of the body (see figs. 5.9, 5.10), and, when cultured, different parts of the epidermis have been shown to require different ecdysteroid exposure times in order to molt (Truman et al., 1974; Riddiford, 1985; Kremen,

1989). This means that the succession of events of the molting cycle, although well correlated with the ecdysteroid titers, may not be causally dependent on the rising and falling phases of the ecdysteroid titer. Ecdysteroids evidently do not act as simple triggers that set the molting cycle in motion, yet exactly why the sequence of events in the molting cycle should require such a prolonged presence of ecdysteroids is not fully understood. The declining phase of the ecdysteroid titer at the end of the pupa-to-adult molt is important because it controls several of the late events of adult development. If the decline of ecdysteroids at the end of adult development of *Manduca* is prevented by means of injections or infusions of 20-hydroxyecdysone, it is possible to delay or prevent programmed cell death of certain larva- and pupa-specific neurones in the central nervous system and of the intersegmental muscles, as well as the development of adult pigmentation and the secretion of the eclosion hormone. By contrast, programmed cell death in the central nervous system and intersegmental muscles can be made to occur prematurely if ecdysteroid titers are caused to decline prematurely by abdominal ligation (Truman and Schwartz, 1982; Truman et al., 1983; Schwartz and Truman, 1983, 1984; Riddiford, 1985).

During a larval molting cycle the ecdysteroid titer has a simple profile consisting of a single fairly symmetrical peak (fig. 5.4). During the molt from the penultimate to the last larval instar this peak lasts about three days in *Locusta* and about one day in *Manduca* (Baehr et al., 1979; Bollenbacher

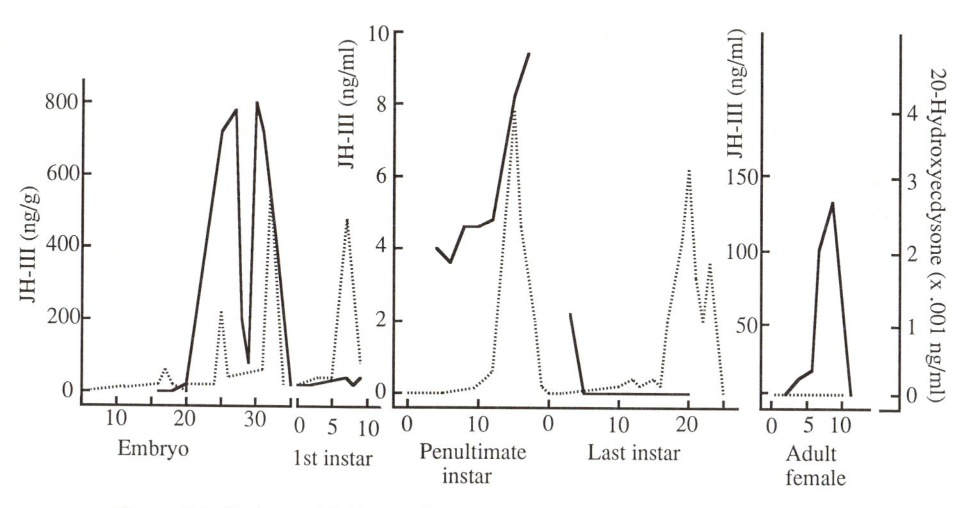

Figure 5.4. Ecdysteroid (dotted lines) and juvenile hormone (solid lines) titers during development in the cockroach *Nauphoeta cinerea*. Ecdysteroid titers are in 20-hydroxyecdysone equivalents, and JH titers are in JH-III equivalents. In the embryo and first larval instar, the JH titer is given per gram live weight, at other times the JH titer is per ml hemolymph (hemolymph volume in ml is about 21% of wet weight in grams). (After Lanzrein et al., 1985.)

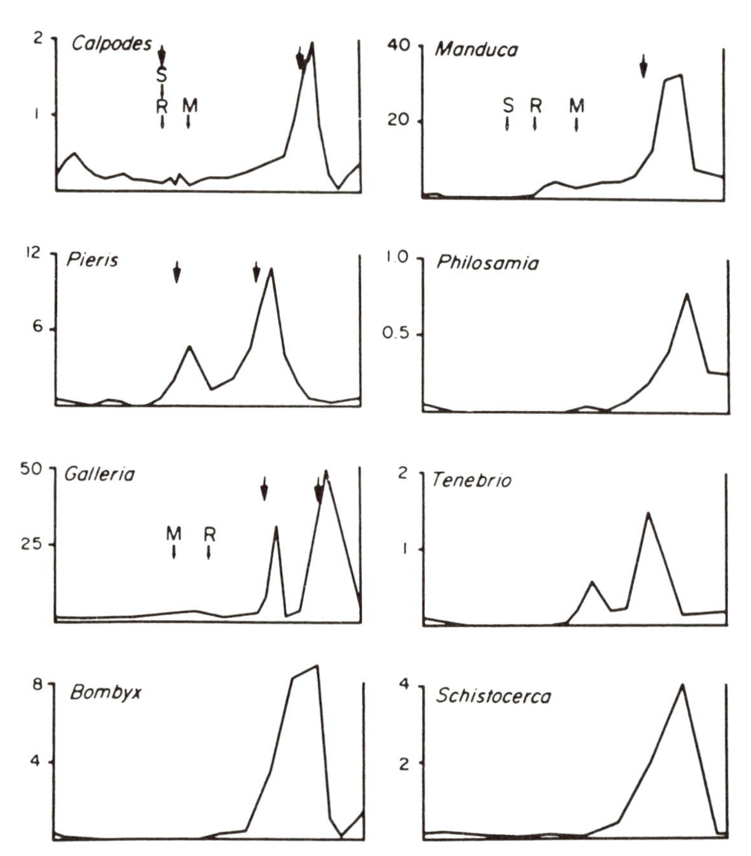

Figure 5.5. Diagrammatic representation of the ecdysteroid titers during the last larval stage of selected insects. Arrows indicate consecutive brain and prothoracic gland critical periods. S, onset of DNA synthesis; R, reprogramming of epidermis; M, onset of mitoses in epidermis. (From Dean et al., 1980. Reprinted with kind permission of Pergamon Press Ltd.)

et al., 1981). Prior to the first metamorphic molt, however, the ecdysteroid profile is more complicated. In several species of Holometabola the molt to the pupal stage is preceded by two peaks of ecdysteroid secretion, and so is the molt to the adult stage in some Hemimetabola (figs. 5.4 and 5.5). In *Rhodnius*, for instance, the ecdysteroid titer rises to a small peak within four hours after the blood meal and remains at a low but significant plateau for 6–8 days (fig. 5.6). This first peak does not initiate the molt but induces a series of preparatory changes in the epidermal cell (Wigglesworth, 1957), which are believed to be required for the subsequent expression of adult-specific genes.

A second, much larger peak of ecdysteroid appears shortly after the head critical period is over (fig. 5.6). This second peak induces apolysis and the

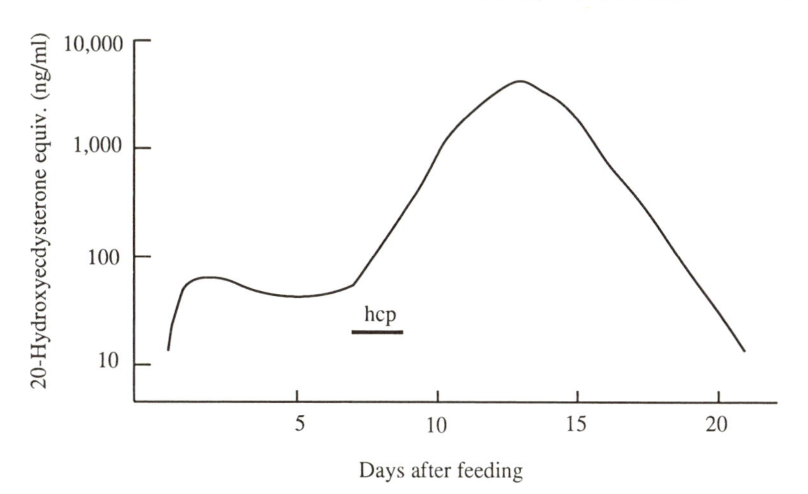

Figure 5.6. Ecdysteroid titers after a blood meal in fifth (final) instar larvae of *Rhodnius prolixus*. Blood meal is taken on day 0, and ecdysis is on day 20; hcp, head critical period, after which the molt is independent of PTTH secretion by the brain. (After Steel et al., 1982.)

onset of the actual molt. The prior "activation" period of the epidermis by the first peak and plateau of ecdysteroid is necessary to switch the developmental fate of the epidermis from a larval commitment to that of the adult. If the timing of the second peak of ecdysteroid is accelerated, by injecting a large dose of 20-hydroxyecdysone before the fifth day after the blood meal, the epidermis manufactures a larval cuticle, whereas 20-hydroxyecdysone injections after the fifth day result in the manufacture of an adult cuticle (Steel et al., 1982). The manner in which the small initial peak and plateau phase of ecdysteroid secretion are regulated is not known. Steel et al. (1982) suggest that an independent early peak of PTTH is responsible for stimulating this secretion. If this is true, then there may be two peaks of PTTH secretion after a blood meal in the last larval instar, with the first one triggered by the blood meal, and the second one following as a programmed response to the first.

The larval-pupal molt in *Manduca* is likewise preceded by two peaks of ecdysteroid secretion (figs. 5.5, 5.7, top). As in *Rhodnius*, the first peak does not cause the molting cycle to begin but is required to switch the state of determination, or commitment, of the epidermis. Prior to the first peak the epidermis is committed to produce larval characters, while after the peak it is committed to produce pupal characters (Truman et al., 1974; Riddiford, 1978, 1981; Riddiford and Hiruma, 1990). The second peak of ecdysteroids, which begins about two days after the first, is much larger and induces the actual molting cycle that produces the pupa. Each peak is stimulated by a different pattern of PTTH secretion. The first peak of ec-

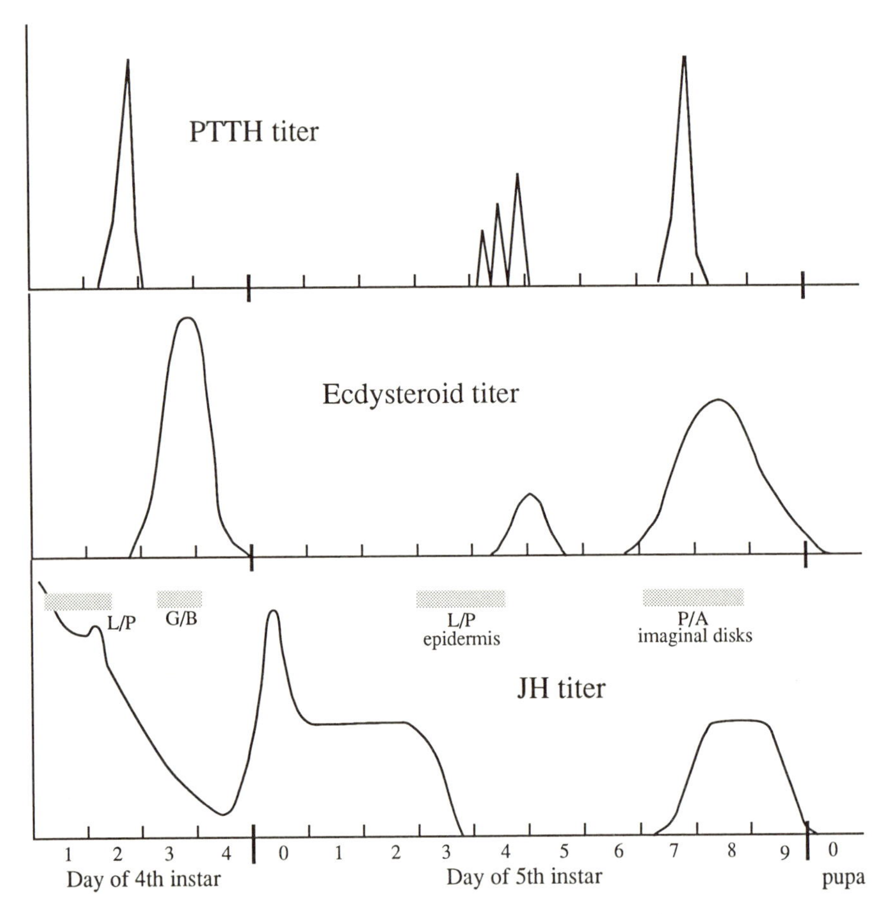

Figure 5.7. Hormone titers during the last two larval instars of *Manduca sexta*. The titers for PTTH, ecdysteroids, and juvenile hormone are shown on a common time scale. Gray bars indicate the approximate timing and duration of the JH-sensitive periods for larval versus pupal determination (L/P), for green versus black larval pigmentation (G/B), and for pupal versus adult determination (P/A). The epidermis and imaginal disks have independent times of commitment to pupal and adult development, as explained in the text. (Based on data from Truman et al., 1973, 1974; Nijhout and Williams, 1974b; Fain and Riddiford, 1975; Riddiford and Truman, 1978; Bollenbacher and Gilbert, 1981; Bollenbacher et al., 1981; Curtis et al., 1984.)

dysteroid is preceded by a short sequence of small pulses of PTTH, while the second peak is preceded by a standard single large pulse of PTTH (fig. 5.7 top).

The switchover to pupal commitment during the larval-pupal molt in *Manduca* requires ecdysteroids, but only occurs in the absence of juvenile hormone (JH) during the last larval instar. Reprogramming of the epidermal cells and some of the imaginal disks is induced by the first peak of ecdysteroid that occurs in the absence of JH. It is accompanied by the loss of mRNAs for larval-specific proteins and the synthesis of new mRNAs and proteins that appear to be required to permanently suppress larval-specific genes (Riddiford, 1982, 1985; Riddiford and Kiely, 1981). In addition, it seems that certain molecular preparations for the expression of pupal-specific genes are also made at this time, although new pupal gene products are not synthesized until after the second peak of ecdysteroids (Kiely and Riddiford, 1985a,b; Riddiford, 1985). These findings suggest that the first peak of ecdysteroids during the last larval instar is the trigger for a complex set of molecular processes that permanently inactivate larval-specific genes and make previously unexpressed pupa-specific genes ready for transcription.

Although the endocrine control of pupal commitment has been best studied in the Lepidoptera, it is now generally believed that in all insects whose metamorphic molt is preceded by two periods of ecdysteroid secretion, the first peak causes reprogramming of the epidermis from a larval to a pupal (or adult in Hemimetabola) commitment, while the second peak induces the molt (see, for instance, Riddiford, 1989, 1992). In the species that do not have a clear double peak of ecdysteroids during their last larval instar the rising phase of ecdysteroid secretion often has a distinct shoulder or hump which may be homologous to the first peak of *Manduca*, and possibly to the early plateau of ecdysteroids in *Rhodnius*.

In addition to reprogramming the epidermis, the first peak of ecdysteroids in last instar larvae of holometabolous insects also causes dramatic changes in the animal's physiology and behavior. The first and most obvious behavioral response is that the animal stops feeding and voids its gut contents. Most larvae then move away from their food source and go in search of an appropriate site for pupation. In the larvae of Lepidoptera and Diptera this activity period is referred to as the "wandering phase." Larvae of Lepidoptera that pupate underground become positively geotropic, climb down their host plant, and burrow into the soil. There they excavate a small pupation chamber and begin to secrete a sticky proteinaceous fluid through their anus that serves to strengthen the walls of the chamber. The larvae of many Lepidoptera that pupate above ground wander in search of an appropriate substrate, and there they undertake an elaborate and stereotyped cocoon-spinning behavior. In larvae of Saturniidae the entire complex sequence of behaviors required to build a peduncle, then an outer and then an inner cocoon (often with escape hatches built in), followed by the impregnation of the inner cocoon with a predator repellent fluid from the Malpighian tubules, is entirely preprogrammed in the central nervous system and triggered by the secretion of ecdysteroids (Van Der Kloot and Williams, 1954; Lounibos, 1975, 1976; Giebultowicz et al., 1980). Among the interesting biochemical changes that take place early during the wandering phase of certain Lepidoptera is the loss of the blue hemolymph protein, insecticyanin, and the synthesis of red ommochromes in the dorsal epidermis (Bückmann, 1959; Riddiford, 1985). This results in a distinctive change in the color of the larva by which the wandering phase can be readily identified even when the animal is temporarily quiescent or experimentally restrained. Wandering, burrowing, and spinning behavior all occur during the period between the two peaks of ecdysteroids. The second peak of ecdysteroids terminates the wandering phase and, with the onset of apolysis, initiates the stage commonly called the *prepupa*, during which the imaginal disks grow and evert, and the cuticle of the pupal stage begins to be deposited. The mechanism by which ecdysteroids cause this elaborate array of molecular and physiological changes is not known.

The last larval instar of holometabolous insects thus differs from all the other instars in the pattern of ecdysteroid secretion and in the physiological response to ecdysteroids. There are two peaks of ecdysteroid secretion instead of one. The first peak of ecdysteroids reprograms the response pattern of the epidermis, which now becomes committed to secrete a pupal instead of a larval cuticle and triggers the sequence of physiological and behavioral events that prepare the animal for pupation. The second peak provokes a normal molting cycle during which the altered commitment of the epidermal cells is manifest. Both the altered pattern of ecdysteroid secretion and the novel physiological responses to ecdysteroids have a common cause, namely the disappearance of JH from the hemolymph. As we will see below, the corpora allata cease to secrete JH during the last larval instar, and the loss of JH is intimately involved in reprogramming the molecular and physiological responses to ecdysteroids.

Ecdysone and Ecdysteroids

Ecdysone was first purified by Butenandt and Karlson (1954) from a half ton of pupae of the silkworm *Bombyx mori*. Two slightly different steroid molecules with molt-inducing activity were identified from these early isolation attempts and named α-ecdysone and β-ecdysone. Their structure (Box 4) was determined by Karlson et al. (1965), and this constituted the first chemical identification of an insect hormone. The hormone that Karlson and his coworkers identified as α-ecdysone (now simply known as ecdysone) is secreted by the prothoracic glands, but it is not the hormone that actually provokes the physiological and developmental effects generally attributed to the molting hormone. Ecdysone is a relatively inactive prohormone that is converted into the much more active form, 20-hydroxyecdysone (the original β-ecdysone, also still occasionally referred to as ecdysterone) by the fat body and epidermal cells. 20-Hydroxyecdysone is now well established as the primary active form of the molting hormone in most insects, though there are exceptions, as we will see below.

Today ecdysone, its analogs, and metabolites are usually identified by chromatography and quantified by means of a radioimmunoassay (Borst and O'Connor, 1972, 1974; Granger and Goodman, 1988; Reum and Koolman, 1989; see chapter 2). The accuracy and sensitivity of an RIA depends critically on the specificity of the antibody that is used. In general, antiecdysone antibodies cross-react to various degrees with a number of ecdysone analogs and several ecdysone metabolites, and it is therefore customary to refer to ecdysone-activity identified by RIA and other nonspecific assays as *ecdysteroids* (Goodwin et al., 1978). Recent studies are uncovering an ever-increasing diversity of natural ecdysteroids in insects, and it has been suggested that some of these may have a role in regulating the physiological events normally attributed to 20-hydroxyecdysone. Thus, unless a molecule has been specifically identified, it is best to refer to all agents with ecdysonelike activity as ecdysteroids rather than ecdysone, as the latter implies a specific molecule (Box 4). More than sixty naturally occurring ecdysteroids have been described in insects and other arthropods (Rees, 1989), and more than one hundred "phytoecdysteroids" have been isolated from plants (Lafont and Horn, 1989). Some of the phytoecdysteroids are identical to known insect ecdysteroids, others are so far known only from plant sources but have a definite biological activity in insects. It is thought that the phytoecdysteroids evolved primarily as a chemical defense mechanism against insect herbivory (Jones and Firn, 1978). The principal ecdysteroids known to be involved in the normal physiology of insects are shown in Box 4.

It was long believed that the prothoracic glands produce only ecdysone, and that this hormone is activated peripherally (in the hemolymph, epidermis, and fat body) by hydroxylation to form the more highly active 20-hydroxyecdysone. Warren et al. (1988) have shown that the prothoracic glands of *Manduca sexta* cultured *in vitro* may actually secrete 3-dehydroecdysone and 2-dehydroecdysone in approximately a 2:1 ratio, though more recent evidence suggests that 3-dehydroecdysone is the predominant (if not the only) ecdysteroid secreted, as it is in many other insects (Howarth et al., 1989; Kiriishi et al., 1990). After they are secreted these compounds are rapidly transformed to ecdysone by enzymes in the hemolymph (Sakurai et al. 1989; Kiriishi et al., 1990). In *Drosophila*, the ring gland

BOX 4

The Ecdysteroids

Structure and Nomenclature: Ecdysteroids are sterol derivatives. The five molecules shown below occur most commonly in insects.

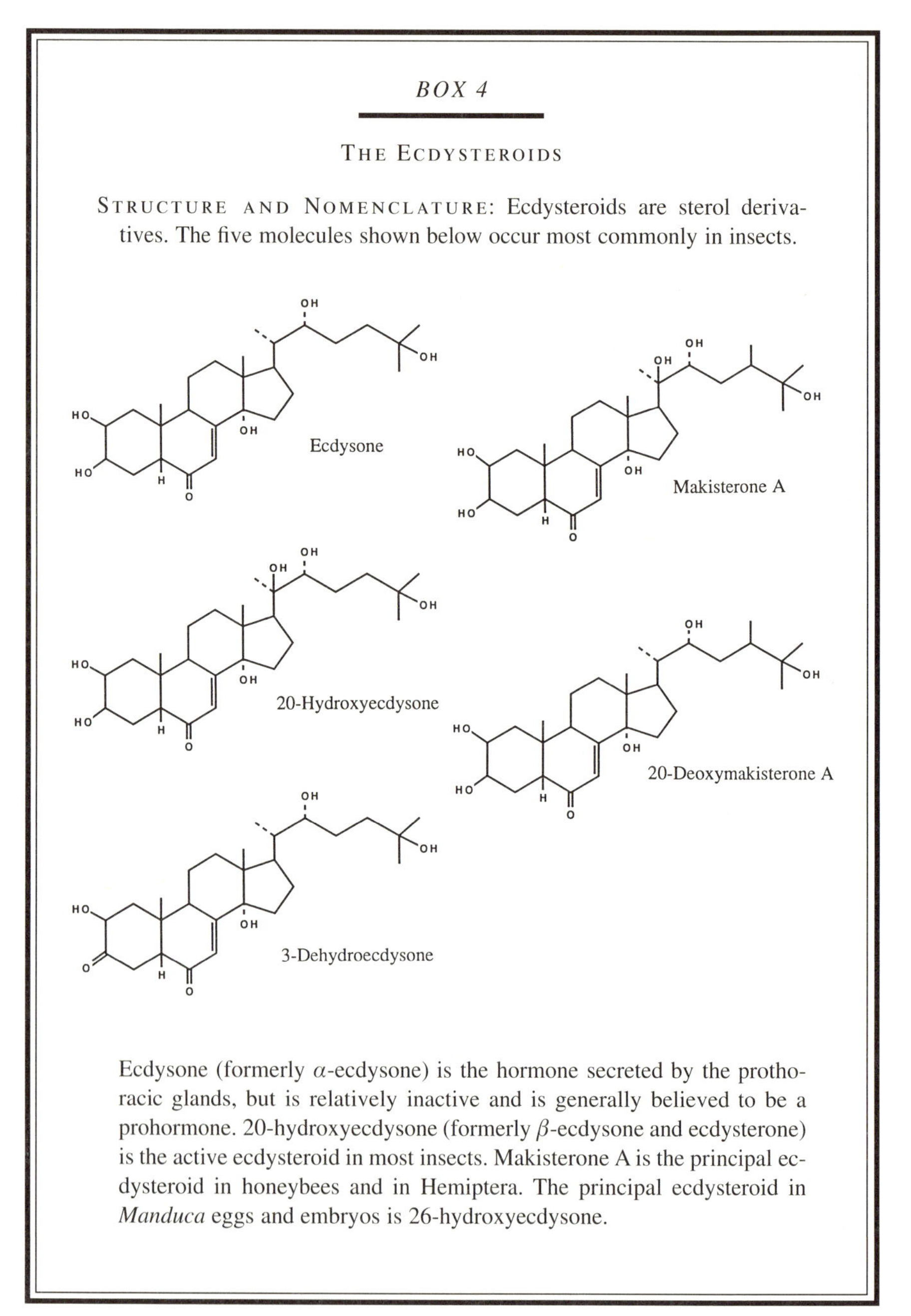

Ecdysone (formerly α-ecdysone) is the hormone secreted by the prothoracic glands, but is relatively inactive and is generally believed to be a prohormone. 20-hydroxyecdysone (formerly β-ecdysone and ecdysterone) is the active ecdysteroid in most insects. Makisterone A is the principal ecdysteroid in honeybees and in Hemiptera. The principal ecdysteroid in *Manduca* eggs and embryos is 26-hydroxyecdysone.

BOX 4 (cont.)

SOURCES: The prohormone ecdysone is the principal molecule that is secreted by the larval and pupal prothoracic glands in most insects studied so far. Prothoracic glands secrete other ecdysteroids as well. The *Drosophila* ring gland in vitro secretes ecdysone and 20-deoxymakisterone A and possibly other ecdysteroids as well. In *Manduca*, the prothoracic glands secrete primarily 3-dehydroecdysone. An enzyme in the hemolymph converts this dehydroecdysteroid to ecdysone. In *Manduca* and other species, circulating ecdysone is transformed into the active hormone, 20-hydroxyecdysone, by the epidermal cells and fat body.

In adult insects the follicle cells of the ovary are the source of ecdysone. The ovaries of *Bombyx* produce four ecdysteroids, in addition to ecdysone and 20-hydroxyecdysone:

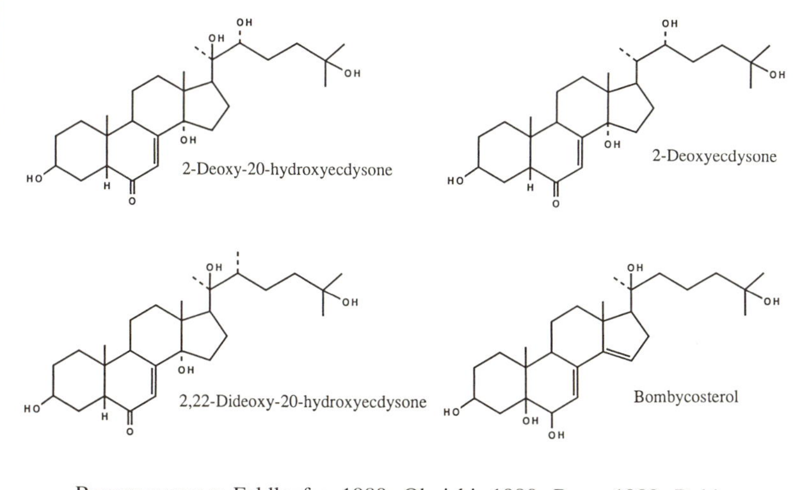

REFERENCES: Feldlaufer, 1989; Ohnishi, 1990; Rees, 1989; Robinson et al., 1991; Warren and Gilbert, 1986; Warren et al., 1988; Watson et al., 1989.

secretes ecdysone as well as its analog 20-deoxymakisterone A (Redfern, 1984; Riddiford, 1992). These findings indicate that the prothoracic glands of insects may be quite variable in the exact ecdysteroids they secrete, and that processing and activation of the prohormones in the hemolymph, fat body, and epidermal cells may be more elaborate than previously thought.

Sterols such as cholesterol constitute the immediate precursors for ecdysteroid synthesis by the prothoracic glands. Insects cannot, however, synthesize sterols from smaller precursors and therefore require a dietary source of sterols (Rees, 1985). Both the synthesis and secretion of ecdysteroids are controlled by the prothoracicotropic hormone (PTTH) of the brain (Bollenbacher et al., 1979). Once secreted into the hemolymph, ecdysteroids are bound to a specific carrier protein. Although the role of the carrier protein is not fully understood, it is believed that it enhances the carrying capacity of the hemolymph for ecdysteroids, and may help the hormones to penetrate cell membranes. The titer of ecdysteroids in the hemolymph is determined by a balance between their synthesis and release from the prothoracic glands and their degradation by various catabolic enzymes in the hemolymph and tissues. There is at present little evidence that the rate of degradation (or excretion) of these hormones is regulated to control the ecdysteroid titer profile. Ecdysteroid catabolism is variable, to be sure, and while it is possible that the rate of catabolism is controlled in a physiologically meaningful way, a regulatory mechanism has not yet been demonstrated. It appears that the overall ecdysteroid titers in the hemolymph are regulated largely by the rate at which they are secreted by the prothoracic glands (Koolman and Karlson, 1985; Smith, 1985; Lehmann and Koolman, 1989). The pattern of ecdysteroid secretion, in turn, depends in large measure on the pattern of PTTH secretion by the brain, and it can therefore be said that the brain's PTTH indirectly regulates the titer of ecdysteroids.

The prothoracic glands secrete ecdysone prior to each molt, and the direct action of 20-hydroxyecdysone on the epidermal cells causes them to undergo apolysis, cell division, digestion of the old cuticle, and secretion of a new cuticle (fig. 4.1). This action of ecdysteroids is the same in all stages in the insect's life cycle, whether the molt is larval or metamorphic. A fairly normal molt can be induced at an inappropriate time in many insects by exogenous ecdysteroids. The hormones can be provided by parabiosis to an animal that is undergoing normal ecdysteroid secretion (Williams, 1952a,b), or by injection. Injections seem to work best when the hormone is introduced steadily by infusion over a fairly prolonged period of time (Nijhout, 1976). The need for a prolonged presence of ecdysteroids is in accord with the finding that natural episodes of ecdysteroid secretion generally last one to three days. In adult development at least, a prolonged presence of ecdysteroids is required for the activation and maintenance of a fairly long sequence of biochemical and physiological processes associated with molting, as we saw above (fig. 5.3). Surprisingly, little is known about the exact sequence of cellular and physiological effects of ecdysteroids during a normal larva-to-larva molt or about the precise role of the different ecdysteroids in each of the successive stages of a molt.

The prothoracic glands degenerate during the pupal stage in holometabolous insects, or during the first few days of the adult stage in hemimetabolous insects. It was long believed that the consequent inability to produce ecdysteroids was responsible for the fact that adult insects do not molt. This is probably not the complete explanation of the absence of adult molting in insects, however. While it is possible to artificially induce adult molting in some species such as *Rhodnius* and *Hyalophora* by parabiosing them to other individuals that are undergoing a molt (Wigglesworth, 1940; Williams, 1963), it has proven to be extraordinarily difficult to get adults of many other species to initiate a molting cycle, even with massive doses of ecdysteroids. Furthermore, it was apparent from many of the early ecdysteroid bioassays that adult insects often have substantial titers of ecdysteroids in their hemolymph, sometimes higher than those produced during larval molting (Karlson and Stamm-Menendez, 1956; Feir and Winkler, 1969; King and Marks, 1974). Yet such animals show no evidence of molting or mitoses in their epidermis. In these cases adult molting must be prevented by mechanisms that somehow render adult epidermal cells incapable of responding to ecdysteroids. The absence of ecdysteroid receptors in adult epidermal cells could provide such a mechanism.

Adult insects lack prothoracic glands, so the source of the high ecdysteroid titers in adult females was something of a mystery. Fallon et al. (1974) and Hagedorn et al. (1975) discovered that the ovaries are the source of the adult ecdysteroids in mosquitoes (*Aedes aegypti*), and that the ecdysteroids play a crucial role in regulating egg development in these animals (see chapter 6). The follicle cells of the ovaries appear to be the secretory tissue for ovarian ecdysteroids (Hagedorn 1985, 1989). Numerous studies in the course of the past decade have shown that ecdysteroids are present in adult female (and to a much lesser degree in male) insects across the entire taxonomic range, from the Apterygota to the Diptera. It is clear that in several species of Diptera (*Aedes*, *Drosophila*, *Musca*, *Calliphora*) ecdysteroids play a role as reproductive hormones that control yolk protein synthesis and egg development (Hagedorn, 1985). Ecdysteroids are also necessary for normal ovarian development in *Bombyx mori* and *Thermobia domestica* (Tsuchida et al., 1987; Bitsch and Bitsch, 1988). The secretion of ovarian ecdysteroids in Diptera is controlled by a neurohormone from the brain called the ovarian ecdysteroidogenic hormone (OEH) (Hagedorn et al., 1979; Lea and Brown, 1989) that is probably unrelated to PTTH. The principal role of ecdysteroids in adult insects is the stimulation of vitellogenin (yolk protein) synthesis by the fat body. As in the case of the prothoracic glands, the ovaries secrete ecdysone which is transformed into 20-hydroxyecdysone by various peripheral tissues, and it is the latter form of the hormone that stimulates vitellogenin synthesis in the fat body.

Ecdysteroids also appear to play a role in embryonic development. Eggs and embryos of many species contain large quantities of ecdysteroids, and discrete periods of ecdysteroid production have been demonstrated in embryos of *Locusta*, *Blaberus*, *Oncopeltus*, and *Nauphoeta*. The embryos of these and many other insects are known to secrete and molt several cuticles while developing within the egg. In *Locusta* each embryonic molt is well correlated with a peak of ecdysteroid production in the embryo, but in other species a correlation between ecdysteroid peaks and embryonic molts is not always detectable (Lageux et al., 1979; Hagedorn, 1985; Hoffmann and Lageux, 1985). In *Locusta migratoria*, *Schistocerca gregaria*, *Bombyx mori*, and *Galleria mellonella* the female incorporates large amounts of ecdysteroids into her eggs. These are believed to exist as inactive conjugates from which active ecdysteroids may be released periodically to control at least some of the embryonic molts. In cases where the concentration of ecdysteroids rises during embryonic development, their source is still unknown (Hoffmann and Lageux, 1985).

Hormonal Control of Ecdysis and Tanning

The molting cycle of insects culminates with the escape of the new stage or instar from the old partially digested cuticle of the previous stage. The general name for this process is "ecdysis," except for the escape of adult holometabolous insects from their pupal cuticle, for which the term "eclosion" is used. Ecdysis and eclosion typically occur only at restricted times of the day and involve the playout of a stereotyped set of behaviors designed to free the animal of its old cuticle and expand and harden the new cuticle. Ecdysis and eclosion are controlled by a neurosecretory hormone, the eclosion hormone (EH). The EH acts directly on the nervous system, where it is a releaser for a largely preprogrammed pattern of nervous and muscular activity (Truman, 1971c, 1985; Truman et al., 1981). Secretion of the EH is, in turn, controlled by physiological clocks and usually occurs at characteristic times in the day/night cycle. The biology of EH, its source, and mechanism of action will be discussed more fully in chapter 9.

Following ecdysis, insects must harden their new cuticle. This sclerotization of the cuticle is controlled by the neurosecretory hormone, *bursicon*. In the higher Diptera, sclerotization of the larval cuticle at the time of puparium formation is controlled by an independent set of neurosecretory hormones, the *pupariation factors*. These two mechanisms of control over sclerotization are discussed separately below.

Bursicon

Bursicon is a neurosecretory hormone that controls tanning (sclerotization) and the mechanical properties of insect cuticle during and after molting. The first indications of a special control over cuticle tanning were found by Fraenkel (1935) in his studies on the control of molting in blowflies (Calliphora). Fraenkel found that hardening and tanning of the cuticle of blowflies could be delayed for as much as 24 hours after emergence of the adults from the puparium if the animal was denied a place to sit and expand its wings. Fraenkel and Hsiao (1962, 1965) subsequently showed that sclerotization in adult blowflies could be prevented indefinitely by placing a ligature around the neck of newly emerged flies. They also determined that neither JH nor ecdysone could stimulate tanning in such neck-ligated flies, and they concluded that cuticle tanning was due to a previously unknown hormone, which they named bursicon (see Box 5). Bursicon has

BOX 5

BURSICON

STRUCTURE AND NOMENCLATURE: Bursicon has proven to be a notoriously difficult molecule to purify; it loses its activity rapidly upon purification, a common problem with peptide hormones. Available data suggest that bursicon is a medium-sized polypeptide with an estimated molecular weight of about 40,000 D, although bursicons with molecular weights as low as 30,000 D and as high as 60,000 D have been reported from some preparations. As noted in the text, it is possible that the activities ascribed to bursicon actually represent the actions of a heterogeneous group of physiologically active peptides that affect the development of various properties of the cuticle.

SOURCE: Bursicon activity has been found in most ganglia of the central nervous system.

REFERENCES: Reynolds, 1983, 1985.

since been shown to control cuticle sclerotization after ecdysis in a number of species of Orthoptera, Hemiptera, Lepidoptera, Diptera, and Coleoptera (Reynolds, 1985). The mechanism by which bursicon controls the sclerotization reactions in the cuticle is not well understood.

Studies on the localization of the source of bursicon have revealed that in blowflies both the brain and the thoracic ganglionic mass (consisting of the fused thoracic and abdominal ganglia in the higher flies) contained bursicon neurosecretory cells. Studies with other insects likewise have shown that bursicon occurs in the brain as well as in the ganglia of the ventral nerve cord. In spite of its widespread distribution throughout the nervous system, it has been shown that the abdominal ganglia are the principal sources of bursicon in most insects and that the hormone is released into the hemolymph from the abdominal perivisceral organs (Reynolds, 1983). In *Manduca sexta* the secretion of bursicon is controlled indirectly by EH (Truman, 1981). Injections of EH that cause premature eclosion behavior also cause a premature secretion of bursicon, which suggests that bursicon release may be part of the preprogrammed series of eclosion "behaviors" released by EH. Some modulation of bursicon release is evidently possible because if an emerging adult *Manduca* is forced to dig for several hours (pupation occurs underground, so there is considerable variation in the amount of time required for animals to dig out), bursicon release is delayed until the animal has found a suitable place to spread its wings (Truman, 1973b, 1981).

Extracts with bursicon activity also provoke an array of physiological effects associated with regulating the mechanical properties of cuticle during molting. Among these effects is the plasticization of the new cuticle just prior to ecdysis in blowflies and locusts and moths. Plasticization is not just an inhibition of sclerotization but is an actual softening of the cuticle, accompanied by an increase in its extensibility that presumably facilitates ecdysis from the old cuticle and postecdysial expansion of the new cuticle. In moths such as *Manduca*, plasticization of the wings is essential in order to enable them to be expanded after adult eclosion (Reynolds, 1977).

In blowflies bursicon has also been shown to stimulate endocuticle deposition after adult eclosion, and cell death of the wing epidermis (Fogal and Fraenkel, 1969; Seligman and Doy, 1973). While both of these developmental processes occur in other insects as well, they have not been shown to be regulated by hormones except in blowflies (Reynolds, 1983).

In view of the diversity of its physiological effects and the fact that it has been difficult to isolate bursicon as a single active molecule, Reynolds (1985) has suggested that the activities usually identified as "bursicon" in different insects or at different stages in the life cycle may not be due to the same hormone. A variety of neuroendocrine factors are probably involved in controlling the mechanical properties of insect cuticles. Isolation and chemical identification of "bursicons" from several different species will be required in order to determine whether it is a universal hormone in insects, or whether it is actually a functional category of hormones.

Pupariation Factors

The unusual metamorphosis of the higher Diptera (Cyclorrhapha) requires a suite of control processes different from those found in other holometabolous insects. The higher Diptera are unique in that they discard their entire thoracic and abdominal epidermis at metamorphosis and build their adult body wall anew from imaginal disks and abdominal histoblasts. In order to contain this near total disintegration of the larval body wall the pupa never ecdyses from the last larval cuticle. Instead, at the end of larval life the cuticle of the last larval instar contracts to a fairly smooth barrel shape and then becomes sclerotized into a hard dark protective case within which pupation and adult development take place. This structure is called the *puparium* and is unique to the higher Diptera. Sclerotization of the puparium is controlled by a bursiconlike molecule called the *puparium tanning factor* (PTF) (Zdarek and Fraenkel, 1969; Zdarek, 1985).

PTF is but one of an array of hormones that has been found to play a role in puparium formation in the higher Diptera. Studies, primarily on the fleshfly *Sarcophaga bullata*, have identified two additional hormones that are involved in the control of puparium formation. They are the *anterior segment retraction factor* (ARF) and the *puparium-immobilizing factor* (PIF). All three factors can be extracted from central nervous system homogenates. They are proteinaceous and are therefore probably neurohormones (Sivasubramanian et al., 1974; Zdarek, 1985).

The ARF is found in the hemolymph during the first few hours of pupariation. It causes the anterior segments of the larva to retract and shapes the larval integument into the barrel-shaped puparium. Injection of an ARF extract into larvae can cause premature cessation of activity followed by retraction of its anterior segments into the puparium shape, but sensitivity of the larva to ARF is limited to a period of a few hours before pupariation would normally take place. The PIF is also found in the hemolymph of larvae during the first few hours after pupariation begins. When injected into larvae that have stopped feeding in anticipation of pupariation, PIF causes the larvae to become immobile for a prolonged period of time. Larvae become increasingly sensitive to PIF as the normal time of pupariation approaches, possibly because of a gradual increase in hormone receptors. The three pupariation factors appear to be secreted virtually simultaneously and control both the behavior of puparium formation (contraction, which rounds up the puparium and increases the internal pressure to produce a smooth envelope, and immobilization, an enforced quiescence until the puparium hardens) and the actual tanning of the old larval cuticle to form the hardened puparium.

Juvenile Hormone (JH)

The juvenile hormones are probably the most versatile hormone in the entire animal kingdom. They play a role in almost every aspect of insect development and reproduction, including metamorphosis, caste determination in the social insects, the regulation of behavior in honeybee colonies, the polyphenisms of aphids and locusts, larval and adult diapause regulation, vitellogenin synthesis, ovarian development, and various aspects of metabolism associated with these functions. The many roles of juvenile hormones will be discussed in some detail in the chapters on reproduction (chapter 6), diapause (chapter 7), polyphenism (chapter 8), and behavior (chapter 9). Here we will limit our discussion to the general endocrinology of these hormones and their role in the control of metamorphosis.

Juvenile hormones are secreted by the corpora allata, which constitute, so far, the only known source of these hormones. The endocrine function of the corpora allata was discovered by Wigglesworth (1934, 1936). Wigglesworth found that when mature (final instar) nymphs of the hemipteran *Rhodnius prolixus* were parabiosed to immature nymphs of an earlier instar, the mature nymph metamorphosed into an adult that retained many of the larval characters of the species. This blood-borne metamorphosis-inhibiting factor was traced to the corpora allata and dubbed the juvenile hormone (JH). Wigglesworth showed that the corpora allata were active in immature nymphs and became inactive in mature, final instar, nymphs. He ascribed the onset of metamorphosis in the mature nymph to the disappearance of JH from the hemolymph. Wigglesworth's early experiments also showed that the corpora allata of *Rhodnius* became active again in the adult.

The chemical structure of JH was first determined by Röller et al. (1967). The hormone proved to be an unusual sesquiterpenoid with an epoxide group near one end and a methyl ester on the other (Box 6). The hormone that Röller and his colleagues identified soon proved to be but one of a series of three naturally occurring juvenile hormones, all with nearly identical chemical structures, differing only in the number of ethyl side chains.

BOX 6

THE JUVENILE HORMONES
(JH)

STRUCTURE AND NOMENCLATURE: Juvenile hormones are terpenoids derived from farnesenic acid. All are methyl esters and have a terminal or subterminal epoxide group. Five structurally related molecular forms occur in insects.

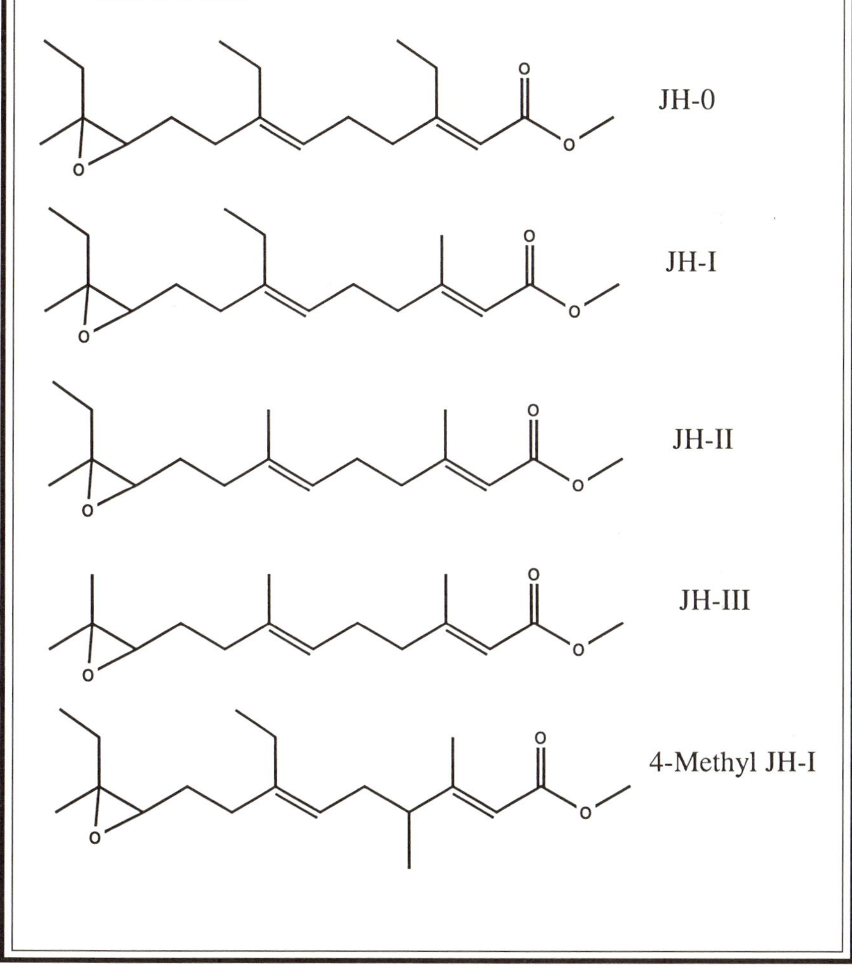

BOX 6 (cont.)

The first JH identified was called JH-I (or C-18 JH, in reference to the 18 carbons in the molecule). The following two JHs to be identified were accordingly named JH-II (C-17 JH) and JH-III (C-16 JH). When a 19-carbon JH was identified from eggs of *Manduca* it was called JH-0 to maintain the parallelism of the two nomenclatural schemes. JH-III is the most widespread of the juvenile hormones and has been found in all orders examined so far. JH-I and JH-II appear to be restricted to the Lepidoptera, as is JH-0. JH-0 and 4-methyl JH-I have been found only in the eggs of Lepidoptera.

JH is a nonpolar lipidlike molecule, yet it has has two mechanisms of action at the cellular level. One is in the fashion of a steroid hormone, binding to high affinity receptors in the nucleus. The other is in the manner of a peptide hormone, binding to a cell surface receptor that activates a phosphatidyl inositol-mediated second-messenger system.

SOURCE: All forms of JH are synthesized and secreted by the corpora allata, which also constitute the only known source for these hormones.

REFERENCES: Henrick et al., 1976; Palli et al., 1991; Schooley and Baker, 1985; Sevala and Davey, 1989; Yamamoto et al., 1988.

Röller's hormone is now named JH-I (formerly C18-JH, in reference to its 18-carbon skeleton), and the other two are called JH-II and JH-III (formerly C17-JH and C16-JH, respectively). Two additional forms of JH, named JH-0 and 4-methyl JH-I, have been found more recently in the eggs of *Manduca sexta*, but nothing is known yet about their function.

Some insects secrete only one of the three forms of JH, others secrete a mixture of two or three of them. Juvenile hormone-III is the only form of JH found in the Orthoptera, though the corpora allata of *Schistocerca gregaria* have been shown to make JH-I when cultured *in vitro*. JH-III is also the principal form of JH found in the Coleoptera, Diptera, Hemiptera, and Hymenoptera. The Lepidoptera appear to be unique in secreting a mixture of JH-I and JH-II. The corpora allata of the tobacco hornworm moth *Manduca sexta*, however, secrete JH-III, in addition to JH-II (Tobe and Stay, 1985). Each of the three forms of JH has a different relative activity when injected into insects of different orders. These differences in activity are purely quantitative, however. There is no evidence that the three forms provoke qualitatively different physiological effects.

The natural JHs are extremely unstable molecules. They are readily degraded by various esterases in the hemolymph and by sunlight, and they also degrade rapidly at room temperature when purified. Their chemical instability makes the JHs difficult to work with experimentally and a considerable effort has gone into developing stable and effective analogs of these hormones (an effort driven in no small measure by the promise of insecticidal potential of such stable compounds; see the last section in this chapter on Third Generation Pesticides). The biological activity of a hormone is believed to depend on the efficiency with which it binds to a receptor, and active analogs of a hormone are those molecules with a 3-dimensional structure that resembles that of the receptor-binding site of the hormone sufficiently to be recognized by the receptor. Studies on the structure-activity relations of JH analogs have shown that the number of carbons in the chain is important (14–16 are optimal), as is the correct configuration around the unsaturated bonds and the position of the methyl (or ethyl) side chains. The methyl ester group at the C-1 position is also important for biological activity, and it is the degradation of this ester that is the first step in the natural inactivation of the hormone. (The structures of three of the most widely used stable JH analogs are shown in figure 5.15 later in this chapter.)

As in the case of ecdysteroids, the juvenile hormones, once secreted, become attached to specialized hemolymph proteins, collectively referred to as the *juvenile hormone binding proteins*. These proteins enhance the solubility of the juvenile hormones in the hemolymph and also protect them from degradation by hemolymph esterases. In some insects the juvenile hormone binding proteins also play an important role in regulating the cellular response to the hormones. They may facilitate entry of the hormones into a cell, and, once inside, they may regulate the way the hormones bind to their receptors. In some cases the hormone receptors only recognize the complex of the hormone and its binding protein, in other cases only the unbound hormone is recognized (Goodman and Chang, 1985).

The synthesis of JH by the corpora allata is controlled by the nervous system, but insects differ widely in the mechanism by which they control their corpora allata (Tobe and Stay, 1985). The corpora allata do not store JH, so that the rate at which JH secretion occurs is determined entirely by its rate of synthesis. In adult *Leptinotarsa decemlineata*, corpora allata whose nervous connections with the brain have been severed function normally, so they appear to be regulated via a blood-borne hormone. In many other insects where corpus allatum regulation has been studied, an intact nervous connection between brain and corpora allata is necessary for normal regulation.

The nerves that connect the corpora allata with the brain contain conventional as well as neurosecretory axons, and brain neurosecretory cells have been implicated in corpus allatum regulation in many species. In larvae of *Manduca sexta* and in adults of *Locusta migratoria* surgical removal of the medial neurosecretory cells of the brain abolishes JH secretion by the corpora allata. This indicates that JH secretion in these species requires stimulation by neurosecretory hormones. These stimulatory hormones are called *allatotropins* (Box 7). The structure of an allatotropin of *Manduca* has been described by Kataoka et al. (1989). An intact nervous connection between the brain and corpora allata is also required for normal regulation of their activity in the moth *Diatraea grandiosella* and the earwig *Anisolabis maritima*, but it is not known whether the regulation occurs via conventional or neurosecretory neurons. In the larvae of the roaches *Leucophaea maderae* and *Diploptera punctata*, and in the bug *Rhodnius prolixus*, however, severance of the nerves between the brain and corpora allata causes them to become permanently active. The corpora allata in these species thus appear to be under inhibitory control either via nerves or via neurosecretions traveling through the nerves that connect the corpora allata to the brain. In *Diploptera*, allatostatin-containing neurons from the lateral neurosecretory cells of the brain have been shown to project directly to the corpora allata (Stay et al., 1992). Inhibitory control of the corpora allata has also been demonstrated in *Galleria*, *Manduca*, and *Leptinotarsa*. Five different neurohormones that inhibit JH synthesis in the corpora allata of adult *Diploptera* have been isolated and identified by Woodhead et al. (1989). They are small polypeptides of 8–13 amino acids (Box 7).

Regulation of the hemolymph titer of JH involves not only control over the synthesis of JH by the corpora allata, but also control of specialized JH-catabolizing enzymes. Even though JH is unstable and degraded rapidly by general esterases in the hemolymph (normal half-lives of JH vary from about 30 minutes to about 4 hours, depending on the species; Tobe and Stay, 1985), a special set of JH-specific esterases has evolved whose function it is to clear the hemolymph of JH at the end of larval life in anticipation of metamorphosis (Hammock, 1985). Clearance of JH from the hemolymph is an essential prerequisite for the initiation of a metamorphic molt. JH-specific esterases are synthesized during the last larval instar at about the time that the corpora allata cease to secrete JH. In the moth *Trichoplusia ni*, there is evidence that JH-esterase synthesis by the fat body is controlled by a hormone from the brain (Jones et al., 1981; Hammock, 1985), but the physiological mechanism that coordinates the inactivation of the corpora allata and the activation of JH-esterase synthesis is not yet understood.

BOX 7

ALLATOTROPINS AND ALLATOSTATINS

STRUCTURE AND NOMENCLATURE: Allatotropins are hormones that stimulate JH secretion by the corpora allata, and allatostatins are hormones that inhibit JH secretion. Allatostatins are also called allatohibins. All allatotropins and allatostatins identified so far are neurosecretory polypeptides. A 13-amino acid allatotropin has been identified from adult *Manduca sexta* and a 20 kD one from larvae of *Galleria mellonella*, respectively. The *Manduca* allatotropin appears to be active only on adult Lepidoptera and has no effect on larvae or on species in other orders. The *Manduca* allatotropin appears to act via an inositol phosphate-mediated second messenger system. The structure of the *Manduca* allatotropin is:

Gly-Phe-Lys-Asn-Val-Glu-Met-Met-Thr-Ala-Arg-Gly-PheNH_2

The *Manduca* allatotropin has no structural similarity to any known peptides. A group of five structurally related allatostatins has been identified from *Diploptera punctata*:

Allatostatin A 1:	Ala-Pro-Ser-Gly-Ala-Gln-Arg-Leu-Tyr-Gly-Phe-Gly-LeuNH_2
Allatostatin A 2:	Gly-Asp-Gly-Arg-Leu-Tyr-Ala-Phe-Gly-LeuHN_2
Allatostatin A 3:	Gly-Gly-Ser-Leu-Tyr-Ser-Phe-Gly-LeuNH_2
Allatostatin A 4:	Asp-Arg-Leu-Tyr-Ser-Phe-Gly-LeuNH_2
Allatostatin B:	Ala-Tyr-Ser-Tyr-Val-Ser-Glu-Tyr-Lys-Arg-Leu-Pro-Val-Tyr-Asn-Phe-Gly-LeuNH_2

SOURCES: The allatotropin of *Galleria* is produced by the medial neurosecretory cells of the brain. Allatostatin 1 in *Diploptera* is produced by the lateral neurosecretory cells which have their neurohemal site in the corpora allata, and by a group of medial neurosecretory cells which have their neurohemal sites within the brain.

REFERENCES: Bogus and Scheller, 1991; Kataoka et al., 1989; Pratt et al., 1989, 1990, 1991; Reagan et al., 1992; Riddiford, 1992; Stay et al., 1992; Woodhead et al., 1989.

Juvenile hormone acts as a regulatory as well as a developmental hormone. Two major regulatory functions are known: in the adults of many insects JH stimulates the synthesis of vitellogenin (yolk protein) by the fat body, and in final instar larvae of Lepidoptera and other holometabolous insects, JH inhibits the secretion of PTTH. The rate of vitellogenin synthesis has been shown to be proportional to the concentration of JH, and continued vitellogenin synthesis appears to require the continued presence of JH (Hagedorn and Kunkel, 1979). The inhibition of PTTH during the last larval instar is designed to inhibit the metamorphic molt until such a time that all JH has been cleared from the hemolymph. As in the case of vitellogenin synthesis, continued inhibition of PTTH release requires the continued presence of JH, suggesting its role as a regulator, not a trigger. This inhibitory action of JH has been secondarily captured as a mechanism to control larval diapause in some insects (chapter 7).

As a developmental hormone, JH controls switches between alternative pathways of development at several points in the life cycle. In its role as a developmental hormone JH does not act in a concentration-dependent manner, as was once widely believed, but acts during discrete critical periods, the juvenile hormone-sensitive periods discussed below. In general, if JH is present during a critical period, no developmental switch takes place, and the current developmental state is maintained. If JH is absent during a critical period, gene expression changes and new developmental processes begin that launch the developing insect on a new developmental pathway. The exact nature of the developmental switch depends on the species and on the time in the life cycle that the critical period occurs; these will be discussed in some detail below and in the chapter on polyphenisms (chapter 8). The continuation of the current developmental state in the presence of JH led Williams (1952b) to refer to JH as the "status quo" hormone.

The Role of Juvenile Hormone in Metamorphosis

The conceptualization of the way in which JH regulates the metamorphosis of insects has gone through a considerable evolution in the course of the past fifty years. The first indication that a hormone from the head controlled metamorphosis was provided by Wigglesworth (1934, 1936), who demonstrated that third and fourth instar larvae of *Rhodnius prolixus*, which were decapitated soon after their head critical period, underwent precocious metamorphosis. When the corpora allata from fourth instar larvae were transplanted into final fifth instar larvae, the latter failed to metamorphose and instead molted to a supernumerary larval instar. Wigglesworth suggested the corpora allata produced an inhibitory hormone, dubbed the juvenile hormone (JH), that suppressed the progression of "normal" development to the adult form. This interpretation was supported by the studies of Williams (1952b, 1956, 1959) on the metamorphosis of *Hyalophora cecropia*. Williams found that when exogenous JH was supplied to pupae just before the onset of adult development, they formed a second pupal cuticle beneath the old one, and he concluded that in the presence of JH, insect tissues simply remained at their current level of differentiation.

Based on an extensive series of corpus allatum transplant experiments in *Galleria mellonella*, Piepho (1942, 1946) developed a very different hypothesis that required two hormones for the control of metamorphosis. Piepho found that when the corpora allata from penultimate and antepenultimate larval instars were implanted into very early final instar larvae, most of them molted to a supernumerary larval instar but a few formed larval-pupal intermediates. When corpora allata were implanted into progressively older final instar larvae, the percentage that molted to perfect supernumerary larvae rapidly diminished; most formed larval-pupal intermediate monsters and a few formed perfect pupae. Finally, when implants were made very late in the final larval instar, few intermediates were formed and most larvae pupated normally. Since the corpora allata in all these experiments were taken from equally young larvae, Piepho proposed that the diminishing effect of transplanted glands was due to the appearance and gradual increase of a pupa-determining substance in the course of the final larval instar. Piepho also found, however, that the corpora allata taken from late final instar larvae and prepupae likewise caused the formation of larval-pupal intermediates when implanted into final instar larvae, though with diminished effectiveness. The corpora allata of adult moths were found to be almost completely unable to prevent normal pupation when implanted into last instar larvae. These experiments suggested that the corpora allata were active throughout the final instar and prepupal stage, but that their activity was considerably diminished compared to that found in the penultimate larval instar. Piepho thus proposed that under normal conditions the last larval stage is characterized by a gradual decrease in corpus allatum hormone (JH) and a complementary increase in the pupa-determining factor. When in subsequent investigations no further evidence could be found for the actual existence of a pupa-inducing factor, this hypothesis about the endocrine control of metamorphosis was modified to propose that larval molting required the presence of a relatively high titer of JH, while the pupal molt required a low titer and the adult molt an even lower one (Piepho, 1950, 1951).

The hypothesis that pupation required a low or intermediate titer of JH received its strongest support from the finding that allatectomy (removal of the corpora allata) of last instar larvae of cecropia moths (*Hyalophora*), tobacco hornworms (*Manduca*), and honeybees caused them to metamorphose, not to perfect pupae, but to monsters with a mixture of pupal and adult characters (Schaller, 1952; Williams, 1961; Kiguchi and Riddiford, 1978). In the absence of the juvenile hormone, certain tissues bypass the pupal stage and metamorphose directly from larva to adult. In *Hyalophora*, all structures that are derived from imaginal disks undergo precocious adult differentiation, while most of the thoracic and abdominal integument pupate normally. Sehnal (1972) has shown that in *Galleria*, at least, the abdominal integument as well as the imaginal disks can be made to bypass the pupal stage. If larvae of *Galleria* are ligated between thorax and abdomen (in effect removing their source of JH) and a period of at least five days is allowed for the natural JH to degrade before inducing them to molt with an injection of ecdysteroids, then the imaginal disks develop their adult characteristics and some of the epidermis develops the scales characteristic of the adult integument. The presence of JH is evidently essential for normal pupation in several species, and these findings support the notion that a low level of JH is required to induce the normal pupal morphology. Furthermore, the fact that in the presumptive absence of JH many tissues transformed directly to their adult form is in accord with the hypothesis that adult differentiation requires the absence of JH.

Most of the experimental work on the hormonal control of molting and metamorphosis done during the 1950s and 1960s gave results consistent with the hypothesis that metamorphosis in holometabolous insects was caused by a successive lowering of the JH titer. This scheme for the endocrine control of metamorphosis, known popularly as the *high-low-no hypothesis* or the *Classical Scheme*, was one of the cornerstones of insect developmental endocrinology for nearly three decades (Schneiderman and Gilbert, 1964; Doane, 1973).

There were, however, several problems with this simple and attractive explanation for metamorphosis (Slama, 1975). It was known from the very earliest experiments on the control of metamorphosis that larvae of *Bombyx mori* metamorphosed to normal pupae in the absence of their corpora allata (Fukuda, 1944). Thus, in this species, at least, normal pupation can occur in the absence of JH . Pupation of the abdominal epidermis of *Hyalophora* that are allatectomized as larvae, likewise, proceeds normally in the absence of JH; only the imaginal disks of this species appear to require JH prior to pupation. The inadequacy of the high-low-no model, however, was revealed when it finally became possible to measure JH titers in the hemolymph and no evidence could be found for intermediate titers of JH at or around the pupal molt. Instead, JH titers declined to undetectable levels during the middle of the final larval instar and, with the exception of a brief but substantial peak near the end of the instar, remained undetectable throughout apolysis and the early stages of pupal cuticle formation.

One of the most complete and best understood JH titer profiles for the period prior to and during metamorphosis is that of the tobacco hornworm, *Manduca sexta* (fig. 5.7, bottom). In this species the JH titer is high but variable during the fourth (penultimate) larval instar. It drops to a low level at about the time of ecdysis to the fifth instar, rises to a brief peak early in the fifth instar, then reaches a brief plateau before declining to an undetectable level at about the middle of the instar. For about a two-day period prior to the pupal ecdysis, the JH titer again rises to a peak and then drops back to an undetectable level during the early period of the pupal stage. The physiological significance of most of these fluctuations in the JH titer are now known and they reveal much about the manner in which the orderly progress of metamorphosis is controlled, as we will see below.

Juvenile Hormone-Sensitive Periods and Developmental Switches

It is now clear that metamorphosis in holometabolous insects is *not* caused by the progressive decline of JH titers. The manner in which JH controls metamorphosis is a little more complicated, but far more interesting, than was believed before. In the course of larval life there exists a series of periods of sensitivity to JH that alternates with much longer periods during which the tissues are largely insensitive to JH (Nijhout and Wheeler, 1982). Evidence for the existence of discrete JH-sensitive periods began to accumulate soon after purified JH and JH analogs became available for experimental use. Numerous studies using a large range of species have shown that JH has an effect only if it is injected or applied before a particular time in the molting cycle. The point in the molting cycle after which exogenous JH no longer produces a physiological or developmental effect marks the end of the JH-sensitive period. This usually occurs a short time after the secretion of ecdysone initiates the next molt. The general developmental "rule" is that if JH is present during a JH-sensitive period, the current developmental state is maintained, whereas if JH is absent, there is a switch in developmental pathways. The actual titer of JH during the JH-sensitive period is immaterial, as long as it is above or below a certain threshold. Different portions of the epidermis have slightly different, partially overlapping, JH-sensitive periods, and this is why a single application of JH often has a mosaic juvenilizing effect.

The onset of a JH-sensitive period appears to be a preprogrammed developmental event that is itself independent of the presence or absence of JH (Nijhout and Wheeler, 1982). During a JH-sensitive period the cells of an insect can become committed to a new developmental fate. The actual switchover to the new developmental pathway generally requires an exposure to ecdysteroids (see above). Often a JH-sensitive period ends when ecdysteroid-stimulated development has progressed to the point at which the actual expression of the JH-sensitive character begins. The actual beginning and end of a JH-sensitive period has so far only been determined in *Rhodnius* (Nijhout, 1983). In the final larval instar of this species, the JH-sensitive period lasts six days, from day 3 to day 9 after a blood meal is taken. Interestingly, JH need not be present during this entire sensitive period. Instead, any individual animal requires only a 2–4-day exposure to JH to be competely juvenilized and it does not matter whether that exposure occurs early or late in the six-day JH-sensitive period. In this case the JH-sensitive period can be regarded as a window during which cells are responsive to the hormone (perhaps because that is the only time the hormone receptors are present in the cells), while the effects of the hormone can be established in a relatively brief period during this window.

In hemimetabolous metamorphosis there is one JH-sensitive period during each larval instar. Generally, JH titers fluctuate to some degree throughout larval life, but during the JH-sensitive periods they are always above threshold. During the last larval instar the corpora allata are turned off and JH falls below threshold during the JH-sensitive period (fig. 5.8, top). At that point, the switch from larval to adult commitment is made. In holometabolous metamorphosis the situation is complicated by the necessity to pass through two different developmental switches, first to the pupal and then to the adult stage. The absence of JH during the sensitive period while the animal is a larva results in the development of pupal characters, while absence of JH during the pupal JH-sensitive period provokes adult development. The sequence larva → pupa → adult appears to be strictly preprogrammed and requires no special graded or progressive stimulus (fig. 5.8, bottom). The experiments outlined above show, however, that some tissues can bypass the pupal stage if JH is entirely absent during the last larval instar. Tissues derived from imaginal disks are evidently capable of making adult structures without previously passing through a pupal morphology. Some provision must therefore be made to suppress premature adult differentiation and ensure normal pupation of those tissues.

In *Manduca* the suppression of premature adult differentiation of imaginal disk tissues is accomplished by the brief peak of JH just prior to pupa-

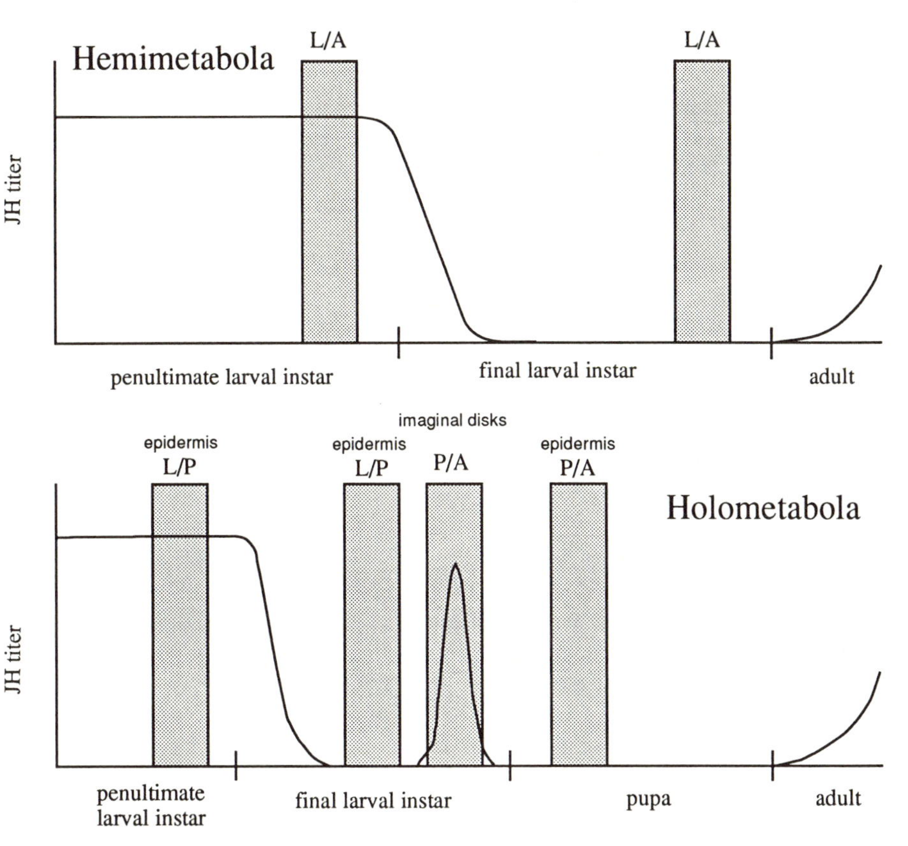

Figure 5.8. Typical juvenile hormone titer profiles during the end of larval life of hemimetabolous and holometabolous insects. Hypothetical JH-sensitive periods are indicated by gray bars. In the Hemimetabola there is a JH-sensitive period near the end of each larval instar, during which the commitment to larval versus adult development is controlled. In the Holometabola there is a JH-sensitive period for larval versus pupal determination near the end of each larval instar, except in the final larval instar. The two JH-sensitive periods in the final larval instar are based on the findings in *Manduca sexta* (fig. 5.7). There is an additional JH-sensitive period for pupal versus adult determination early in the pupal stage, at the outset of adult development.

tion (figs. 5.7, 5.8, bottom). If this peak is eliminated, by neck ligation or by allatectomy, some or all of the imaginal disk integument will bypass the pupal stage (Kiguchi and Riddiford, 1978). This peak of JH coincides with a JH-sensitive period during which the pupa/adult switchover of the imaginal disks is controlled. Thus there are two successive JH-sensitive periods during the last larval instar. The first one controls a larva/pupa developmental switch of the general integument. The second one controls a pupa/adult developmental switch of the integument derived from the imaginal disks (fig. 5.8, bottom). If JH is present during either of these periods, the current developmental state is maintained, whereas if JH is absent, reprogramming to the next developmental state takes place. Normal metamorphosis requires the absence of JH during the first JH-sensitive period, which causes a switch to pupal determination and the presence of JH during the second JH-sensitive period, which stabilizes the pupal-determined state of imaginal disk structures and prevents their commitment to adult development (fig. 5.8, bottom). This scenario explains the requirement for the corpora allata during pupation because, if they are removed, there will be no JH during the second JH-sensitive period and switchover to adult determination will take place (Nijhout and Wheeler, 1982). These observations imply that the imaginal disk tissues are committed to pupal development early in the last larval instar, probably well before the general epidermis becomes committed, and require some sort of restraint late in the instar, when ecdysteroids are secreted, to prevent premature adult commitment. It is interesting in this regard that in *Manduca* the imaginal disks are the only tissues that contain the enzyme JH methyl transferase, which converts JH acid released by the corpora allata to active JH (Sparagana et al., 1985). Possibly this affords a mechanism that ensures the presence of sufficient JH to stabilize the pupal state of the imaginal disks at this critical time in development.

Commitment to adult development does not take place until after pupation. In insects that develop without pupal diapause there is a prolonged period of ecdysteroid secretion during the pupal stage that begins within a few days after pupation and initiates adult development. Adult commitment of the general integument, as well as that of the imaginal disks, occurs during the early phase of this period of ecdysteroid secretion. Hence, the application or injection of exogenous JH early during the pupal stage can cause development of a normal or nearly normal second pupa (Williams, 1959). In the higher Diptera, application of JH at the time of head eversion can still prevent normal adult differentiation of the abdominal histoblasts (Bhaskaran, 1972; Postlethwait, 1974).

In chapter 8, when we discuss hormone-controlled polyphenisms we will see that there are several additional JH-sensitive periods at various points in the life of many insects. In each of those cases, the absence of JH causes a developmental switch to take place. It is believed that these developmental switches always involve the inactivation of some of the genes that are currently active, while causing the expression of new genes (Willis, 1986). The activities of the new gene products then account, at least in part, for the appearance of new biochemical, physiological, and morphological characteristics.

It should be noted in conclusion that pupal or adult commitment of epidermal cells during a JH-sensitive period is not an all-or-none process. Several covert steps are involved and the commitment of different characters of the epidermal cells can be dissociated. Lawrence (1969) and Willis et al. (1982) have shown that when larvae of *Oncopeltus fasciatus* or *Pyrrhocoris apterus* are treated with threshold doses of JH they develop patches of cuticle that have a larval pigmentation but adult surface sculpturing. Epidermal cells can thus simultaneously express larval and adult characteristics. Composite cuticles with larval bristles and pupal surface morphology have also been obtained in several species of holometabolous insects (Willis et al., 1982). Such composite responses indicate that during normal metamorphosis different characters require different exposures to ecdysteroids in the absence of JH to change their commitment. It may therefore be possible to subdivide JH-sensitive periods in epidermal cells into subperiods that deal with determination of different structures, each with their own patterns of gene activation and inactivation. Sequential gene activation upon exposure to ecdysteroids in the absence of JH has been well studied at the descriptive level (e.g., Lepesant and Richards, 1989; Lezzi and Richards, 1989); such studies must now be integrated into the causal analysis of morphological sequences in metamorphosis.

Inhibition of PTTH Secretion by JH

In addition to its ability to control the choice of developmental pathways, JH also has several direct physiological effects. Among these are the stimulation of vitellogenin synthesis in adult insects (see chapter 6) and the inhibition of PTTH secretion in larvae. The inhibitory effect of JH on PTTH secretion is a phenomenon restricted to the final larval instar, and has so far been demonstrated only in holometabolous insects. In final instar larvae of *Manduca* the JH titer remains at a high plateau during the first half of the instar (fig. 5.7, bottom). The decline in JH in the middle of the instar coincides with the attainment of the critical weight (Nijhout and Williams, 1974a,b; Nijhout, 1981). In a normal animal, PTTH secretion does not occur until JH has completely disappeared from the hemolymph, 24–48 hours after the critical weight is reached. The secretion of PTTH in the final larval instar can be delayed indefinitely by injections or topical aplications of JH (Nijhout and Williams, 1974b; Rountree and Bollenbacher, 1986). It is now clear that when a larva reaches its critical weight its corpora allata are shut off. If a larva is prevented from attaining its critical weight by starvation, its corpora allata remain active indefinitely and the next molt is postponed. But when a larva is starved after it has reached its crtitical weight, its corpora allata remain shut off and pupation takes place at the normal time. The timing of PTTH secretion, and therefore the onset of metamorphosis, appears to be controlled entirely by the mechanism that shuts off the corpora allata and clears the hormone from the hemolymph (Nijhout and Williams, 1974b; Safranek et al., 1980; Rountree and Bollenbacher, 1986).

The inhibitory effect of JH on PTTH secretion is restricted to the final larval instar. In earlier instars the JH titer is high during PTTH secretion, and injections of additional JH have no effect on the timing of PTTH release or on molting. During the final larval instar, however, even a small amount of JH prior to the pupal molt can prevent the metamorphosis of some tissues and results in poorly viable larval-pupal intermediates. The inhibition of PTTH secretion by JH during the final larval instar therefore probably evolved as a safety mechanism designed to prevent the premature onset of the metamorphic molt as long as any JH is still present. This safety mechanism has been adapted by several species of insects to control other aspects of their development. Diapause during the final larval instar in Lepidoptera, for instance, is due to the persistence of high titers of JH (see chapter 6). The control of soldier caste determination in ants also makes use of the inhibitory effect of JH on PTTH secretion by delaying pupation of presumptive soldier larvae, so they grow larger during the final larval instar than presumptive worker larvae (see chapter 8).

Ecdysiotropic Effects of JH

We have just seen that in the last larval instar JH inhibits the secretion of PTTH and therefore has an indirect inhibitory effect on the secretion of ecdysteroids. Early studies in insect endocrinology suggested that at other times JH may also stimulate ecdysteroid secretion because implantation of active corpora allata could stimulate brainless pupae of Lepidoptera to initiate normal adult development. Injections of crude extracts of JH into brainless pupae likewise stimulated adult development, suggesting that JH could either stimulate the prothoracic glands to secrete ecdysteroids or could itself mimic the action of ecdysteroids (Gilbert and Schneiderman, 1959; Smith, 1985). Hiruma et al. (1978) and Hiruma (1980) showed that topical applications of a JH analog on isolated abdomens of *Mamestra brassicae* could stimulate these abdomens to molt only if a prothoracic gland was first implanted. Hiruma's experiments thus suggest that JH causes molting by stimulating ecdysteroid secretion by the prothoracic glands. Yet, while these in vivo experiments are consistent with the hypothesis that JH has a prothoracicotropic function, they do not prove that JH acts directly on the prothoracic glands. In order to prove this, it would be necessary to demonstrate that isolated glands can be activated by JH in vitro. Gruetzmacher and her coworkers have provided this critical test and have shown that while topical application of JH can stimulate neck-ligated larvae of *Manduca* to molt it is impossible to stimulate isolated prothoracic glands of *Manduca* to secrete ecdysteroids in vitro with JH analogs (Gruetzmacher at al., 1984a,b; Smith, 1985). Thus JH does not seem to act directly on the prothoracic glands, and the physiological significance of the apparent ecdysiotropic effect of JH remains unclear.

The Coordination of Physiological and Endocrine Events during Molting and Metamorphosis

We are now ready to integrate several features of the physiological control of molting and metamorphosis discussed in chapter 4 into what we have just learned about the endocrine events during molting and metamorphosis. We will also add a few points of special interest not mentioned before. The correlations and causal relations among various endocrine and physiological effects are best understood in *Manduca sexta* (Bollenbacher, 1988; Gilbert, 1989), and this species will serve as our exemplar.

Figure 5.7 illustrates, on a common time scale, the profiles of the various hormones and of the JH-sensitive periods during the last two larval stages and the pupa. During the fourth (penultimate) larval instar, the JH titer declines, gradually at first and then more rapidly, until a low point at the end of the instar. There is a JH-sensitive period that ends just before the rapid decline phase that controls a potential reprogramming to pupal development, should JH be absent. Secretion of PTTH takes place just about the time the rapid decline phase of JH begins and is followed by a period of ecdysteroid secretion lasting about a day and a half. Apolysis occurs shortly after the onset of ecdysteroid secretion and ecdysis to the fifth instar takes place about two days later. There is one additional brief JH-sensitive period near the time that the JH titer reaches its lowest point some 20–30 hours before ecdysis. This JH-sensitive period controls a color polymorphism switch. If JH is below a threshold at that time, the fifth instar larva will be black, whereas if JH is above threshold the fifth instar will be green, like all prior instars. In *Manduca quinquemaculata* such black larvae develop when reared under cool temperatures, and in *M. sexta* they can be easily induced in the laboratory by allatectomy after the first JH-sensitive period is passed (Hudson, 1966; Truman et al., 1973).

Early in the fifth (final) larval instar there is a brief peak of JH followed by a more prolonged plateau. When the larva reaches its critical weight, the corpora allata are shut off and the JH titer begins to decline. Inactivation of the corpora allata is followed shortly by a rise in the level of JH-specific esterases in the hemolymph which rapidly inactivate the remaining JH (Hammock, 1985). While the JH titer is still high during the fifth instar, the secretion of PTTH is strongly inhibited and feeding and growth continue. The inhibition of PTTH by JH during the last larval instar is a safety mechanism that ensures that no molt will take place while there is still some JH present. When JH disappears, this inhibition is relieved and PTTH secretion can begin. After that point the actual timing of PTTH secretion is determined by the photoperiod. PTTH secretion occurs only during a particular time period called a *photoperiodic gate* (in fifth instar larvae of *Manduca* this gate occurs during the dark phase of the photoperiod; Truman, 1972; Truman and Riddiford, 1974a). The PTTH is secreted during the first photoperiodic gate *after* JH has disappeared from the hemolymph (Nijhout, 1981). As figure 5.7 (top) shows, the secretion of PTTH at this time occurs as a series of short bursts, though the significance of this secretory pattern is not understood. A small peak of ecdysteroids follows PTTH secretion. This is the peak that stimulates the switchover from larval to pupal commitment in the epidermis and imaginal disks (Riddiford, 1985; Kremen and Nijhout, 1989).

The final larval instar is thus characterized by a new kind of endocrine interaction, namely the inhibition of PTTH secretion by JH (Nijhout and Williams, 1974b; Rountree and Bollenbacher, 1986). It turns out that during the early portion of the final larval instar the prothoracic glands are relatively unresponsive to artificial stimulation by PTTH, and competence of the glands to respond to PTTH depends on the disappearance of JH (Watson et al., 1987; Watson and Bollenbacher,1988). The declining titer of JH after the critical weight is attained thus disinhibits PTTH secretion and activates prothoracic gland competence. The mechanisms by which the endocrine system is apprised of the fact that the animal is in the final larval instar, and by which these new JH-inhibitory interactions are regulated, are as yet poorly understood. We only know that this mechanism is activated during the first larval instar that exceeds a sharply defined threshold size for metamorphosis (Nijhout, 1981).

The actual cause of the switch to pupal commitment is complex. It requires both the absence of JH during the first JH-sensitive period and the secretion of ecdysteroids. Exactly what happens during the JH-sensitive period and the ecdysteroid peak to alter the developmental program is not yet known. It is possible that during the JH-sensitive period the larva-specific genes are actually repressed while something happens to pupa-specific genes that makes them capable of expression. The role of ecdysteroids, then, is to trigger the actual expression of the pupa-specific genes. There is evidence that different tissues gain pupal commitment at different times during the JH-sensitive period, and with different thresholds of sensitivity to the hormone (see the next section, on Spatial Patterns in Molting and Metamorphosis).

When the onset of the molt is accelerated by infusion of 20-hydroxyecdysone early in the final instar before the critical size is attained, a normal larval molt ensues, except that some of the head imaginal disks produce pupal instead of larval cuticle (Nijhout and Wheeler, 1982). Such results indicate that some imaginal disks have already begun their pupal commitment at the outset of the final instar, while JH titers are still high. In contrast to the epidermis, pupal commitment in the imaginal disks does not appear to require exposure to ecdysteroids and may be stimulated simply by the decline of JH from its high peak early in the last larval instar.

About two days after the first small peak of ecdysteroids in the fifth instar there is a large release of PTTH, followed immediately by a large peak of ecdysteroid secretion (fig. 5.7). Apolysis and the molt to the pupal stage begins within a few hours after the onset of ecdysteroid secretion. At the same time the JH titer rises to a level equivalent to that of the earlier plateau phase. This JH peak lasts about as long as the period of ecdysteroid secretion. The JH-sensitive period for the switch to adult development occurs at this time. Since JH is normally present, the pupal status quo is maintained. Ecdysis to the pupal stage takes place about three days after this last secretion of PTTH. By this time the JH titer has fallen below a detectable level again.

About 48 hours after the pupal ecdysis PTTH is secreted again, and this is now followed by a very long period of ecdysteroid secretion that lasts about two weeks. Apolysis and the onset of adult development once more coincide with the onset of ecdysteroid secretion. There is also one more JH-sensitive period during the first few days after pupation. If JH should be present again at this time, the pupal stage is repeated. If not, the switch to adult commitment takes place at that time. The prolonged period of ecdysteroid secretion is apparently necessary to stimulate the long and diverse processes of adult development within the pupal skin. The final stages of adult development, such as pigmentation of the scales and the programmed cell death of some pupal muscles and neurons, are cued by the declining titer of ecdysteroids.

At the end of adult development, when the new adult moth is fully formed and pigmented within the old pupal cuticle, eclosion hormone (EH) is released in response to the declining titer of ecdysteroids at that time (Truman, 1981). The EH triggers the stereotyped behavior of eclosion and also causes plasticization of the cuticle of the wings. Finally, after the adult has eclosed, bursicon is secreted. The secretion of bursicon is probably a programmed consequence of the prior secretion of EH, though it can be modified by sensory input. *Manduca* and *Calliphora*, for instance, can delay the secretion of bursicon for an hour or more until they have found a suitable resting place for cuticle expansion and tanning (Truman, 1981, 1985). Bursicon initially causes a further plasticization of the wing cuticle, which makes the wings easy to expand, and after expansion of the wings causes them to sclerotize and harden.

The physiological control of molting and metamorphosis in *Manduca* thus consists of a complex interaction of physiological and endocrine events. There are at least two different kinds of size-monitoring mechanisms (the threshold size for metamorphosis and the critical weight for inactivation of the corpora allata) that control the onset of precise temporal patterns of hormone secretion and inactivation. The pattern of hormone secretion, in turn, is superimposed on an independent temporal pattern of JH-sensitive periods during which dramatic changes in gene expression can take place. The genetic and developmental switches that occur during a JH-sensitive period are generally triggered or activated by ecdysteroids. Progress to the next developmental stage requires the absence of JH during a JH-sensitive period.

Spatial Patterns in Molting and Metamorphosis

The ecdysteroids and juvenile hormones circulate in the hemolymph throughout the body so that all tissues are exposed to the same concentrations of these hormones and for the same periods of time. Yet the epidermal cells do not all respond equally or synchronously to ecdysteroids and JH (Truman et al., 1974). Not all parts of the epidermis require ecdysteroids for the same period of time to achieve a normal molt. During a larval molt in *Manduca sexta*, the epidermal cells around the prolegs of each abdominal segment require the shortest period of exposure to ecdysteroids, while the cells along a narrow diagonal strip across the center of the segment require the longest period of exposure (fig. 5.9). During the pupal molt of *Manduca*, the pattern by which epidermal cells become independent of ecdysteroids is completely different (fig. 5.10). Now narrow strips along the anterior and posterior border of the segment and the cells around the spiracle require the least exposure to ecdysteroids while the cells around the proleg require the longest. In both kinds of molts, epidermal cells become independent from ecdysteroids in a complex spatial pattern that sweeps across each segment of the body over a period of 4–8 hours.

The commitment to pupal development also occurs at different times in various locations of a segment (Truman et al., 1974). Commitment to pupation can be tested by applications of JH or a JH analog. If JH is applied during the middle of the commitment process, a larval-pupal intermediate

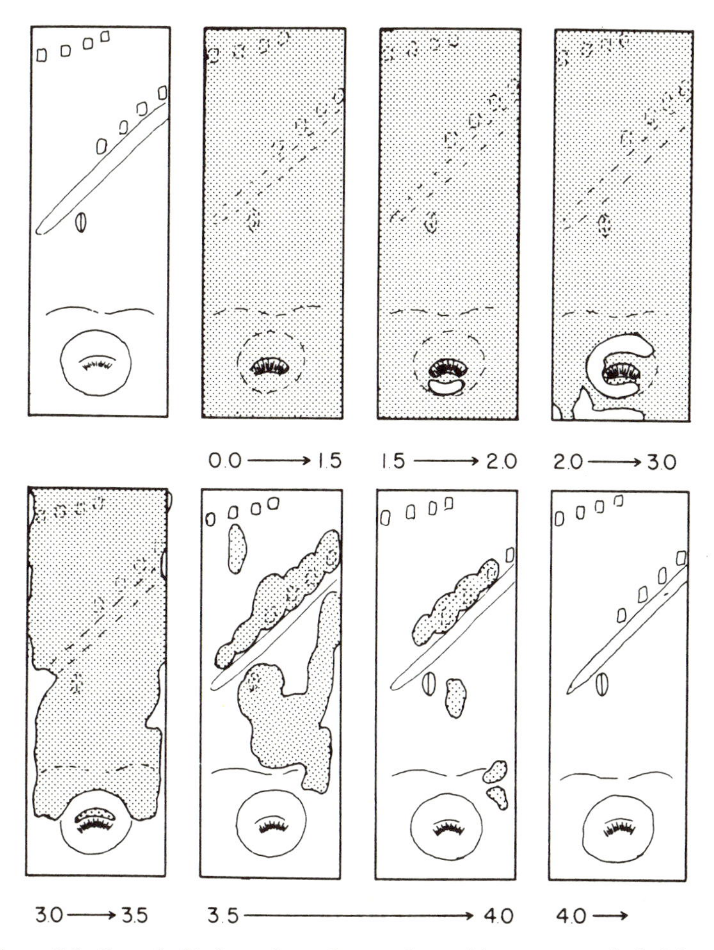

Figure 5.9. Spread of independence from ecdysteroids across a typical abdominal segment of a fourth (penultimate) instar larva of *Manduca sexta*, during a larval-larval molt. Stippled areas are unmolted epidermis. Numbers below each figure give the time (in hours) required for the entire progression across the segment. (From Truman et al., 1974. Reprinted with permission of *Dev. Biol.* and Academic Press.)

animal results whose integument is a mosaic of patches of larval and pupal cuticles. In *Manduca* the entire process takes about five hours and progresses across each abdominal segment in a pattern that is similar, but not identical, to the pattern by which the epidermis becomes independent from ecdysteroids during a larval molt (fig. 5.11).

The spatial pattern of pupal commitment is not the same in all species. In *Precis coenia* pupal commitment starts from a spot on the dorsal midline

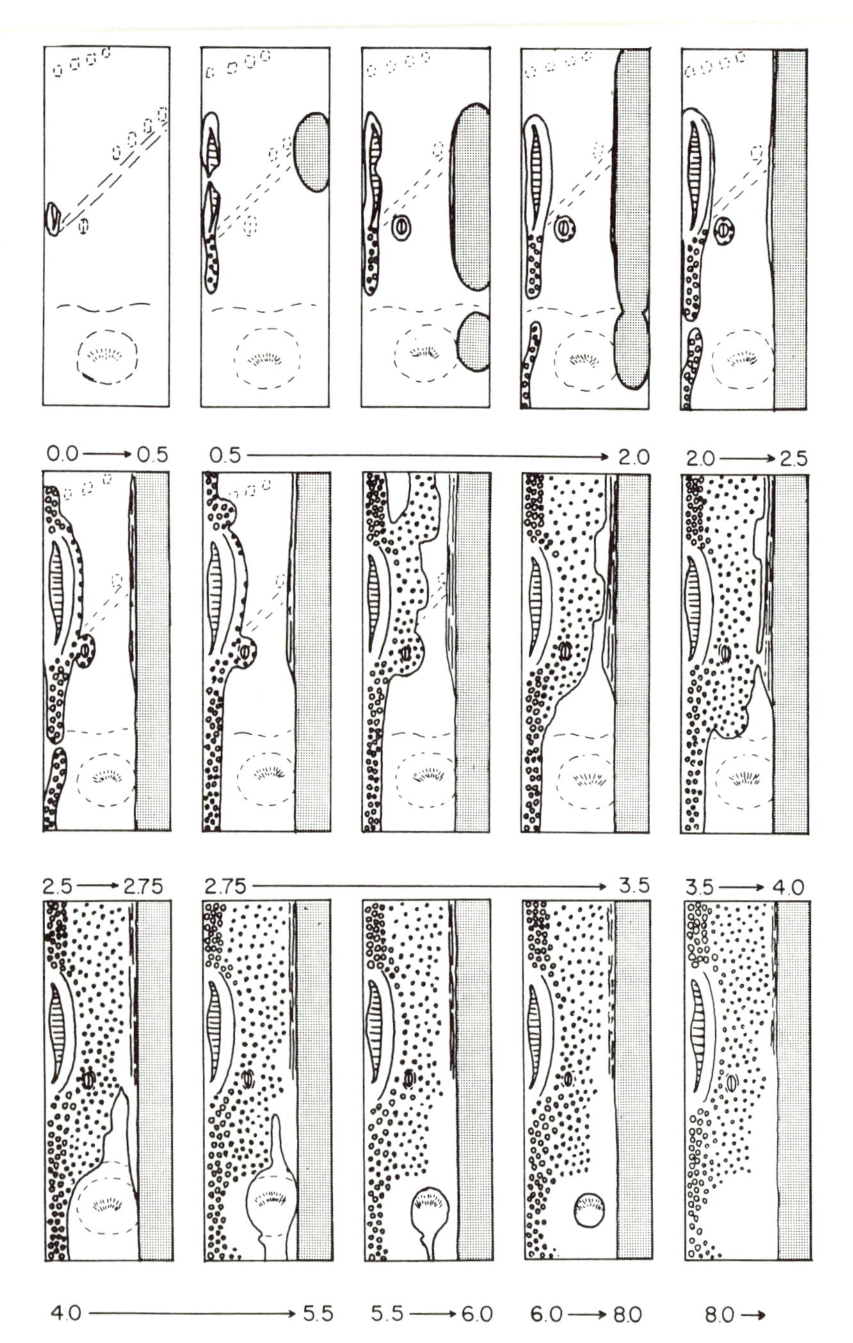

Figure 5.10. Spread of independence from ecdysteroids across a typical abdominal segment of a fifth (final) instar larva of *Manduca sexta*, during the larval-pupal molt. Each drawing shows the area of pupal cuticle (heavily dotted) and pupal intersegmental membrane (light stippling) that will develop if ecdysteroids are withheld at progressively later times by ligation between thorax and abdomen. Numbers below each figure give the time (in hours) required for the entire progression across the segment. (From Truman et al., 1974. Reprinted with permission of *Dev. Biol.* and Academic Press.)

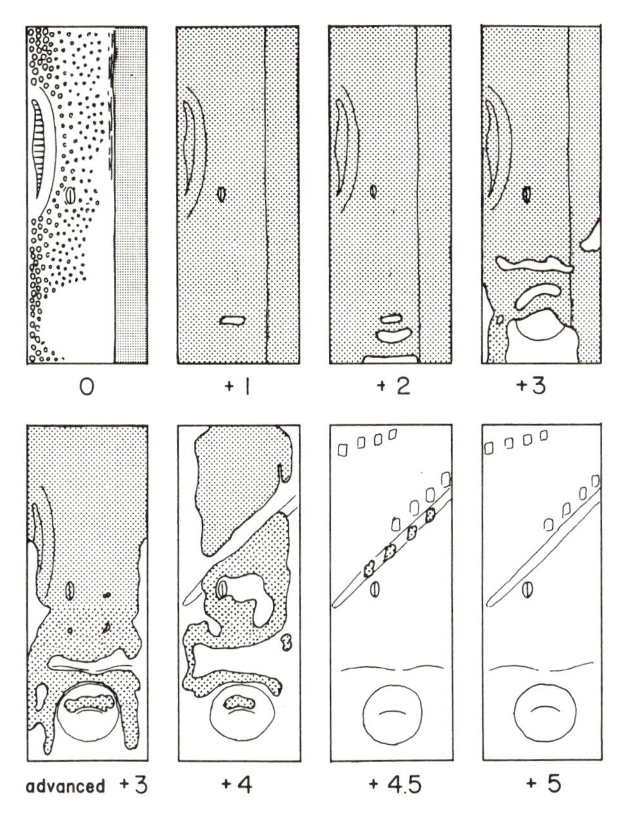

Figure 5.11. Progression of response to exogenous JH in an abdominal segment of *Manduca sexta*. These mosaics were produced by topical application of a JH analog at progressively later times during the period in which pupal commitment occurs. Numbers below each figure are an arbitrary scoring system used to quantify the response to JH. Except for the score of 0, stippled areas indicate pupal cuticle; in all other panels white areas are larval cuticle. (From Truman et al., 1974. Reprinted with permission of *Dev. Biol.* and Academic Press.)

of each thoracic segment and progresses laterally from there. Kremen (1989) has found that if the initial cells in one segment are killed by cautery, pupal commitment in the whole segment is initially inhibited. Commitment in such segments then starts from two lateral initiation points and eventually spreads across the entire segment. Furthermore, if a narrow strip of cells next to the initial cells is killed by cautery it acts as a barrier to the spread of commitment. Pupal commitment can be restricted to a small region of the dorsal thorax by surrounding the initial cells with a ring of cauterized cells (fig. 5.12). Evidently the spatial pattern of pupal commitment in *Precis* is due to the transmission of a signal that originates from a small group of initial cells on the dorsal midline (and two groups on the lateral sides) and passes from cell to cell.

A. Controls

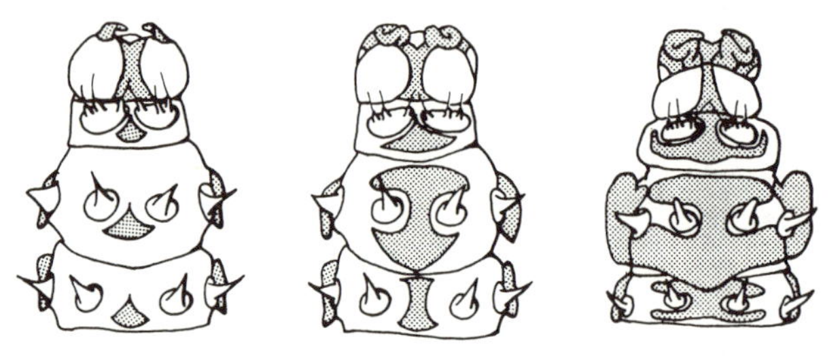
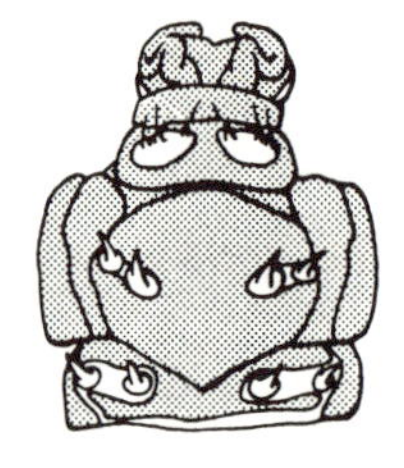

B. Mesothoracic midline cautery

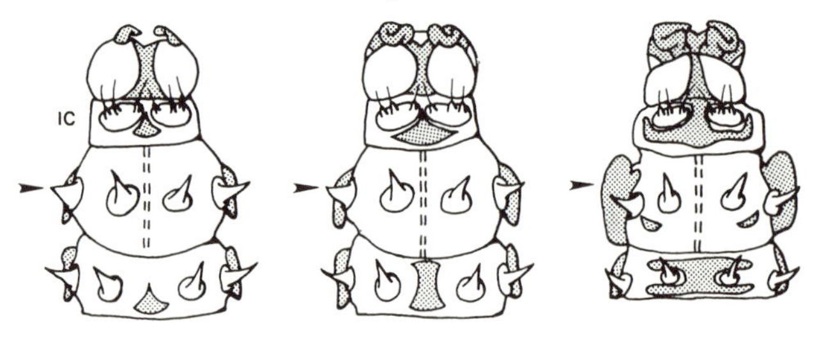

C. Mesothoracic off-center cautery

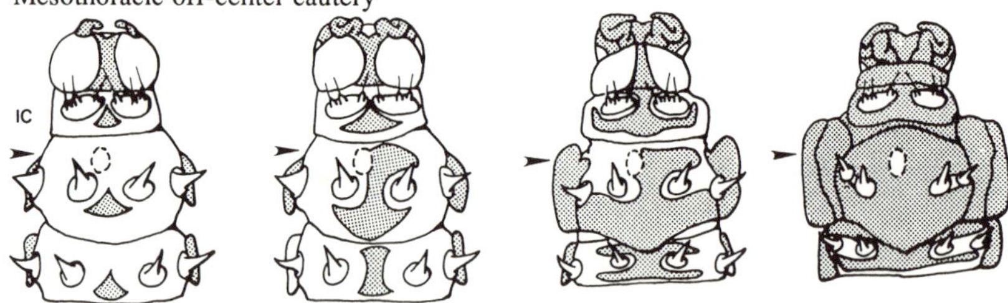

D. Mesothoracic circular cautery

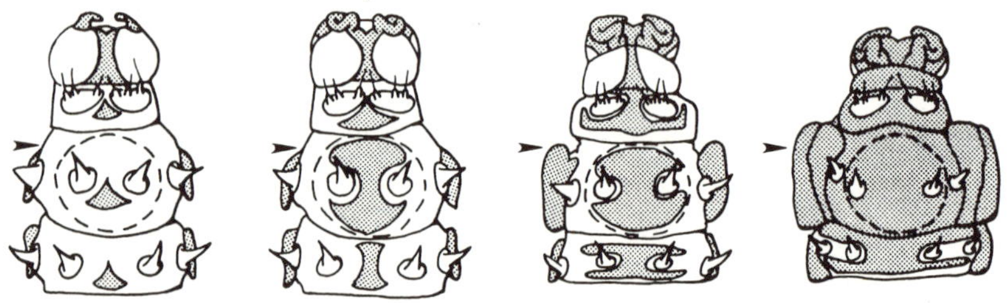

Figure 5.12. Progression of pupal determination on the dorsal thorax of *Precis coenia*, under normal conditions and after various types of cauteries of the epidermis, as explained in the text. Stippled areas represent pupal cuticle; white areas are larval cuticle. The progression of pupal commitment is tested by application of exogenous JH at various times during the period of pupal commitment. (From Kremen, 1989. Reprinted with permission of *Dev. Biol.* and Academic Press.)

Pupal commitment in *Precis* thus requires not only the absence of JH but also a special cellular signal. The nature of this cellular signal is still unknown. Several interesting questions emerge from these findings. First, how do the initial cells become activated? Are they stimulated by the disappearance of JH, or by the first peak of ecdysteroids in the absence of JH? Second, is the cellular signal alone sufficient for pupal commitment, or do the epidermal cells respond to the signal only in the absence of JH, or only in the presence of ecdysteroids?

Delayed Effects of JH

The effects of JH, like those of ecdysteroids, are generally immediate because these hormones act as triggers or reprogramming switches for a variety of physiological and developmental events. There are, however, a few instances in which exposure to JH reprograms events that do not occur until weeks or months later. When JH is applied to eggs of *Hyalophora cecropia*, embryonic development may be disrupted, but individuals that continue to develop normally later metamorphose into pupae that retain patches of larval integument (Riddiford and Williams, 1967; Willis, 1969). More dramatic cases of delayed effects of JH are found in the Hemiptera. When JH is applied to the embryonating eggs of *Oncopeltus fasciatus* or *Pyrrhocoris apterus*, embryonic and larval development proceed quite normally but metamorphosis can be severely disrupted. Many individuals undergo one or more supernumerary larval molts and eventually die as giant larvae, while others molt into unviable larval-adult intermediates (Riddiford, 1970). These effects are identical to those obtained when JH is applied to normal larvae at the beginning of the last larval instar. When the corpora allata of animals treated with JH as embryos are removed at the beginning of the final (fifth) larval instar they undergo normal metamorphosis. These results suggest that exposure of embryos to JH either alters the normal programming of their corpora allata so that they do not cease to secrete JH at the end of larval life, or somehow alters the response pattern to JH of future epidermal cells.

Willis and Lawrence (1970) suggested that these delayed effects of JH might instead be due to the persistence of JH. They showed that small pieces of integument taken from larvae of *Oncopeltus* treated with JH as embryos, and grafted onto untreated larvae, showed a slight retention of larval characters when their host underwent metamorphosis. In following up these experiments, Riddiford and Truman (1972) showed that such effects could be due to JH contamination from excreta and old larval skins left in the rearing containers. Larval hemipterans may be unusually sensitive to JH contamination of their environment. Riddiford and Truman (1972) also showed by bioassay that corpora allata taken from untreated animals at the outset of the fifth instar were inactive, while those from treated larvae were at least partially active. It is unlikely, however, that the early application of JH somehow alters the corpora allata themselves. Instead, it is probably the allatotropic activity of the brain that is altered so that at the end of larval life the brain either fails to inhibit the corpora allata through a failure to secrete allatostatin or a failure to stop secreting allatotropin.

While these studies clearly describe laboratory artifacts, it appears that reprogramming of endocrine activity by an embryonic exposure to JH plays a role in the normal developmental physiology of several species of insects. As we will see in chapter 8, there are several polyphenisms in which there is maternal control over the developmental fate of a larva. For instance, queen determination in some species of ants, the green pigmentation in the solitary phase of migratory locusts, and sexual/parthenogenetic polymorphism in aphids are controlled by a maternal effect that can be mimicked by application of JH to the eggs. Under natural conditions the mother may control the fate of her offspring by the amount of JH she incorporates into her eggs.

Endocrine Interactions between Hosts and Parasites

Molting and metamorphosis in many species of insect that are endoparasites or parasitoids of other insects are synchronized with those of their host. This suggests that host and parasite may use shared cues to control their development. Some parasites apparently can cue in on the hormonal cycles of their host, but it is also clear that several species of parasites have evolved mechanisms to modify the timing of developmental events in their host to their own advantage (Beckage, 1985).

A requirement for the host's hormones has been demonstrated in *Biosteres longicaudatus*, a parasite of the fly *Anastrepa suspensa*. When a parasitized fly larva is ligated around the middle of the body, only the parasites in the anterior compartment (containing the host's endocrine centers) molt, while those in the posterior compartment grow but apparently cannot molt (Lawrence, 1982). When an isolated abdomen is parabiosed to an unparasitized pupa the parasites are able to molt again, presumably stimulated by the ecdysteroids secreted during adult development by the parabiotic partner. *Biosteres* larvae evidently need ecdysteroids from their host to be able to molt. A different requirement for host hormones has been described in the tachinid fly *Gonia cinerescens*, which is a parasitoid of noctuid moths and can be reared in the laboratory on *Galleria mellonella*. First instar larvae of this parasitoid live in the muscles of their host but do not begin to feed and grow until the host reaches the last larval instar. Breaking of this larval "dormancy" and the onset of feeding in the parasitoid are stimulated by the peak of ecdysteroids that initiates pupal commitment in the host. Activation of the parasitoid can be prevented by injections of JH (Baronio and Sehnal, 1980). Evidently the parasitoid is activated by the first pulse of ecdysteroids in the host that occurs in the absence of JH.

Other species of parasites disturb and alter the pattern of endocrine secretion of their host. Often the effect is to inhibit metamorphosis so that the host is maintained in its larval stage (Beckage, 1985). The parasitic wasp *Apanteles congregatus*, a parasite of *Manduca sexta*, can alter the metabolism of JH in its host. In parasitized larvae of *Manduca* the titers of JH do not decline fully at the end of larval life, and thus the secretion of PTTH is inhibited (Beckage and Riddiford, 1982). The reason for the persistence of JH appears to be a suppression of the rise of JH-esterases in the hemolymph. Normally the level of JH-esterase increases a hundredfold and this rise serves to eliminate all JH at the end of larval life. In parasitized larvae JH esterase activity completely disappears, and JH cannot be effectively cleared from the hemolymph and tissues. Exactly how *Apanteles* regulates the JH esterases of its host is still unknown. In addition, the parasites also cause a decrease in the activity of 20-monooxygenase, the enzyme that converts ecdysone to its active form, 20-hydroxyecdysone. The consequent inability of host larvae to activate their molting hormone may be an additional factor contributing to their inability to metamorphose when parasitized (Beckage, 1985).

Insect Hormones and Third-Generation Pesticides

Juvenile hormone must be completely absent during the JH-sensitive periods of the last larval instar in order for the normal developmental switches associated with metamorphosis to occur. Due to its lipid nature, JH penetrates readily through the cuticle, and a small dose of JH topically applied at the right stage can prevent metamorphosis of the underlying integument, resulting in a small patch of larval cuticle on the body of an otherwise normal pupa or adult insect (fig. 5.13). The ability of exogenous JH, applied directly to the cuticle, to induce abnormal metamorphosis is the basis for most bioassay methods for JH and its analogs. When an insect is exposed to a larger dose of juvenile hormone late in its larval life, metamorphosis usually goes awry. While it is theoretically possible to make a final instar larva undergo a supernumerary (additional) larval molt by applying JH, in practice such hormonal interventions seldom work perfectly, and a nonviable monstrous larval/pupal or larval/adult mosaic usually results. Thus exposure to JH at an inappropriate time during metamorphosis is almost always lethal.

The possibility that topically applied JH might be useful as a selective insect control agent was first suggested by the finding that larvae of the bug

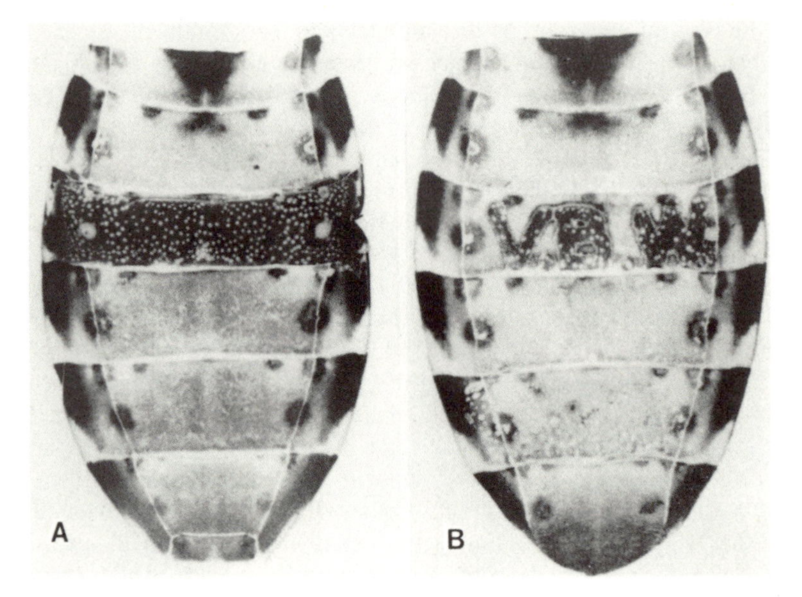

Figure 5.13. Localized response to topical application of JH in *Rhodnius prolixus*. (A) JH applied to a single segment prevent metamorphosis of only that segment (larval cuticle of *Rhodnius* is dark and bears tubercles, adult cuticle is lighter and smooth). (B) Finely localized response to JH is seen where a small amount of the hormone was streaked to form the initials VBW (for Vincent B. Wigglesworth). (From Wigglesworth, 1959. Reprinted with permission of Cornell University Press.)

Pyrrhocoris apterus were extraordinarily sensitive to a juvenile hormone analog found naturally in some paper products. Slama and Williams (1966) found that paper products derived from certain American coniferous trees contained a substance that inhibited metamorphosis of any larval *Pyrrhocoris* that was reared in contact with that paper. This substance was first called "paper factor" and later identified as *juvabione*, a structural analog of juvenile hormone (fig. 5.14). Curiously, juvabione has no effect whatsoever on any other kind of insect; its JH-like activity is restricted to members of the family Pyrrhocoridae (Hemiptera) (Williams and Slama, 1966). This

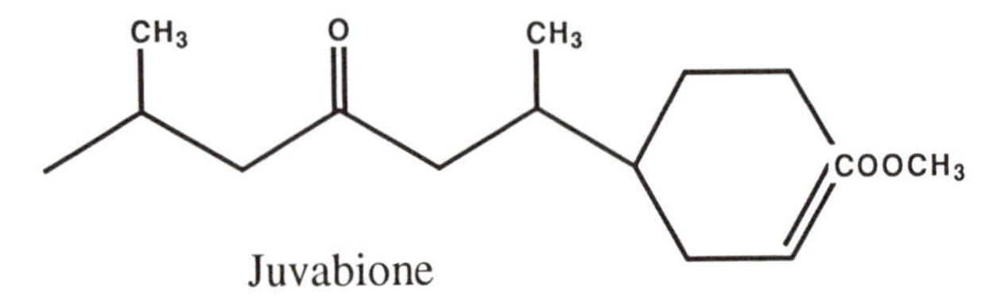

Figure 5.14. Chemical structure of the "paper factor," or juvabione.

highly selective action of the first JH analog to be discovered immediately raised the hopes that JHs in general might be species-specific and that they might form a basis for the development of very selective pesticides. The use of a natural hormone as a control agent would have the additional benefit that the insect would be unable to develop resistance to it (for instance by evolving selective catabolic enzymes for the exogenous hormone) without compromising its own normal endocrine physiology. Williams (1967) coined the term "third-generation pesticide" for this use of an insect's own hormones as a means of insect control (the "first-generation pesticides" are inorganic substances like sulfur and arsenates, used as pesticides since antiquity, and some natural products such as nicotine and pyrethrum, used more recently; the "second-generation pesticides" are the many synthetic organic insecticides mostly developed since World War II).

These high hopes for the use of JHs as selective control agents were dashed by the finding that there are only three natural forms of JH in all the insects, none of which proved to be particularly selective at the genus or species level. Furthermore, it soon became clear that natural JHs were extremely unstable in the environment and extraordinarily difficult, and thus prohibitively expensive, to synthesize in quantity. The idea that hormones might be used, if not as species-specific agents then at least as insect-specific agents, persisted, however, and considerable effort has gone into the development of JH analogs that are both active and stable in the environment. A great many JH analogs have been produced over the years. Figure 5.15 shows the analogs most commonly used in experimental work, and Sehnal (1983) shows the structural formulas for a great many others. Several JH analogs are now commercially available and licensed as insecticides. As pesticides, the JH analogs are effective in disrupting metamorphosis of many insects, and in species that use JH in the regulation of their reproductive cycle, JH-based pesticides can also disrupt reproduction and effectively act as birth control devices (Retnakaran et al., 1985). The earlier expectation that insects might not be able to develop resistance to hormonal pesticides also proved to be unfounded. It has been shown, in both laboratory and field populations, that insects develop a resistance to JH analogs just as rapidly as they do to any other type of insecticide (Hammock, 1985; Hammock et al., 1977; Plapp and Vinson, 1973). While it has proven possible to design JH analogs that are more active in some orders than in others, it seems that insect hormones are subject to the same problems of nonspecificity and resistance as any of the other insecticides.

Perhaps the biggest problem in the use of JH as an insect control agent is that its effectiveness is restricted to a relatively narrow temporal window at the end of the final larval instar. JH is normally present at a high level during the earlier larval instars and exogenous JH has little or no disruptive effect during that time. Unfortunately, it is during these larval stages that

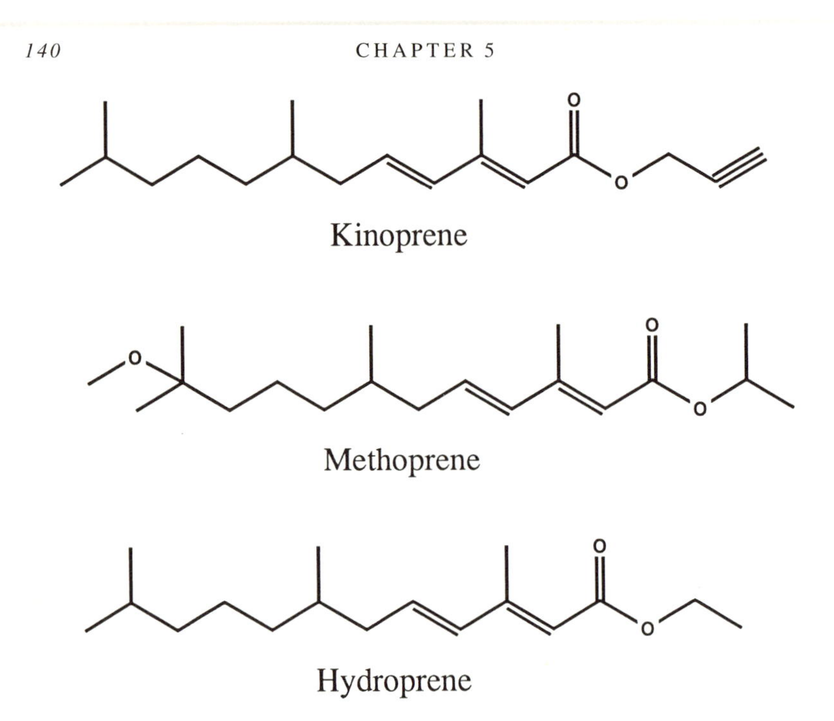

Figure 5.15. Three of the most commonly used JH analogs. In the literature kinoprene is also known as Enstar and ZR 777; methoprene is also known as Altosid and ZR 515, and hydroprene is also known as ZR 512.

many insect pests do much of their damage. A control strategy using JH analogs must therefore depend on disruption of the life cycle, not on acute lethal effects. The narrowness of the window of effectiveness and the short half-life of JH analogs in the environment, therefore, pose problems for their widespread use as general control agents since it would be necessary to apply the analogs repeatedly to ensure exposure of all the sensitive stages in an asynchronous population of pest insects.

A possible way to circumvent the short time window for JH activity might be provided if a means could be found to stop JH secretion prematurely. To this end, Bowers (1976, 1985) has undertaken a directed search for anti-juvenile hormones from natural sources. He was able to isolate two compounds from the common ornamental bedding plant *Ageratum houstonianum*, with the desired properties. When second instar larvae of the milkweed bug *Oncopeltus fasciatus* were exposed to an extract of *Ageratum*, they first underwent two normal larval molts and then metamorphosed prematurely into miniature adults. The active principles in this extract were isolated and were named precocene I and II (fig. 5.16). The precocenes proved to be inhibitors of the secretion of JH by the corpora allata because they act as specific cytotoxic agents to cells of the corpora allata (Pener

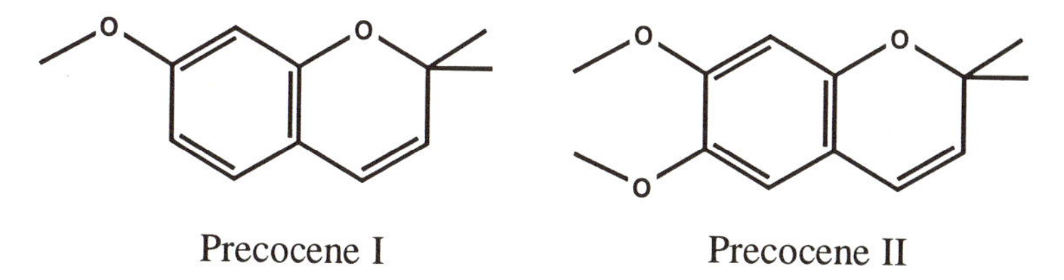

Figure 5.16. The structures of two antijuvenile hormone agents, the precocenes.

et al., 1978; Bowers, 1985). The purified compounds have been shown to provoke premature metamorphosis in a variety of hemimetabolous insects, but they have little or no effect on the development of holometabolous insects. Treatment of a variety of adult insects with the precocenes also inhibits ovarian development, presumably by inhibiting the JH-dependent steps in the ovarian cycle. The precocenes are used extensively in experimental studies on the role of JH in development and reproduction, but have not yet been applied to the problem of insect control.

None of the other insect hormones are particularly promising as insecticides. Ecdysteroids are found abundantly as natural products in plants, particularly in certain ferns. Many of these phyto-ecdysteroids are far more potent than natural 20-hydroxyecdysone, and this suggests that plants may be using these compounds as a defense mechanism against herbivory by insects. A protective or otherwise adaptive function of phyto-ecdysteroids has not yet been critically demonstrated, however. Attempts to use ecdysone analogs as pesticides have been generally unsuccessful.

One interesting way in which insect hormones are involved in the action of pesticides was found by Maddrell and Reynolds (1972), who showed that several neurotoxic insecticides cause the release of diuretic hormone and therefore cause death by dehydration. Insecticides are believed to cause the uncontrolled release of other neurohormones as well, and may thus exert at least some of their toxic effect through the disruption of hormone-mediated metabolism.

In recent years insect physiologists have become interested in exploiting recombinant DNA technology and genetic engineering to disrupt endocrine processes in pest insects (Keeley and Hayes, 1987; Hammock et al., 1990). Genes for neuropeptides, or for the enzymes that metabolize hormones, can be readily inserted into a baculovirus expression vector. Infection with such an engineered baculovirus can then be expected to disrupt specific endocrine events.

CHAPTER 6

REPRODUCTION

THE REPRODUCTIVE cycles of insects are regulated with great precision, and hormones play a role in every case investigated so far. Interestingly, the ecdysteroids and juvenile hormones, which earlier in life control molting and metamorphosis, are the principal hormones that regulate adult reproduction. The physiological effects of these hormones are very different in the adult. Adult insects do not molt and do not grow, nor do they change form, and the hormones that controlled these processes in the larval stage are now used to control the synthesis of yolk protein, the maturation of ovaries, and the development of eggs. Because the prothoracic glands degenerate at metamorphosis, it was long thought that adult insects could not produce ecdysteroids. It is now clear that they do, and in high concentrations, because new tissues of the adult take over the function of ecdysteroid synthesis.

There has been an enormous evolutionary radiation in the morphology of reproductive systems and the physiology of reproductive processes in the insects, with the consequence that no two species are quite alike. In this chapter I will give a brief outline of the general features that have emerged about the morphology and physiology of insect reproduction, and will follow that with several more detailed case histories that illustrate the richness, diversity, and complexity of the endocrine and physiological mechanisms by which insects regulate their reproduction. The last part of this chapter will deal with several additional aspects of reproductive biology, such as sex determination and an interesting biological effect of vertebrate hormones on insect reproduction.

Morphology of the Reproductive Systems of Insects

Females

Female insects have paired ovaries. Each ovary is composed of several ovarioles that lie loose in the body cavity. The ovarioles are joined at their distal end by an apical filament that is connected to the body wall, and proximally by their common oviduct. The number of ovarioles per ovary is species-specific and varies from one (in some beetles) to more than a thousand (in the queens of the higher termites). Typically the ovary contains four to six ovarioles. Each ovariole has a germarium at its distal tip, where oogonia undergo meiosis and where oocytes are produced. Each oocyte becomes surrounded by a single layer of follicle cells that in the early stages of oocyte growth controls the incorporation of vitellogenin (yolk lipoprotein), and later secretes the egg shell around the mature egg. The oocytes travel down the ovariole in single file and grow by the accumulation of vitellogenin. When an egg is fully grown an impermeable proteinaceous egg shell is secreted around it and it enters the common oviduct.

The ovaries of various groups of insects differ in the way provisions are made for nourishment of the oocytes (fig. 6.1A). The ovarioles of Thysanura, Ephemeroptera, Odonata, Plecoptera, Orthoptera, and Isoptera have only follicle cells around their oocytes, and are called *panoistic ovarioles*. In most other orders the oocyte is connected to a cluster of nurse cells that supply it with proteins, nucleic acids, and ribosomes. This type of ovary is called *meroistic*. The nurse cells are sister cells of the oocyte that arise by mitosis and remain connected to it by cytoplasmic strands. In the Hemiptera, Homoptera, and some Coleoptera the nurse cells are collected in a cluster at the distal end of the ovariole (a telotrophic ovariole). In the other orders the nurse cells remain broadly attached to the oocyte and travel down the ovariole with it, jointly surrounded by the follicle cells (a

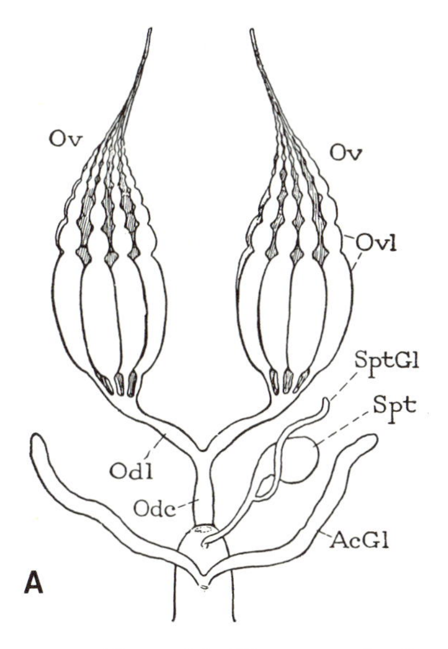

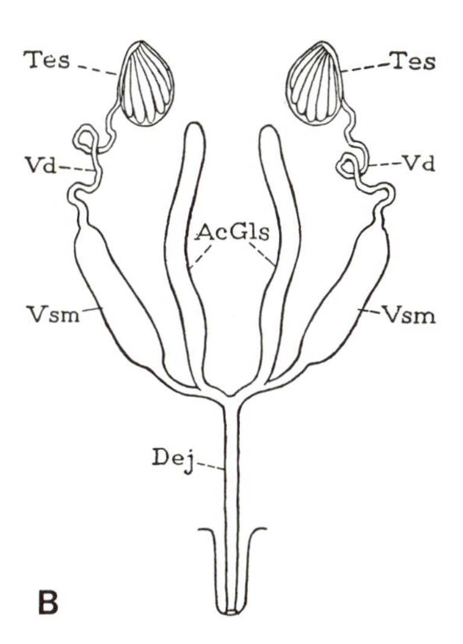

Figure 6.1. Diagrammatic view of the generalized reproductive systems of insects. (A) female; (B) male. AcGl, accessory gland; Dej, ejaculatory duct; Odc, common oviduct; Odl, oviduct; Ov, ovary; Ovl, ovariole; Spt, spermatheca; SptGl, spermathecal gland; Tes, testis; Vd, vas deferens; Vsm, seminal vesicle. (From Snodgrass, 1935. Reprinted with permission of Cornell University Press.)

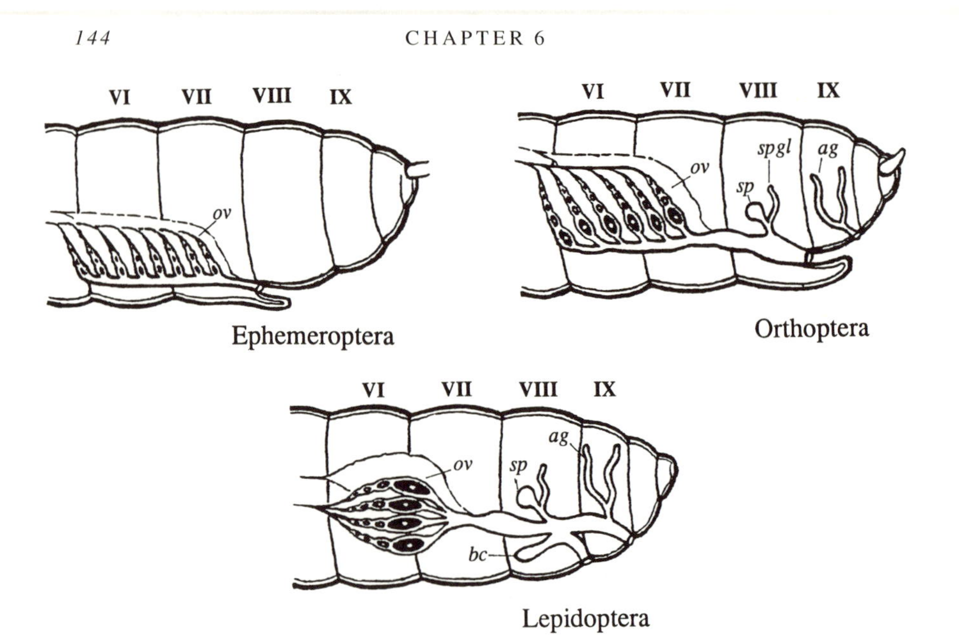

Figure 6.2. Lateral view of the female reproductive systems of three different insects with various degrees of elaboration of the ducts and accessory glands. Roman numerals indicate segment numbers. Genital openings can occur behind segments VII, VIII, and IX, depending on the taxonomic group; *ag*, accessory glands; *bc*, bursa copulatrix; *ov*, ovaries; *sp*, spermatheca; *spgl*, spermathecal gland. (Redrawn after Snodgrass, 1935, and Weber, 1954.

polytrophic ovary). An oocyte with its surrounding follicle cells (and its nurse cells in the case of polytrophic ovaries) is referred to simply as a follicle.

Insect eggs are fertilized within the body of the female and in most species the female can store sperm for a prolonged period of time in a spermatheca. The extreme case is that of social Hymenoptera, where the queen can live for ten years or more and produce thousands of offspring each year, all on stored sperm acquired from inseminations soon after eclosion. Eggs are fertilized in the common oviduct as they pass the spermathecal duct. The sperm enters the egg through a small pore in the egg shell, the micropyle. In most species the eggs are deposited singly or in small groups, and may be glued to a substrate by secretions of the accessory glands. Some species, particularly in the Orthoptera and Blattaria, lay large batches of eggs encased in an ootheca for protection. The ootheca is made of tanned proteins that are also produced by the accessory glands. Figure 6.2 illustrates three types of anatomical arrangements of the female reproductive organs and ducts as they have evolved in different groups of insects.

Males

Male insects generally have paired testes (fig. 6.1B), although these are fused into a single median testis in some species. Each testis is composed of a number of follicles, each with its own germarium, analogous to the ovarioles of females except that the testicular follicles are generally encased in a common sheath, making the testis a fairly compact body. Within each follicle, groups of spermatocytes are surrounded by a membrane to form a spermatocyst. All spermatocytes within the cyst develop synchronously and in some species are ejaculated as a group. There is some uncertainty as to whether the sperm in a single spermatocyst are genetically identical or not (Hannah-Alava, 1965; Engelmann, 1970). Spermatogenesis in short-lived insects takes place entirely during the larval and pupal stage, while in long-lived insects it may continue for some time into adult life.

The male accessory glands produce a fluid that surrounds and nourishes the sperm and serves as a vehicle for ejaculation. In most hemimetabolous insects (except the Homoptera) and in the Lepidoptera, males package their sperm into a spermatophore, a tough proteinaceous capsule secreted by the accessory glands. In some insects the male accessory glands produce substances that alter the behavior of the inseminated female, making her refractory to further attempts at insemination and thus rendering her effectively monogamous (see chapter 9).

General Features of Insect Reproductive Physiology

Insects can be divided into two groups depending on how their egg production is controlled. Insects with short-lived adults and those that do not feed as adults generally develop their eggs from reserves accumulated during the larval stage. In *Bombyx mori* and *Hyalophora cecropia*, for instance, eggs start to mature during adult development, while the animal is still within the pupal skin, and the adult emerges with a full complement of eggs that are ready to be laid as soon as the female has mated. Insects with longer-lived adults usually do not begin to mature eggs until after adult eclosion, and may produce several batches of eggs in the course of their lifetime. Egg production in such insects has, where it has been examined, been found to be under hormonal control, though the actual mechanisms of regulation are extremely diverse. Each of the diverse control schemes that have evolved among the insects are adaptations to a particular life-style and reproductive strategy, as we will see below.

Most insects produce eggs continuously and lay them singly or in small clusters. But some insects, such as mosquitoes and other blood feeding insects, and most Orthoptera and Blattaria, produce eggs in discrete batches. Control over egg production by these two methods is necessarily different. In batch producers there is usually a specific trigger such as a blood meal, or oviposition of a prior batch of eggs, that starts egg maturation in all ovarioles simultaneously. In continuous producers there are no unique triggers, and eggs are produced in continuous succession. In *Manduca* and other Lepidoptera, egg maturation occurs continuously in each ovariole and a hundred or more eggs in all stages of maturation can be found along its length. Mosquitoes, by contrast, develop a single egg synchronously in each of their many ovarioles. Development of the next oocyte in each ovariole is delayed until the previous egg is deposited. In *Drosophila* there are a few immature eggs in each ovariole in a state of slow or arrested development and the ovarioles produce single mature eggs in a staggered fashion (King, 1970).

The process of egg development in an ovariole can be divided into three stages. First, the oocyte undergoes a period of *previtellogenic growth* during which it accumulates a variety of proteins, carbohydrates, and other nutrients. Then follows a period of *vitellogenesis*, during which the oocyte specifically accumulates vitellogenin from the hemolymph. The final stage is *choriogenesis*, during which the proteinaceous egg shell (chorion) is secreted around the oocyte. In some species choriogenesis is preceded by a brief period of hydration during which the oocyte swells significantly. All three phases of oocyte maturation require a special activity of the follicle cells. During previtellogenic growth and vitellogenesis the follicle cells control the access of proteins to the oocyte. Proteins that enter the oocyte do not pass through the follicle cells, but through the intercellular space between them. These intercellular spaces are particularly evident during vitellogenesis and arise by a special shrinkage and deformation of the follicle cells. Vitellogenin must pass through the intercellular spaces before it can enter the oocyte, so the follicle cells can effectively regulate the timing and rate of vitellogenesis (vitellogenin uptake) by the oocyte. Selectivity of protein uptake appears to reside entirely in the membrane of the oocyte itself. Uptake is by pinocytosis and presumably involves specialized surface molecules that recognize specific dissolved proteins. The chorion is a direct secretory product of the follicle cells. An imprint of the follicle cells is usually visible as a fine microstructural sculpturing of the chorion surface and serves as a species-specific taxonomic character in many insects.

In the majority of insects the fat body is the only source of vitellogenin, the lipoprotein that makes up most of the yolk of insect eggs. In the Diptera the follicle cells of the ovary also produce a small amount of vitellogenin. The source of amino acids for the synthesis of vitellogenin is diverse. Some insects must take protein meals before they can mature a batch of eggs, while others can produce eggs on any source of food and presumably synthesize amino acids from a variety of precursors via intermediary metabolism in the fat body. Female insects that receive a spermatophore from the male digest the spermatophore and use the liberated amino acids for the synthesis of vitellogenin. During times of stress and prior to adult diapause, insects can resorb partially matured eggs and store or reuse the protein for other purposes.

Hormonal Control of Reproduction in Females

The progress of egg maturation can be controlled at various points and, depending on the species, one or several processes may be regulated independently. The best-documented control points in the reproductive cycle are the following:

- Vitellogenin (yolk protein) synthesis by fat body
- Separation of a new follicle from the germarium
- Previtellogenic growth of the oocyte
- Vitellogenesis (yolk protein uptake by oocyte)

Ecdysteroids, juvenile hormones, and various neuroendocrine hormones have been shown to control these processes, but species differ greatly in which hormones they use for control and in the number of processes regulated. In a few species the synthesis of vitellogenin is not under external control, but in most it is stimulated by JH. A notable exception are the Diptera, in which vitellogenin synthesis is stimulated by ecdysteroids. Uptake of the vitellogenin by the oocyte is, likewise, uncontrolled in some species, stimulated by JH in others, and seems to require a neurosecretory hormone in *Locusta* and *Danaus* (Raabe, 1986). Even though the control of ovarian development has been studied in dozens of species ranging across the entire taxonomic range of the insects, few if any general rules have emerged that could lead us to deduce a basic or ancestral scheme of ovarian regulation. Even the reproductive cycle of *Thermobia* (see below), an apterygote and thus one of the most primitive insects, appears to be derived and adapted to its peculiar life cycle in which molts and reproductive cycles alternate. The most widespread regulatory mechanism in egg maturation is the stimulation of vitellogenin synthesis by JH (Engelmann, 1983; Koeppe et al., 1985). This occurs in all orders so far examined, from the Apterygota to the Diptera, and in some species it appears to afford the only mechanism of control over ovarian maturation. Insofar as prevalence across a broad taxonomic range may be taken as an indication of primitiveness, perhaps the regulation of vitellogenin synthesis by JH constitutes the basic scheme from which all other regulatory mechanisms have evolved.

Research over the past fifteen years has revealed an increasing number of species, mostly among the Diptera, that use ecdysteroids in one aspect or another of the control of ovarian maturation. Since the prothoracic glands degenerate during metamorphosis, there has to be a different source of ecdysteroids in the adult insect. In all cases studied so far it is the ovary itself, and specifically the follicle cells, that synthesize and secrete ecdysteroids in adult females.

Below I will outline five case histories that illustrate the scope and diversity of physiological control of female reproductive cycles in insects. It should be noted, though, that many of the regulatory steps in the ovarian control pathways I will discuss are still incompletely understood and continue to be the subject of considerable current research.

Thermobia domestica (Thysanura)

The firebrat, *Thermobia domestica*, is one of the Apterygota and thus represents one of the most ancient lineages of insects. Unlike other insects, *Thermobia* does not lose its prothoracic glands during metamorphosis and continues to molt periodically throughout its adult life. Females produce one batch of eggs during each molting cycle, and there is a well-developed coordination between the molting cycle and the reproductive cycle (Watson, 1964; Rohdendorf and Watson, 1969; Bitsch, 1985). When a female *Thermobia* molts, it sheds the lining of its spermatheca and loses its store of sperm. A female must therefore mate after each molt if it is to reproduce. Mating usually takes place within five days after ecdysis and the act of mating triggers the next reproductive cycle (fig. 6.3). Immediately after mating, the JH titer in the hemolymph rises and vitellogenin synthesis begins (Bitsch et al. 1985; Rousset et al., 1987). The rise in JH also stimu-

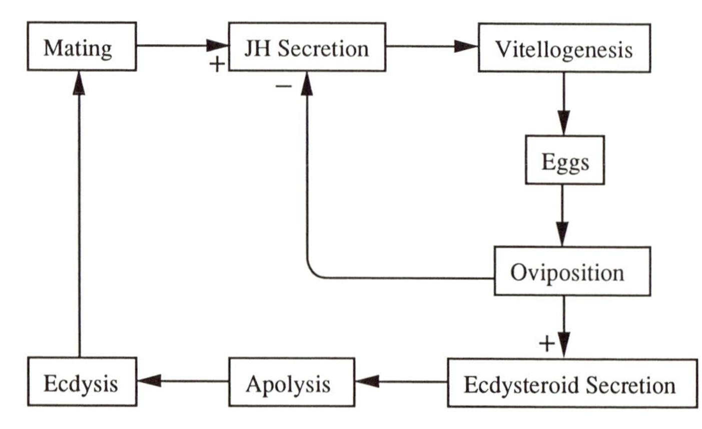

Figure 6.3. Control diagram of a reproductive and adult molting cycle in the firebrat *Thermobia domestica*.

lates vitellogenesis in the terminal follicle of each ovariole. The terminal oocytes grow to full size over a period of 2–4 days. The eggs are then oviposited and the molt to the next stage takes place exactly 4–5 days later. Oviposition appears to trigger the onset of the next molting cycle because soon after oviposition the JH titer declines and the ecdysteroid titer begins to rise.

The intermolt period of *Thermobia* can thus be divided into three phases. The first phase is between ecdysis and mating. The length of this phase is variable, and if it is too long, then the terminal oocytes degenerate and the female eventually initiates a molting cycle without maturing a batch of eggs (Bitsch, 1985). The second phase is between mating and oviposition. This phase lasts 2–4 days and is characterized by an elevated titer of JH that stimulates both vitellogenin synthesis and vitellogenesis. The third phase is between oviposition and the next ecdysis. It is characterized by a declining titer of JH and an elevated titer of ecdysteroids and lasts 4–5 days. Each adult molt therefore takes place in the absence of JH, even though the reproductive cycle is dependent on the presence of JH. The actual length of the intermolt period in *Thermobia* is thus determined by the sum of these three phases. Since the duration of the second and third phases is fixed, it is the timing of mating that determines the actual duration of each adult instar.

Diploptera punctata (Blattaria)

Diploptera punctata is a viviparous roach. Its eggs are enclosed in a thin ootheca which is retained in the brood chamber, the terminal portion of the female's reproductive tract. The control of the ovarian cycle of *Diploptera* is designed to deal with the need to alternate the production of a batch of eggs with a prolonged retention of developing embryos within the brood chamber. In *Diploptera* both vitellogenin synthesis by the fat body and vitellogenesis in the ovary depend on JH. Only the terminal oocyte in each ovariole undergoes vitellogenesis at any one time, so that a nearly synchronous batch of eggs is produced during each ovarian cycle.

The secretion of JH by the corpora allata is actively inhibited in virgin females as well as in "pregnant" females, but by different physiological mechanisms (fig. 6.4). In virgin females the inhibition is via nerves; when the nerves between the brain and corpora allata are cut the latter become active and begin to secrete JH and egg production begins (Engelmann, 1959). Normally, JH secretion and egg development begin immediately after mating, stimulated by deposition of a spermatophore in the bursa copulatrix. The effect seems to be purely mechanical, because glass beads placed in the bursa likewise induce the initiation of egg development (Roth and Stay, 1961). If the ventral nerve cord is cut before or shortly after

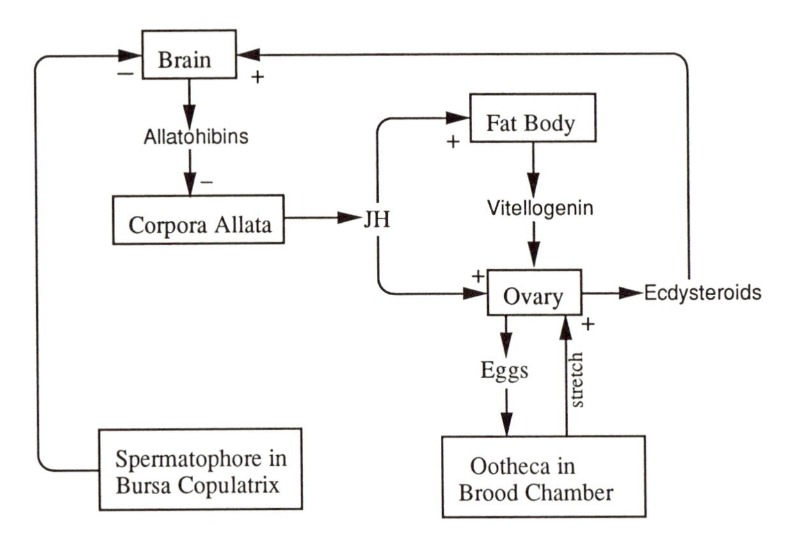

Figure 6.4. Control diagram of the reproductive cycle in the cockroach *Diploptera punctata*.

mating, JH secretion fails to occur. Thus stretch of the bursa after mating stimulates the brain via the ventral nerve cord and this somehow disinhibits the corpora allata.

Inhibition of the corpora allata during pregnancy occurs via a complex endocrine pathway. While the ootheca is in the brood chamber the ovaries are stimulated to secrete ecdysteroids. The pathway by which the ovaries are stimulated is not yet known but it may involve stretch receptors in the bursa which stimulate the secretion of a hormone, possibly related to the ovarian ecdysteroidogenic hormone of mosquitoes (see below). Ecdysteroids inhibit secretion of JH by the corpora allata. Injected 20-hydroxyecdysone likewise inhibits the secretion of JH, but does not inhibit JH secretion by isolated corpora allata cultured in vitro, so its effect must be indirect. Ecdysteroids evidently act on the brain which, in turn, inhibits the corpora allata via an allatostatic neurosecretory hormone (Stay et al., 1980; Rankin and Stay, 1987). When the eggs hatch and the hatchlings are deposited, the inhibition of the corpora allata is relieved and a new cycle of egg development begins.

Locusta migratoria (Orthoptera)

Locusta has a relatively simple control over its ovarian cycle. Like most other orthopterans, *Locusta* produces its eggs in batches. Both vitellogenin synthesis and vitellogenesis are stimulated by JH. Secretion of JH is controlled by the brain, but unlike the case of *Diploptera*, severance of the

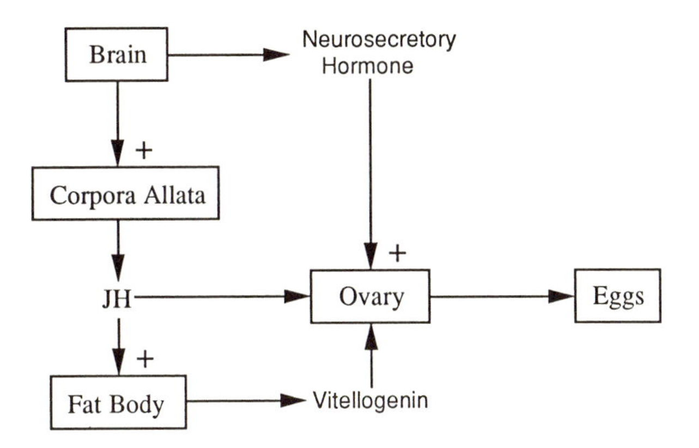

Figure 6.5. Control diagram of the reproductive cycle of *Locusta migratoria*.

nervous connection between brain and corpora allata depresses JH synthesis and vitellogenesis. This implies that in *Locusta* JH secretion during oogenesis is stimulated by the brain. The presence of an allatotropin has been demonstrated by Ferenz and Diehl (1983) and appears to stimulate the corpora allata directly through the nerve (Tobe et al., 1982). Secretion of JH begins autonomously shortly after emergence of the adult, and this induces maturation of the first batch of eggs. Allatectomy abolishes vitellogenesis and either the reimplantation of *corpora allata* or the application of exogenous JH restores vitellogenesis and oocyte development (Roussel, 1978).

Whereas vitellogenin synthesis only requires stimulation by JH, vitellogenesis in *Locusta* appears to require both JH and a neurosecretory hormone from the brain (fig. 6.5). McCaffery (1976) has shown that removal of the medial neurosecretory cells from the brain by cautery caused the cessation of vitellogenesis, and that implantation of corpora allata could not reverse this inhibition. The relative roles of JH and the putative neurohormone in regulating the uptake of vitellogenin is not clear at present. The concentration of vitellogenin in the hemolymph rises and falls with successive cycles of egg maturation, but how this cyclical behavior is controlled is not known.

Rhodnius prolixus (Hemiptera)

Adult females of the bloodsucking bug, *Rhodnius prolixus*, produce a batch of eggs each time they take a blood meal. The blood meal serves two functions in the control of reproduction. First, it provides a rich source of protein whose amino acids are used in the manufacture of vitellogenin. Second, the stretching of the abdomen by the blood meal provides the physical stimulus that initiates the endocrine events associated with oogen-

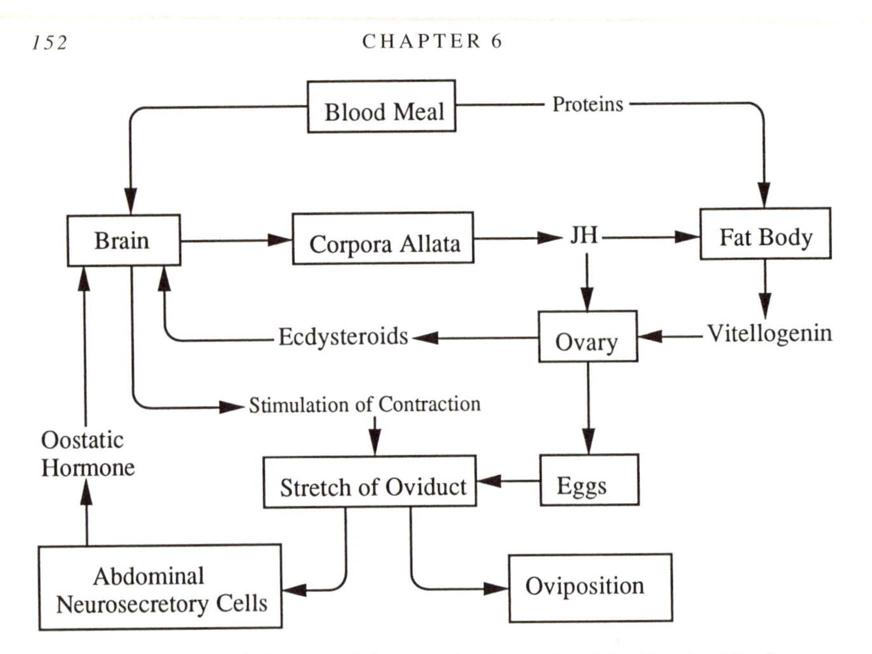

Figure 6.6. Control diagram of the reproductive cycle of the bloodsucking bug *Rhodnius prolixus*.

esis. Abdominal stretch by the blood meal activates JH secretion by the corpora allata, presumably through stimulation of the brain's allatotropic system (Wigglesworth, 1936). The JH stimulates both vitellogenin synthesis by the fat body and vitellogenesis by the ovaries. The main function of JH in vitellogenesis, here as in other species, is to cause an increase in the intercellular spaces among the follicle cells. This may be accomplished by altering the ionic balance within the follicular cells, which causes them to lose fluid (Abu-Hakima and Davey, 1977). Vitellogenin and other hemolymph proteins are believed to diffuse passively through these intercellular spaces. The membrane of the oocyte then takes up the vitellogenin selectively. JH simply regulates the access of vitellogenin to the oocyte membrane; the actual uptake of the vitellogenin does not appear to require JH.

While the terminal eggs in each ovariole are undergoing vitellogenesis and growth, they inhibit vitellogenesis in the remaining oocytes via an oostatic hormone (fig. 6.6). This hormone is secreted by a special neurosecretory organ in the abdominal nerves that is stimulated by stretch of oocytes in the oviduct. The oostatic hormone does not inhibit JH secretion but appears to act on the follicle cells to antagonize the action of JH (Davey and Huebner, 1974). During vitellogenesis the ovaries of *Rhodnius* begin to produce ecdysone, whose production, in turn, causes the brain to secrete a neurosecretory hormone that stimulates muscular contractions in the oviduct. These contractions eventually lead to the expulsion of the eggs from the ovary (Davey, 1967; Kriger and Davey, 1982). The next cycle of oogenesis then starts when the female once again takes a blood meal.

Aedes aegypti (Diptera)

In the yellow-fever mosquito *Aedes aegypti*, the production of a batch of eggs is triggered by a blood meal. As in the case of *Rhodnius*, the digestive products of the blood meal are used in the synthesis of vitellogenin by the fat body, and abdominal stretch triggers the endocrine events associated with egg maturation. But the sequence of endocrine events in *Aedes* is very different and much more complex than that in *Rhodnius* (fig. 6.7). In *Aedes* secretion of JH begins autonomously soon after emergence of the adult. At this time each ovariole contains one undeveloped oocyte called the primary follicle. The initial rise of JH acts on the ovaries and stimulates previtellogenic growth of the primary follicle. During this period the oocytes grow slightly in size and then enter a period of arrested development called the resting phase. JH also acts on the fat body and induces a set of biochemical changes that make the fat body competent to synthesize vitellogenin (Gwadz and Spielman, 1973; Flanagan and Hagedorn, 1977).

Both the early rise of JH and some stimulus associated with mating alter the behavior of a female so that she goes in search of a blood meal. When a female takes a blood meal, abdominal stretch activates the brain, probably via the ventral nerve cord, and causes the secretion of a neurosecretory hormone, the *ovarian ecdysteroidogenic hormone* (OEH; Box 8; Lea, 1972), also known as the *egg development neurosecretory hormone* (EDNH). The peak of OEH is relatively brief and acts as a stimulus that causes the follicle cells of the ovary to secrete ecdysone (figs. 6.7, 6.8). The prior action of JH on the ovary at the time of eclosion is necessary in order to make it competent to respond to OEH (Shapiro and Hagedorn, 1982). As in the case of ecdysone secretion by the prothoracic glands, the ovaries secrete the relatively inactive prohormone which is converted to the active 20-hydroxyecdysone in the fat body. The 20-hydroxyecdysone then acts back on the fat body and stimulates it to secrete vitellogenin (fig. 6.7), and here too the prior action of JH has been shown to be essential to make the fat body competent to respond to 20-hydroxyecdysone (Flanagan and Hagedorn, 1977). The appearance of vitellogenin in the hemolymph activates the resting primary follicles, which then initiate vitellogenic growth. The ecdysteroid peak stimulated by OEH also has a second important function. It causes the separation of a new follicle from the germarium (Beckemeyer and Lea, 1980). This is the secondary follicle that will respond to the next blood meal.

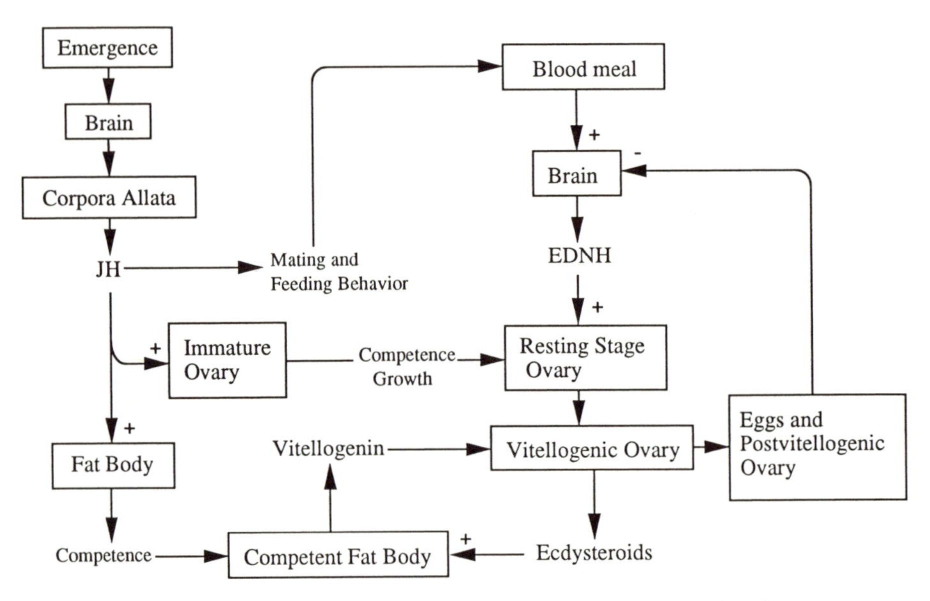

Figure 6.7. Control diagram of the reproductive cycle in the yellow-fever mosquito *Aedes aegypti*. (After Hagedorn, 1985.)

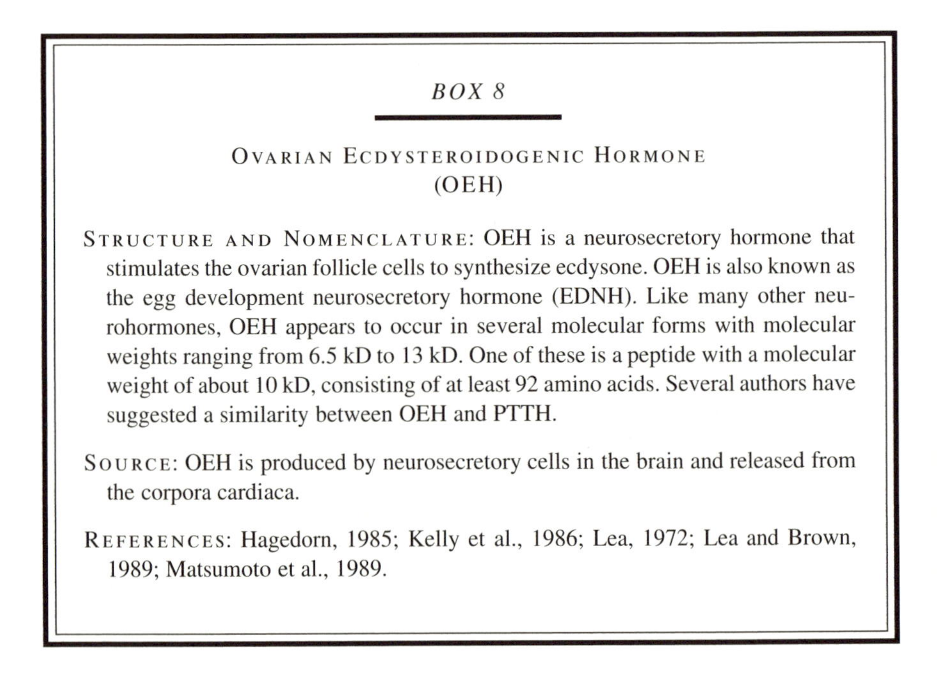

BOX 8

Ovarian Ecdysteroidogenic Hormone (OEH)

Structure and Nomenclature: OEH is a neurosecretory hormone that stimulates the ovarian follicle cells to synthesize ecdysone. OEH is also known as the egg development neurosecretory hormone (EDNH). Like many other neurohormones, OEH appears to occur in several molecular forms with molecular weights ranging from 6.5 kD to 13 kD. One of these is a peptide with a molecular weight of about 10 kD, consisting of at least 92 amino acids. Several authors have suggested a similarity between OEH and PTTH.

Source: OEH is produced by neurosecretory cells in the brain and released from the corpora cardiaca.

References: Hagedorn, 1985; Kelly et al., 1986; Lea, 1972; Lea and Brown, 1989; Matsumoto et al., 1989.

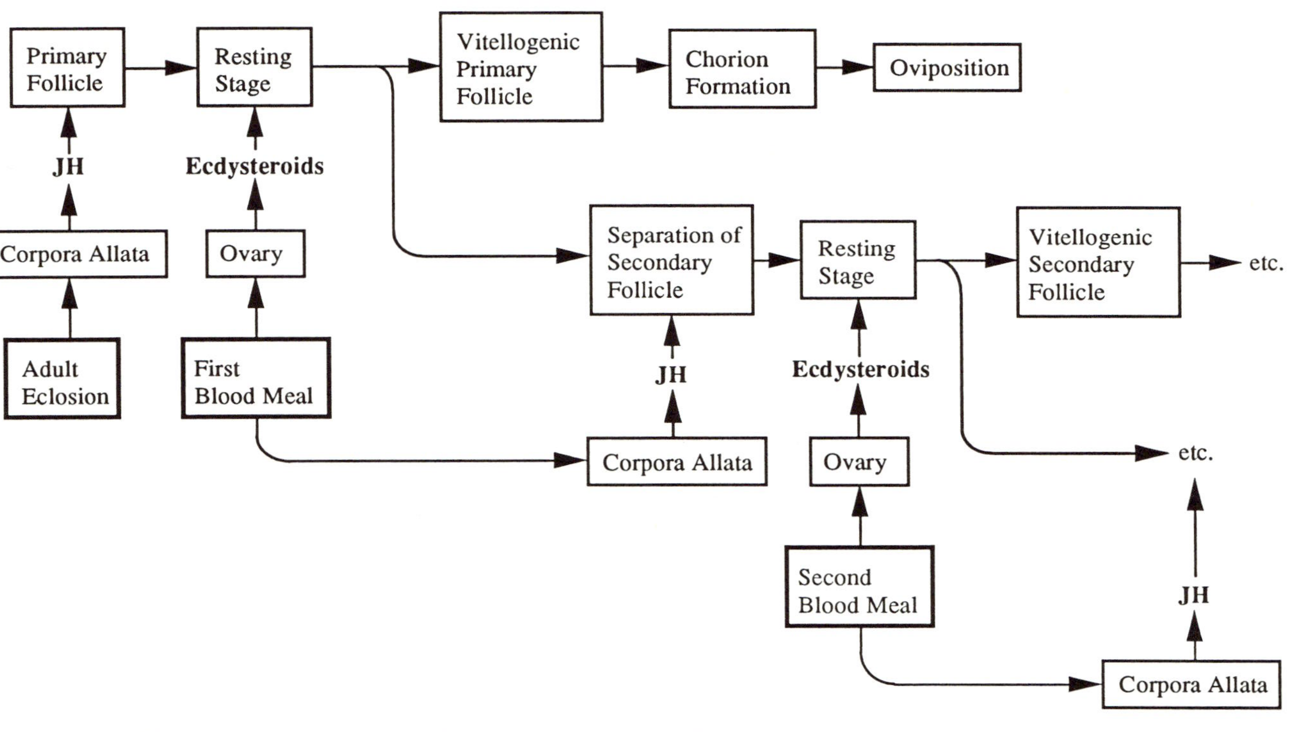

Figure 6.8. Diagrammatic representation of the sequential action of JH and ecdysteroids in controlling the maturation of eggs (follicles) in the mosquito *Aedes aegypti*. Secretion of JH after eclosion stimulates the primary follicle to grow to the resting stage. A blood meal stimulates secretion of ecdysteroids which initiate vitellogenesis and the separation of the secondary follicle in each ovariole. On the second day after the blood meal, the corpora allata are reactivated and the JH stimulates the new secondary follicle to grow to the resting stage. The next blood meal starts the cycle over again. (After Hagedorn, 1985.)

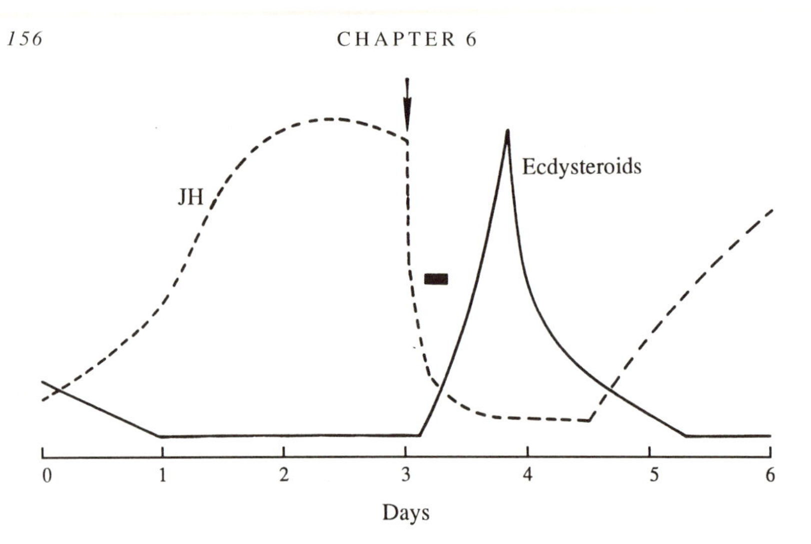

Figure 6.9. Diagrammatic representation of the JH and ecdysteroid titers in an adult female mosquito for the first six days after emergence, and assuming a blood meal is taken on day 3 (arrow). The black bar indicates the presumed period of OEH secretion. (From Shapiro et al., 1986. Reprinted with kind permission of Pergamon Press Ltd.)

The blood meal also induces the appearance of JH esterases and a decline in the JH titer (fig. 6.9). About two days after the blood meal, while the oocytes are growing, the JH titer begins to rise once more. The rising JH titer causes the next undeveloped oocyte (the secondary follicle) in each ovariole to undergo previtellogenic growth (Feinsod and Spielman, 1980; Shapiro et al., 1986). The next set of oocytes is thus prepared for the second blood meal, which is usually taken soon after the first batch of eggs is laid. There is one final control mechanism in the reproductive cycle of *Aedes* that prevents the secretion of OEH while there are still growing eggs in the ovary (Meola and Lea, 1972). The nature and origin of this OEH inhibitor is still unclear, but it acts via a blood-borne factor and provides a safety mechanism that prevents the development of the next batch of oocytes should the female take a blood meal before the first batch is laid (Hagedorn, 1985).

Hormonal Control of Reproduction in Males

Male insects do not produce gametes cyclically, but appear to be always ready to mate. Insects with short-lived adults begin spermatogenesis during the late larval or pupal stage. In these species, spermatogenesis is stimulated by ecdysteroids, probably 20-hydroxyecdysone (Kambysellis and Williams, 1972). Insects with long-lived adults usually do not initiate spermatogenesis until after metamorphosis and there is some evidence that in these species, too, spermatogenesis is stimulated by ecdysteroids. The source of ecdysteroids in adult males appears to be the testis itself (Hagedorn, 1985). Loeb et al. (1982) have shown by tissue culture experiments that in *Heliothis virescens* the cellular sheath of the testis is the actual source of the hormone. A number of investigators have studied the possible effect of JH on spermatogenesis with mixed success. Koeppe et al. (1985) note that in all apparently successful cases of spermatogenesis acceleration by JH, the animals also broke diapause, so the apparent effects of JH may be confounded by the simultaneous endogenous secretion of ecdysteroids (see chapter 5).

Other aspects of male reproductive biology can also be affected by hormones. Males of *Schistocerca* undergo a period of sexual maturation after metamorphosis during which the pigmentation of their legs and abdomen changes dramatically (Norris, 1954). These changes are accompanied by the onset of sexual activity. Removal of the corpora allata from a male inhibits its sexual maturation and mating activity, and implantation of these glands restores the male's ability to mate (Loher, 1960; Pener, 1967).

Hormonal Control of Accessory Gland Function

In females the accessory glands (also called colleterial glands) secrete the glue that attaches the eggs to a substrate, or the proteins and enzymes from which oothecae are made. The best evidence for hormonal regulation of female accessory gland function comes from the roach, *Periplaneta americana*. This species, like many orthopteroids, has a pair of dissimilar accessory glands. The right gland is specialized to secrete the enzyme, beta-glucosidase. The left gland secretes a complex mixture of structural proteins, the oothecins, a high concentration of protocatechuic acid glucoside, and the enzyme phenoloxidase. In the oviduct the eggs are first enveloped by secretions form the right gland, and then pass the duct of the left gland. When secretions of the two glands become mixed, the beta-glucosidase hydrolyzes the glucoside to release free protocatechuic acid which is then oxidized by the phenoloxidase to a highly reactive quinone. This quinone, in turn, reacts with the oothecins and crosslinks them to form an extremely tough tanned case around the eggs (Brunet, 1952; Brunet and Kent, 1955; Davey, 1985b; Koeppe et al., 1985). The synthesis of all three components of the left gland secretion are stimulated by JH and may be coordinated with JH-stimulated vitellogenin synthesis (Koeppe et al., 1985). Interestingly, the right gland is unaffected by JH.

In the tsetse, *Glossina austeni*, the accessory glands of the female act as "milk" glands from which the larva feeds while it grows in utero. Secretions of these glands form the sole food for the entire larval stage. This size of the accessory glands diminishes when flies are allatectomized and increases again when they are injected with JH. The activity of the milk gland may therefore require stimulation by JH (Ejezie and Davey, 1974, 1976, 1977; Davey, 1985b).

Male accessory glands produce the fluid vehicle for the sperm. In some species they also produce specific sperm activation substances and the proteinaceous secretions from which spermatophores are made. Allatectomy results in a diminution of the accessory glands in males of *Rhodnius*, *Schistocerca*, and *Locusta* (Wigglesworth, 1936; Loher, 1960; Girardie and Vogel, 1966), but not in roaches (Engelmann, 1970). In other insects such as *Danaus* (Herman, 1975), *Aedes* (Ramalingam and Craig, 1977), and *Oncopeltus* (Davey, 1985a; Koeppe et al., 1985), JH deprivation results in the cessation of secretory activity in the accessory glands. It is not yet clear exactly at what level in the control pathway JH influences accessory gland development and function in these species (Koeppe et al., 1985).

Hormones and Sex Determination

The secondary sexual characters of insects (the external genitalia and sexually dimorphic features such as antennae, color patterns, pheromone glands, horns, or body size) are under strict genetic control. Sex hormones play no role in the differentiation of these characteristics and that is why sexual mosaics, such as gynandromorphs, made up of patches of distinctly female and distinctly male tissues, are possible and, in fact, quite common in some species of insects. There exists to date but a single instance of a sex-determining hormone in insects. This was discovered by Naisse (1966, 1969) in the firefly *Lampyris noctiluca*. In this species a special tissue at the tip of the testes produces a hormone that controls the differentiation of secondary sexual characters as well as the sex of the gonad. When a male larva is castrated by removal of both the testes and the endocrine tissue, the larva develops into a normal-looking but sterile female adult. When a testis is transplanted into a normal female larva, her own ovaries will develop as testes and she develops as a normal morphological male. Since female appears to be the default sex in this species, and the testes induce male development, the hormone they produce was dubbed the androgenic (male-producing) hormone. The androgenic hormone has not yet been isolated and characterized, but male-inducing hormones are well known in the Crustacea in which they have been tentatively identified as steroids (Highnam and Hill, 1977).

The endocrine activity of the testicular tissue appears to be controlled by the brain, probably via a neurosecretory hormone (Naisse, 1969). It is curious that a sex-inducing hormone should exist in such an advanced insect as a firefly while being absent in the more primitive groups. One would have to assume either a unique evolution of such a hormone in *Lampyris*, which seems unlikely if an elaborate neuroendocrine control also exists, or that androgenic hormones are more widespread but not easily recognized due to the overriding importance of cell-autonomous genetic sex-determination mechanisms in many insects. It is worth noting that there is circumstantial evidence for diffusible sex-determining factors in the wasp *Bracon hebetor* (= *Habrobracon juglandis*). In this species males are either haploid or homozygous at a sex-determining locus, while females are diploid and heterozygous. Whiting et al. (1934) reported finding male-male gynandromorphs that were homozygous for different alleles at the sex locus and were thus composed of a mosaic of male tissues that were genetically different. They found that in many such individuals a narrow strip of female cells formed at the interface between the two genetically different male regions, and proposed that this was due to complementation via a diffusible substance. If such a substance indeed exists and can diffuse between cells, it must pass through gap junctions and is thus unlikely to be a polypeptide. Whether such a substance could in other circumstances be regarded as a sex hormone remains to be determined.

Rabbit Fleas and Rabbit Hormones

A unique kind of endocrine interaction is known to occur between certain species of rabbit fleas and their host in which reproduction of the insect parasite is cued by the reproductive hormones of its vertebrate host. In the rabbit fleas *Spillopsyllus cuniculi* and *Cediopsylla simplex*, ovarian maturation occurs only when the flea feeds on pregnant doe rabbits or newborn young (Rothschild and Ford, 1973). Ovarian maturation in these fleas can be stimulated by exogenous application of vertebrate estrogen and corticosteroids. Exogenous applications of JH likewise stimulate ovarian maturation, which suggests that vertebrate steroid hormones may act to stimulate JH secretion in the flea. Interestingly, application of progesterone, the hormone of pregnancy, has exactly the opposite effect and inhibits ovarian maturation in the flea. The rabbit hormones also stimulate the fleas to increase their rate of feeding. (Rothschild et al., 1970; Rothschild, 1975). With increased feeding rates the defecation rate also increases and this results in the accumulation of protein-rich fecal droppings, the primary food source of larval fleas, in the rabbit's nest. Finally, mating behavior and copulation in these fleas occurs only in the presence of newborn young. The stimulus to mating appears to be a volatile substance, yet unidentified, produced by the nestling rabbits (Mead-Briggs and Vaughan, 1969; Rothschild and Ford, 1973; Rothschild, 1975). Thus the entire life cycle of these fleas, including ovarian maturation, the provision of a larval food supply, mating, and ultimately the production of a new generation of fleas, is coordinated with the reproduction of their host.

CHAPTER 7

DIAPAUSE

DIAPAUSE is a period of arrested development that has evolved as an adaptation to survive seasonally recurring adverse conditions. In temperate zones, diapause is a strategy for overwintering. In tropical regions diapause is usually called *estivation* and provides a way of surviving seasonal periods of drought and scarcity of food. During diapause insects become quiescent, their metabolic rate becomes greatly reduced, and they develop various physiological and biochemical adaptations to desiccation and cold-tolerance (Tauber et al., 1986). Most insects that are capable of diapause do so only once in their life, always at a species-specific point in their life cycle; very few species are capable of diapausing at more than one stage.

Depending on the species, diapause is either genetically controlled and obligatory, or it is facultative and induced by specific environmental cues. Insects with an obligatory diapause always enter diapause when they reach a specific point in their life cycle and therefore have only one generation per year (they are called *univoltine*). Examples of univoltine species are the giant silkmoths (Saturniidae) and tent caterpillars (*Malacosoma*), which diapause in the pupal and egg stage, respectively. Insects that have a facultative diapause generally (but not necessarily) have more than one generation a year (they are *bivoltine* or *multivoltine*). Such insects enter diapause only when they are exposed to a very specific set of environmental conditions that herald the approach of an unfavorable season. Whether obligatory or facultative, and irrespective of the stage at which it occurs, diapause seems to be universally controlled by the endocrine system.

It is important to note that diapause is not a response to, or a result of, the onset of unfavorable conditions. Many insects, for instance, become inactive when the weather gets too cold or when they are deprived of food. This type of quiescence is called *torpor* and does not involve the manifold physiological adaptations of true diapause. Diapause occurs in response to token stimuli in the environment that predict but are not coincident with seasonal change. Therefore diapause always begins well before the actual onset of unfavorable conditions and usually persists for a period of time after favorable conditions return.

Environmental Induction of Diapause

Insects that live in the temperate zone almost always use cues from the photoperiod (the relative lengths of day and night) to predict the change in seasons (Danilevskii, 1965; Beck, 1980; Tauber et al., 1986). As long as day lengths remain above a certain value, most temperate-zone insects continue normal development. But if day lengths fall below that critical value, diapause will be induced. The transition between noninducing and diapause-inducing day lengths is usually very sharp (fig. 7.1A), generally occurring over a span of less than one hour. The day length at which exactly 50% of the population enters diapause is called the *critical day length.* Insects that enter diapause when day lengths fall below the critical day length are called long-day insects. There are also short-day insects; these continue to develop normally when day lengths are short and enter diapause when exposed to day lengths longer than a particular critical value (fig. 7.1B).

The exact value of the critical day length depends on the species and on the geographic region from which the population is taken (fig. 7.2). In a given species, the critical day length increases on the average by about 1.5 hours for every 5° latitude (or about 800 km) away from the equator, or for every 1,000 m in altitude on a mountain at a given latitude (Danilevskii, 1965; Tauber et al., 1986). The critical day length is under genetic control and is inherited as a quantitative character. Danilevskii (1965) has shown that in the moth *Acronycta*, for instance, hybrids between individuals from different geographic areas have critical day lengths in between those of the two parental populations. When such hybrids are back-crossed to either of the parental populations, their progeny have a critical day length about halfway between those of the parent and hybrid. This kind of multifactorial inheritance ensures that a continuous gradient of critical day lengths can be maintained within large contiguous populations of a species.

Each species of insect has a distinct photosensitive stage during which it must experience an inductive photoperiod if diapause is to occur (fig. 7.3). The photosensitive stage usually occurs well in advance of the stage at which diapause takes place, although the timing and duration of the photosensitive stage can vary widely, even among closely related insects (Denlinger, 1985). Many insects that diapause as pupae must experience the appropriate inductive photoperiod during the latter half of their larval life, but in *Sarcophaga crassipalpis* pupal diapause is induced by short day lengths during the latter half of embryonic life (Denlinger, 1971). By contrast, the closely related species *Sarcophaga argyrostoma* requires short days through nearly its entire larval life if it is to enter diapause (Saunders, 1980). One of the longest delays between photosensitive stage and dia-

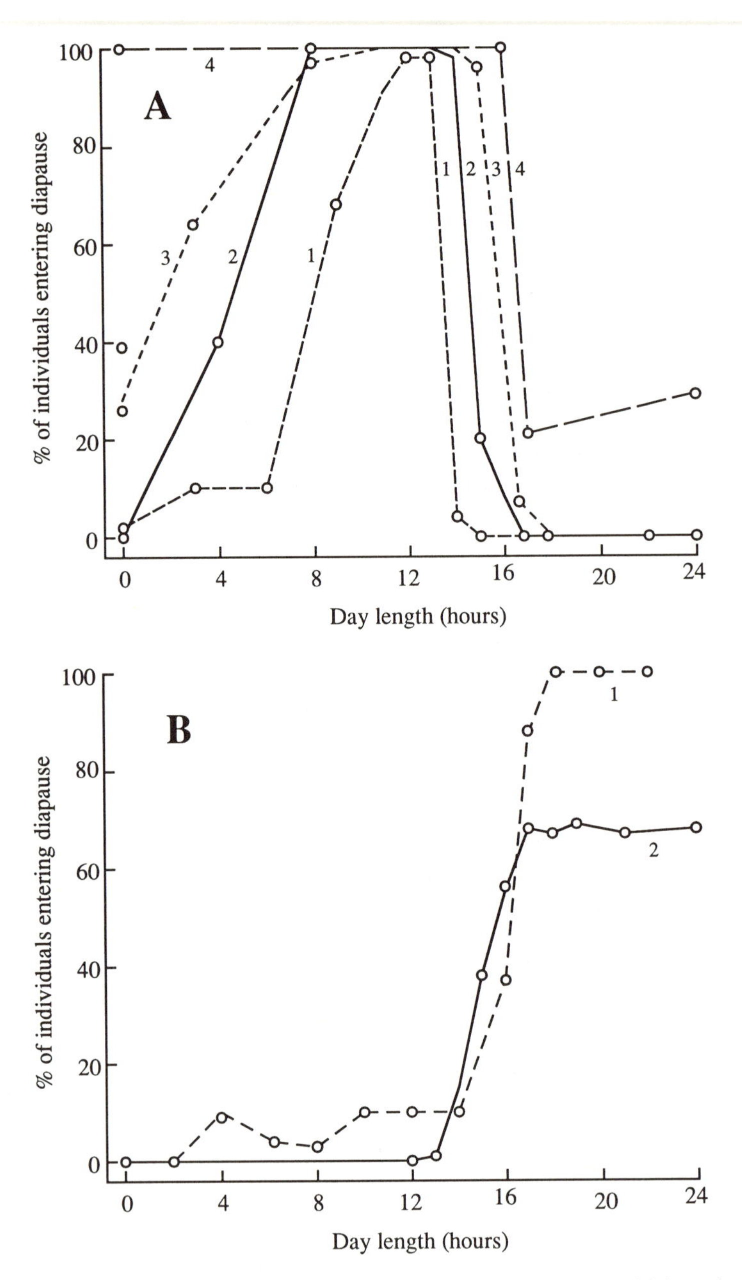

Figure 7.1. (A) The photoperiodic response of short-day insects, which are induced to diapause when the daylight hours rise above a particular critical value (1, *Stenocranus minutus*; 2, *Bombyx mori*). (B) The photoperiodic response of long-day insects, which are induced to diapause when the daylight hours fall below a particular critical value (1, *Laspeyresia molesta*; 2, *Pieris brassicae*; 3, *Acronycta rumicis*; 4, *Leptinotarsa decemlineata*). The decline of the induction curves at day lengths less than about 8 hours probably has no physiological or ecological significance but is a dynamic consequence of the biological clock mechanism that measures time. (After Danilevskii, 1965.)

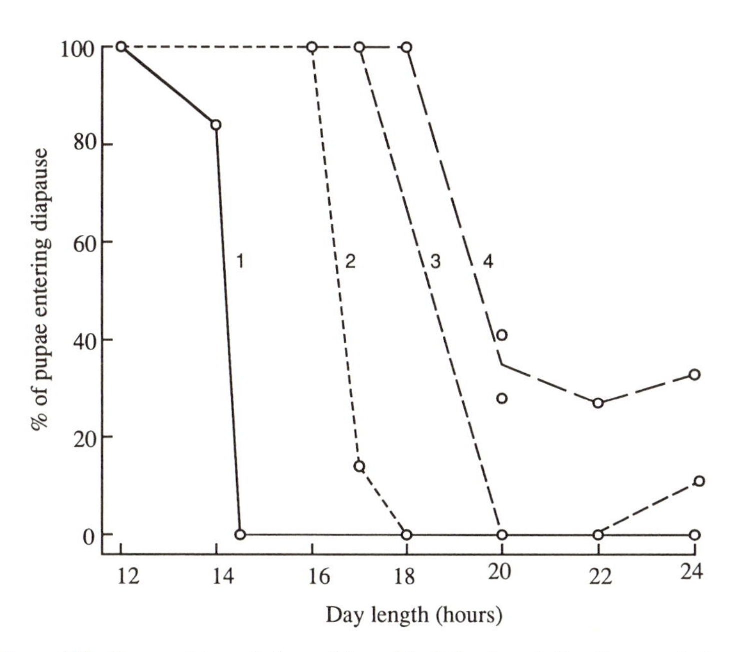

Figure 7.2. Geographic variation of the critical day length for diapause induction in the moth *Acronycta rumicis*. The curves describe the response of populations taken from different latitudes: 1, Abkhazian (43° N); 2, Belgorod (50° N); 3, Vitebsk (55° N); 4, St. Petersburg (60° N). (Redrawn from Danilevskii, 1965.)

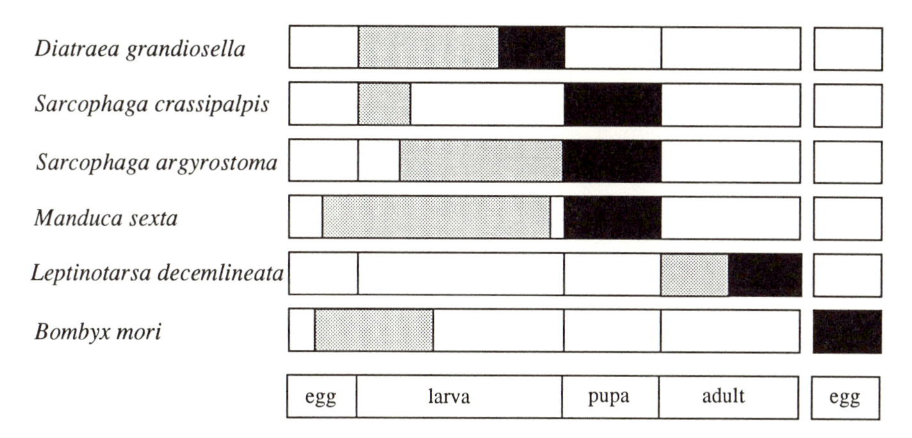

Figure 7.3. Photoperiodic induction of diapause occurs during a sensitive period that precedes the actual onset of diapause by many days or weeks. This diagram gives the sensitive periods for induction (gray) and the period of diapause (black) for six species. (Based on Denlinger, 1985.)

pausing stage is found in *Bombyx*. Here the photoperiod experienced by a female embryo determines whether her offspring will enter diapause as embryos (also see the section on diapause hormone below).

As a rule, insects need to experience a fairly long consecutive series of inductive photoperiods in order to become programmed to enter diapause. Long day lengths during the photosensitive period can cancel the inducing effect of a previous series of short day lengths. Larvae of *Sarcophaga argyrostoma*, for instance, require 13–14 days of short day lengths in order for 50% of the population to enter diapause. As they experience fewer than fourteen consecutive short days, a progressively smaller proportion of the population will enter diapause. Larvae of *Acronycta* require at least eleven consecutive short days for 50% diapause induction. Reversal of the diapause program generally requires fewer days of a noninductive than induction of diapause. Larvae of *Manduca sexta*, for instance, require a sequence of 5–10 short days to enter diapause, but larvae that are reared continually under short-day conditions require only a sequence of three long days during the last larval instar to reverse the diapause program in 96% of the individuals (Bell et al., 1975; Bowen et al., 1984b). In a unique series of experiments Bowen et al. (1984b) showed that in larvae of *Manduca* the entire photoreceptor and response mechanism for diapause induction resides in the brain. They showed that brain-retrocerebral complexes taken from larvae raised on short days could be cultured in vitro. When such complexes were cultured for three days under long-day conditions, they prevented pupal diapause when implanted into short-day larvae. By contrast, when the complexes were cultured under short-day conditions and implanted into short-day larvae, the recipients underwent normal pupal diapause. Evidently, three long days can reverse the diapause program of the brain, even when experienced by isolated brains in vitro. In order to do this the brain must be able to receive the photoperiod, integrate and store this information, and express it some nine days later (after the larva molts to the pupal stage) as an altered pattern of PTTH release.

Temperature is also an important environmental cue that is used in a variety of ways in the regulation of diapause. Thermal periods can sometimes substitute for photoperiods as cues for day length (Beck, 1983). When a thermal period coincides with the photoperiod (so that nights are cold and days are warm), it often has a synergistic effect, so that fewer daily cycles are needed to induce a particular degree of diapause (Beck, 1980). A constant low temperature usually enhances the response to short day lengths, while a constant high temperature often abolishes an insect's sensitivity to photoperiod. One of the ways in which low temperature acts to enhance diapause is to slow down development so that the insect experiences a larger number of inductive photoperiod cycles (Beck, 1980;

Denlinger, 1985). Another way is by changing the critical day length. In *Acronycta*, for instance, the critical day length goes up with decreasing temperature, so that it becomes easier to induce diapause at low temperatures even with moderately short day lengths (Danilevskii, 1965).

In some species, the quality and quantity of food available has been shown to either induce diapause or modify the response to photoperiod and temperature cues. In the lacewing *Chrysopa carnea* and the mosquito *Wyeomyia smithii*, the percentage of individuals entering diapause can be modified over a considerable range by varying the amount of food they are given. In tropical areas, where photoperiod and temperature often differ little from season to season, the change in food quality due to the maturation and senescence of host plants probably provides the most important cue for predicting the dry season. Such changes in food quality appear to be the primary stimuli for diapause induction in several tropical stem-boring moths of the genus *Chilo* (Tauber et al., 1986).

Endocrine Control of Diapause

When insects encounter diapause-inducing environmental conditions during their sensitive period, their physiology and metabolism undergo subtle changes. Often their feeding and growth rates are altered and an increasing portion of the food they take in is channeled away from growth or reproduction and toward storage and accumulation of reserves in the fat body (Tauber et al., 1986). Finally, when the stage in the life cycle is reached at which the species normally goes into diapause, metabolism drops precipitously and development comes to a virtual standstill.

In all species where the physiology of diapause has been studied, it has been shown that the developmental standstill is controlled by one or more hormones. Usually diapause can be attributed to a specific endocrine deficiency, and in many cases diapause can be terminated by simply supplying the hormone exogenously. The exact nature of the endocrine involvement in diapause, however, depends critically on the stage at which diapause occurs. Diapause in the egg, larval, pupal, or adult stages each has a unique physiological character and each is regulated by different combinations of hormones, as we will see below.

Embryonic Diapause

Insects that overwinter as eggs can come to a developmental standstill at various stages in embryonic development. Embryos of the silkmoth *Bombyx mori* enter diapause just prior to segmentation, while those of the tent caterpillar *Malacosoma americana* and the gypsy moth *Lymantria dispar* diapause as fully formed larvae ready to hatch. Developmental arrest at nearly every stage in embryogenesis from blastoderm to hatchling larvae has been documented in different species of Hemiptera (Cobben, 1968). Embryonic diapause can be obligatory, as it is in *Malacosoma* and in many grasshoppers and mantids, or it can be facultative, as in *Lymantria* and in the polyvoltine races of *Bombyx*.

The control of embryonic diapause has been best studied in the Chinese silkworm *Bombyx mori*. Some races of *Bombyx* have an obligatory embryonic diapause, and thus have only one generation per year; they are univoltine. The embryos of other races of *Bombyx*, however, undergo a facultative diapause, and such races can have more than one generation each year. Whether or not an embryo from such a race will enter diapause depends on the photoperiod that its mother experienced while she herself was an embryo. When a mother experiences long day lengths and warm temperatures as an embryo, the eggs she produces as an adult will undergo diapause early during embryonic development. When she experiences short day lengths and cool temperatures as an embryo, her eggs will develop normally without diapause. As a consequence, mothers that grow up in the spring, under short day lengths, lay their eggs in the summer, and these eggs develop normally. While developing, these summer embryos experience long day lengths, and when fully grown a few months later (in the fall) the adult females that result will lay eggs that will diapause. These eggs overwinter in diapause and initiate development in the spring, and the cycle starts over again. As a consequence of this facultative diapause scheme, and the long delay between photoperiodic induction and actual diapause, these races of *Bombyx* are bivoltine (have two generations per year).

The endocrine link between maternal photoperiod and diapause of the offspring was discovered by Fukuda (1951b, 1952) and Hasegawa (1951, 1952, 1957). They found that if the brain of a female *Bombyx* destined to produce diapausing eggs is removed immediately after pupation, she will lay a mixture of diapausing and nondiapausing eggs. By contrast, when the subesophageal ganglion is removed, the female lays only nondiapausing eggs. Reimplantation of the subesophageal ganglion into the abdomen results in females that lay a mixture of diapausing and nondiapausing eggs. Transplanting the subesophageal ganglion from a diapause-producing female into the pupa of a nondiapause-producing female causes the latter to lay a mixture of diapausing and nondiapausing eggs. Transplants of the brain have no such effect. The production of diapausing eggs appears to depend on the presence of the subesophageal ganglion and its normal connection to the brain. Fukuda and Hasegawa interpreted these experiments to show that production of diapause hormone by the subesophageal ganglion is inhibited by nerves from the brain. Severing the connection between the brain and subesophageal ganglion abolishes this negative control and results in a slow leakage of diapause hormone from the subesophageal ganglion. Extracts from the subesophageal ganglion, injected into non-diapause-producing pupae, cause them to lay diapausing eggs when they emerge as females. It appears then that long-day photoperiods experienced as embryos programs females of *Bombyx mori* to secrete diapause hormone during their late pupal stage, when they are maturing their eggs. (See Box 9.)

BOX 9

Diapause Hormone

Structure and Nomenclature: Diapause hormone is a neurosecretory hormone that controls egg diapause in the silk moth *Bombyx mori*. Partial purifications of diapause hormone suggest that it exists as at least two active components, DH-A and DH-B, with molecular weights of approximately 3300 D and 2000 D, respectively. The structure of diapause hormone has been reported by Imai et al. (1991) as a peptide of 24 amino acids with the following sequence:

Thr-Asp-Met-Lys-Asp-Glu-Ser-Asp-Arg-Gly-Ala-His-Ser-Glu-Arg-Gly-Ala-Leu-Cys-Phe-Gly-Pro-Arg-LeuNH_2

The terminal amino acid sequence is similar to that of the pheromone biosynthesis activating peptides (see chapter 9 and Box 10) and several myotropic peptides (see chapter 3).

Source: Diapause hormone activity has been reported in several species of moths. In *Bombyx* it is produced by a single pair of neurosecretory cells in the subesophageal ganglion. Male brains are particularly rich sources of diapause hormone, even though it has no known role in males.

References: Denlinger, 1985; Fukuda, 1951a, b; Hasegawa, 1952, 1957; Imai et al., 1991; Isobe et al., 1976.

The diapause hormone acts on developing eggs in the ovaries and somehow causes development to be arrested a few hours after oviposition, immediately after formation of the cephalic lobe, when the embryo consists of about 12,000 cells (Kitazawa et al., 1963). The diapause hormone acts directly on developing oocytes. Each oocyte goes through a sensitive period around the time it reaches a size of about 500 μg. Exposure to diapause hormone much before or much after this time does not provoke the typical biochemical responses associated with diapause induction (Yamashita and Hasegawa, 1970). Thus the production of a continuous sequence of diapausing eggs requires a continuous exposure to diapause hormone.

In addition to changing the embryo's developmental program, diapause hormone also has several physiological effects on the ovary. Diapause hormone causes the ovary to take up 3-hydroxykynurenine from the hemolymph, which it converts into a reddish brown ommochrome. This ommochrome is incorporated into the serosa of the egg and gives diapause eggs of *Bombyx* their characteristic dark coloration (Yamashita and Hasegawa, 1964; Sonobe and Ohnishi, 1970). Diapause hormone also stimulates the accumulation of glycogen and ecdysteroids in the ovary, and a reduction in its content of cyclic GMP (Yamashita and Yaginuma, 1991).

The mechanism of action of diapause hormone is not known, but its presence in the egg is correlated with several changes in carbohydrate and yolk metabolism. If the accumulation of glycogen in the ovaries of diapause egg-producing females is inhibited by means of the trehalase inhibitor, validoxylamine, their eggs fail to enter diapause (Takeda et al., 1988; Yamashita and Yaginuma, 1991). It may be that the inability of the embryo to use stored nutrients of the egg is at least in part responsible for the developmental arrest, since embryos from diapausing eggs can be induced to develop by simply removing them from the egg and placing them in a tissue-culture medium (Takami, 1959). Immersing the egg in warm dilute hydrochloric acid within 24 hours after oviposition also breaks diapause, presumably by altering the permeability of the chorion, though the exact mechanism of this effect is not understood at present (Watanabe, 1935; Kai and Nishi, 1976).

Embryonic diapause in *Orgyia antiqua* (Lymantriidae) is also due to a hormone produced by the subesophageal ganglion of the mother. But it is clear that not all types of embryonic diapause depend on maternal cues. In insects that diapause late in embryonic development, such as *Lymantria*, the photoperiod-sensitive stage is early in embryonic development and no maternal influence is detectable.

The cause of developmental standstill during embryonic development in other species is not yet understood. Changes have been observed in the ecdysone concentration of diapausing eggs and in the contents of neurosecretory cells of developing embryos of some species (Denlinger, 1985); but while such changes are correlated with diapause it is not clear whether they are involved in the control of diapause. In species that diapause in the later stages of embryonic development it is possible that a deficiency of ecdysone or some other hormone could arrest an embryo during an embryonic molt, or during another hormone-dependent embryonic event. It is also possible that a deficiency in a hatching hormone, which might act much like the eclosion hormone, could account for diapause at the hatchling stage in species like *Malacosoma.*

Larval Diapause

Larval diapause probably occurs in members of most of the insect orders. Larval diapause can occur in any of the larval instars. Fritillaries (*Speyeria* spp.) enter an obligatory diapause as first instar larvae, immediately after hatching from the egg. The mosquito *Wyeomyia smithii* usually diapauses in the penultimate larval instar, while many Lepidoptera diapause as fully grown final instar larvae. Among the most extreme forms of diapause is the four-year developmental arrest that occurs in the early larval instars of the seventeen-year cicadas (*Magicicada*) (Lloyd and White, 1976; Lloyd et al., 1983).

Diapause in the larval stage is generally more difficult to characterize than that in other stages, because diapausing larvae often remain active and sometimes continue to feed intermittently as well. Stationary molting is also known to occur during diapause, but such larvae never grow, nor do they progress in their development. Larval diapause can sometimes be mistaken for torpor (and vice versa). In most cases the relative unwillingness of diapausing larvae to feed, their evident inability to grow and develop at a normal rate, and the restriction of this syndrome to a single larval instar are good diagnostic characters for diapause.

The physiological control of larval diapause has been studied most thoroughly in the final instar of Lepidoptera. Here larval diapause is due in large part to the inhibition of PTTH and ecdysteroid secretion, which prevents continuation of development to the pupal stage. In *Chilo suppressalis* and *Diatraea grandiosella*, the inhibition of PTTH is due to a continued elevated titer of JH. In these species the corpora allata of larvae destined to diapause do not turn off in the middle of the final larval instar but remain active and maintain a high concentration of JH throughout the period of diapause. Two types of experiments have demonstrated that this elevated titer of JH is the principal cause of the diapause syndrome. First, nondiapausing larvae of both species can be induced to enter a diapauselike state by applications of JH during the early portion of the final larval instar. Second, allatectomy of diapausing larvae causes them to pupate, and pupation of such allatectomized larvae can be prevented by treatment with JH (Yagi and Fukaya, 1974; Yin and Chippendale, 1976, 1979). Diapausing larvae of *Chilo* and *Diatraea* occasionally molt during diapause. These molts are not accompanied by growth, but the melanic pattern of the larval cuticle is greatly reduced so that diapausing larvae after their first molt are much paler than normal nondiapausing final instar larvae (Chippendale and Yin, 1976; Yin and Chippendale, 1974, 1976). Stationary molts can also be induced by injections of 20-hydroxyecdysone into diapausing larvae of both species.

The initial suppression of PTTH secretion during larval diapause is probably part of the general adaptation that inhibits PTTH secretion while there is still JH present during the final larval instar (see chapter 5). But this inhibition seems to decay gradually so that irregular and infrequent stationary molts occur. It appears that in the course of the evolution of larval diapause, the JH-induced inhibition of PTTH secretion has become elaborated into the diapause syndrome, which includes the cessation of feeding and growth, and the alteration of behavior and metabolism. The end result of this evolutionary process is that diapause becomes the normal response to an elevated JH titer in the latter part of the final instar. This is why applications of JH induce the normal diapause syndrome in final instar larvae of *Diatraea* and *Chilo*, but have no such effect in species that have not evolved the capacity to diapause in the last larval instar.

Diapause in *Diatraea* and *Chilo* persists as long as the JH titer remains elevated. The larval diapause of *Ostrinia nubilalis*, on the other hand, does not depend on the continued presence of JH. In this species, the persistence of an elevated JH titer is the cause of the initiation of diapause. But if diapausing larvae of *Ostrinia* are injected with ecdysone, they pupate instead of undergoing a larval molt, which indicates that JH gradually disappears during diapause. It is interesting to note in this regard that diapausing larvae of *Ostrinia* do not molt. Exactly what maintains larval diapause of this species is not known (Yagi and Akaike, 1976; Chippendale and Yin, 1979; Bean and Beck, 1980).

Pupal Diapause

Pupal diapause is particularly common in the Lepidoptera and higher Diptera. Pupal diapause can be obligatory, as it is in *Hyalophora* and many other Saturniidae, or it can be environmentally induced and facultative, as it is in *Manduca*, many butterflies, and Diptera. The immediate cause of pupal diapause is the failure of the brain to secrete the prothoracicotropic hormone (PTTH) and the failure of the prothoracic glands to secrete the molting hormone, ecdysone. Under normal development PTTH and ecdysone secretion occur within a few days after pupation in most species. But in pupae that are to enter diapause, the secretion of PTTH does not take place. In the absence of ecdysteroids, adult development cannot begin and the animal's metabolism gradually declines to the level characteristic of the diapausing pupa. In *Manduca* it appears that pupal diapause is maintained by two changes in the endocrine system: the suppression of PTTH secretion by the brain, and a change in the physiology of the prothoracic gland that somehow prevents it from responding normally to PTTH. Cultured prothoracic glands taken from diapausing *Manduca* pupae produce markedly less ecdysteroids when stimulated by PTTH *in vitro* than do glands taken from nondiapausing pupae (Bowen et al., 1984a; Smith et al., 1986). Interestingly, the diapausing glands synthesize significantly more cAMP upon stimulation by PTTH than do nondiapausing glands. This finding suggests that the refractoriness of diapausing prothoracic glands to PTTH is not due to a change in PTTH receptors or second messenger machinery, but must reside somewhere in the steroidogenic pathways beyond the point at which cAMP acts. The change in PTTH-responsiveness of the prothoracic gland during diapause appears to be due to a factor secreted from the brain, because simple extirpation of the brain of a nondiapausing pupa diminishes the PTTH-responsiveness of its glands when tested *in vitro* (Smith et al., 1986). Whether this brain factor is PTTH itself, or a new kind of hormone, remains to be established. It is conceivable that it is simply PTTH, and that this hormone normally maintains the competence of the prothoracic glands by some continuous low level of secretion.

Diapausing pupae can be stimulated to initiate development at any time by implantation of an "active" brain or by a simple injection of 20-hydroxyecdysone. Thus even though the induction and maintenance of diapause may involve a prolonged preparatory period and complex changes in physiology, secretion of ecdysteroids is a sufficient stimulus for the termination of diapause. Under normal conditions the brain of a diapausing pupa of *Hyalophora* is activated by exposure to cold for a period of several months. Pupae that are kept at warm temperatures will remain in diapause until they die. If pupae of *Hyalophora* are exposed to a relatively brief period of cold, and then returned to warm temperatures, only a few animals will break diapause, and they do so asynchronously over a long period of time. On the other hand, if they are exposed to a prolonged period of cold, almost all individuals break diapause within a few days after they are returned to warm temperatures. In species with a facultative diapause the trigger for the termination of diapause is often much less clear-cut. Diapausing pupae of *Manduca sexta* and *Papilio polyxenes*, for instance, have no absolute requirement for an exposure to cold. Their diapause can be broken efficiently and predictably by chilling, but these pupae will also break diapause if they are maintained at a constant high temperature, albeit after a long and unpredictable period of time.

In general, the longer an animal has been maintained in diapause, the quicker it will resume development after being returned to favorable conditions. Thus even while metabolism is slowed and overt development has come to a standstill, something happens to the animal's physiology during diapause that enables it to reactivate its endocrine system. This process is called *diapause development*. The nature of diapause development is unclear at present. It could involve changes in both the central nervous system and prothoracic glands. In *Hyalophora*, removal of the brain at any time before adult development starts places the animal into a permanent state of diapause, but in *Manduca sexta*, *Acronycta rumicis*, *Antheraea polyphemus*, and *Pieris rapae*, pupae can break diapause and initiate adult development in the absence of a brain, provided they have been maintained in diapause for a long enough period of time (Denlinger, 1985). This means that in *Hyalophora* diapause development is possibly a process that occurs in the neuroendocrine system, while in the last four species it may be a process restricted to the prothoracic glands. In *Heliothis zea* the control of diapause appears to reside entirely in the prothoracic glands (Meola and Adkisson, 1977). In this species the brain secretes PTTH soon after pupation but the prothoracic glands do not become activated as long as the temperature remains below 21°C. Diapausing pupae kept at or below 21°C develop normally if injected with ecdysone. Thus diapause is due to the failure of the prothoracic glands to secrete ecdysone despite the release of PTTH (possibly involving a blockage in the ecdysteroidogenic pathway analogous to that found in *Manduca*). An early exposure to PTTH is essential to potentiate the ability of the prothoracic glands to produce ecdysteroids at the end of diapause. After overwintering, pupae of *Heliothis* begin development almost immediately after they are brought to a temperature of 27°C. Apparently diapause in this species is controlled entirely by the inhibitory effects of low temperature on the prothoracic glands.

Adult Diapause

Diapause in the adult stage is characterized by the cessation of egg development, resorption of any eggs that are in the process of development, an increase in size of the fat body, and changes in metabolism and behavior. Many diapausing adults become negatively phototropic and overwinter under bark or fallen logs. In *Leptinotarsa decemlineata* the flight muscles are largely resorbed and the recovered proteins, together with those from the oocytes, are used to build up storage reserves in the fat body. Some insects such as the monarch butterfly *Danaus plexippus* and the milkweed bug *Oncopeltus fasciatus*, by contrast, undertake an extensive migration at the onset of adult diapause (Johnson, 1969; Dingle, 1985), as will be discussed in chapter 9.

The one common feature of adult diapause in all insects in which it has been studied is the cessation of reproduction and regression of the ovaries. This observation gave the first clue that hormones associated with reproduction must decline during adult diapause. Since JH is the best known and most widespread hormone that insects use to regulate their reproduction (see chapter 6), most studies on adult diapause have focused on the role of JH. In all species that have been studied so far, it is indeed the absence of JH that is responsible for the adult diapause syndrome, and adult diapause can be reversed in most of them by simple application of exogenous JH.

The endocrine control of adult diapause has been best studied in the Colorado potato beetle *Leptinotarsa decemlineata*. When reproductively active adult *Leptinotarsa* that are maintained under non-diapause-inducing conditions (long day lengths and warm temperature) are allatectomized, their ovaries regress, they become negatively geotropic, they dig into the soil, their flight muscles begin to degenerate, and their metabolism declines. All these behavioral and physiological changes are identical to those observed if the animals are switched to short-day diapause-inducing conditions (De Wilde and De Boer, 1961, 1969). Conversely, when diapausing adult females of *Leptinotarsa* are treated with JH, they become active, begin to feed again, and egg development begins. With very high doses of JH such females even oviposit a number of eggs (Schooneveld et al., 1977). But if JH-treated females are kept under diapause-inducing conditions, their physiology and behavior soon revert to those typical of diapause (fig. 7.4). The JH titers in the hemolymph of *Leptinotarsa* under diapause and non-diapause-inducing conditions have been measured (fig. 7.5). Under long-day conditions the JH titer rises soon after adult emergence, while under short-day conditions it rises briefly and then drops to an undetectable level as the animal enters diapause. When diapausing animals are exposed to long day lengths once more, the JH titer begins to rise almost immediately and ovarian development resumes within a few days.

Adult reproductive diapause of several species of Dytiscidae, *Locusta migratoria*, and *Pyrrhocoris apterus* can also be reversed by applications of JH or implantation of active corpora allata (Joly, 1945; Darjo, 1976; Hodkova, 1977). The dry-season reproductive diapause of tropical insects,

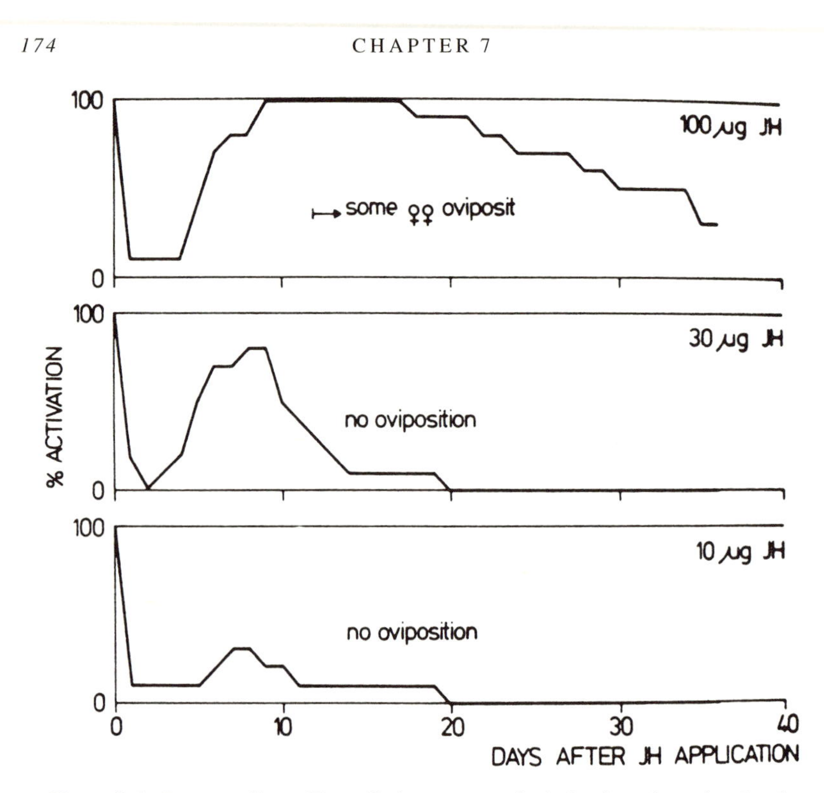

Figure 7.4. Dosage effect of juvenile hormone on the induction of ovarian development in diapausing females of *Leptinotarsa decemlineata*. (From Schooneveld et al., 1977. Reprinted with kind permission of Pergamon Press Ltd.)

known as *estivation*, is likewise due to the absence of JH and can be reversed by applications of exogenous JH. Development of the diapause syndrome, the complex of behavioral and physiological changes that attend entry into diapause, can be induced by allatectomy of the adult, but only in species that normally diapause as adults. In other species, allatectomy may cause regression of the ovaries, because JH is necessary for vitellogenin synthesis or vitellogenesis (see chapter 5), but such species exhibit none of the other symptoms of diapause. The complex behavioral and physiological characteristics of adult diapause are therefore evolved adaptations, with JH (or, rather, with its absence) as the proximate controlling stimulus.

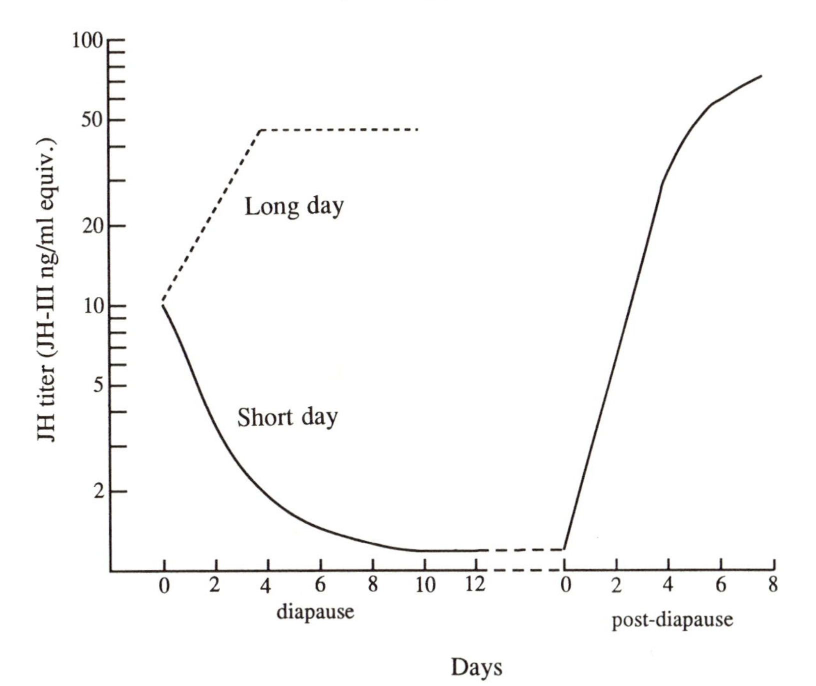

Figure 7.5. Juvenile hormone titers in diapausing and nondiapausing female adults of *Leptinotarsa decemlineata*. Animals reared under long-day conditions (dashed line) have JH titers that rise soon after adult eclosion (day 0). Animals reared under short days (solid line) enter diapause, and their JH titers fall soon after adult eclosion. When females break diapause their JH titer immediately rises. (Redrawn from De Kort, 1981.)

CHAPTER 8

POLYPHENISMS

POLYPHENISM is the occurrence of several distinct phenotypes or forms in a given species, each of which develops facultatively in response to some cue from the internal or external environment. Polyphenisms are formally distinguished from polymorphisms by the fact that in the latter the alternative phenotypes are generally due to genetic differences among individuals in a population, while in the former the alternative phenotypes are developmental variants that are genetically identical. Polyphenisms can arise as sequential steps in a developmental pathway, or through the choice of alternative developmental pathways at some point in the life cycle. Generally, the alternative phenotypes are believed to be adapted to different environmental conditions or life-styles. The best example of a sequential polyphenism is the metamorphosis of insects. Here an individual passes through two or more morphologically distinct forms, each adapted for a specific set of functions. Larvae are specialized for feeding and efficient growth, while adults are specialized for dispersal and reproduction. Larval stages can, in turn, exhibit sequential polyphenism, as in the case of the blister beetles (Meloidae) where the early larval instars are active, slender triungulin larvae, while the later instars are sedentary grubs. An example of an alternative polyphenism is the queen/worker/soldier caste system in ants, in which the alternative phenotypes are morphologically and behaviorally distinct (and adapted for different functions in the colony) but genetically identical.

We have already seen that during larval life, insects must make a "choice" of developmental pathways each time they molt. Each molting cycle can lead to either a larval molt or a metamorphic molt, depending on whether or not the corpora allata have ceased to secrete JH. Normally this occurs only after a particular critical size is reached, but the point is that metamorphosis *could* occur at any molt, even the first one. At the metamorphic molt, many insects actually have more than one developmental choice to make. Many ants and termites, for instance, can metamorphose into workers, soldiers, or queens. All eggs in those species are totipotent (that is, they can develop into any of the three castes), and whether they develop into one caste or another is determined by a complex interaction between their environment and their endocrine system. It appears at this time that all polyphenisms in insects come about by hormonally controlled switches in developmental pathways (Nijhout and Wheeler, 1982).

Insect polyphenisms, then, can be divided into the following categories:

1. Sequential polyphenisms
   a. heteromorphosis
   b. metamorphosis
   c. hypermetamorphosis
2. Alternative polyphenisms
   a. chromatic adaptation
   b. seasonal polyphenism
   c. caste polyphenism (in social insects)
   d. phase polyphenism (in locusts, aphids, and moths)

We have already discussed the control mechanisms of sequential polyphenism in chapters 4 and 5. Below we will examine the nature and physiological control of various alternative polyphenisms.

Chromatic Adaptations

Some insects can change their color to approximate that of their background. These chromatic adaptations are almost always developmental rather than physiological. Only a few insects are able to change their color physiologically by controlling the migration of pigment granules in their epidermis (Atzler, 1930; Veron et al., 1974; Filshie et al., 1975). The stick insect *Carausius morosus* is capable of altering its coloration, and since its epidermal cells are not innervated, the control over color change may be humoral (Raabe, 1966). A specific hormone has not yet been identified, but injection of exogenous neurohormone D can stimulate this color change and may thus mimic the action of the native hormone. The blue/black color adaptation in the dragonfly *Austrolestes annulosus* is controlled primarily by temperature (as it is in other dragonflies), but also appears to require a humoral factor from the terminal abdominal ganglion (Veron, 1973).

Probably the best studied cases of chromatic changes are those which occur at the end of the final larval instar in many Lepidoptera. The onset of the wandering phase, when a larva goes in search of a suitable place in which to pupate, is often accompanied by a change in pigmentation due to the accumulation of ommochromes in the dorsal epidermis. In caterpillars of the European puss moth *Cerura vinula*, this change is particularly dramatic as they develop a brick-red saddle-shaped pattern on the dorsal side of the otherwise green caterpillar. The control over this color shift has been studied by Bückmann (1959, 1974). Ommochrome synthesis in this case is stimulated by the first pulse of ecdysteroid secretion that occurs in the absence of JH. This pulse of ecdysteroids also initiates the wandering phase which, in many Lepidoptera, culminates in the spinning of the cocoon (see chapters 4 and 5).

In most insects, however, chromatic changes are fixed in the integument and require a molt to become manifest. Many Orthoptera have a green/brown polyphenism that is correlated with seasonal plant growth and senescence. In the temperate zone, grasshoppers capable of such color diphenism are green in the summer and brown in the fall and winter, while in the tropics they are green in the wet season and brown in the dry season. Photoperiod and humidity are the principal environmental stimuli that cue the switch. In a few species this switch has been shown to be controlled by JH. Implantation of corpora allata and the application of JH cause brown larvae to become intermediates or green in the next instar (Rowell, 1967, 1971). Nothing appears to be known yet about when in each larval instar the JH-sensitive period of this transformation occurs, nor how it is related to the JH-sensitive period for metamorphosis.

The coloration of butterfly pupae is often adapted to the environment in which pupation takes place. In many species rough or dark-colored substrates and dim light induce the formation of a dark-colored pupa, while smooth, pale-colored surfaces and bright light induce the formation of light-colored pupae (Angersbach and Kayser, 1971; Bückmann, 1960; Koch and Bückmann, 1984; Kayser-Wegmann, 1975). When prepupae are ligated between thorax and abdomen, the anterior portion of the pupa undergoes a normal chromatic adaptation to its substrate, but the posterior portion does not. In general, the abdomens of such ligated pupae either develop the lighter color form or develop a color intermediate between the two extreme forms, which suggests that an endocrine factor is required for development of the dark color form.

Pupae of *Pieris brassicae* have a green/gray polyphenism that is under dual endocrine control (Kayser-Wegmann, 1975). Prepupae have a brief light-sensitive period extending from five to twelve hours after the larvae stop feeding. If they experience yellow light during this period they develop into green pupae, but if they experience blue light they develop into dark gray pupae. Complete darkness during this sensitive period results in intermediate gray animals. Intermediate gray coloration also results if prepupae are ligated behind their prothorax before the light-sensitive period. Carefully timed ligation experiments at various times throughout the light-sensitive period have shown that a melanization-inhibiting hormone is secreted under yellow light, and a melanization-stimulating hormone under blue light. Both hormones are believed to be produced by the prothoracic ganglion. In the related species *Pieris rapae*, it has been shown that development of the green pigment, which resides in the epidermal cells, requires the presence of JH. Green pigment synthesis can be induced in pupae destined to become dark by injection of JH, but a specific JH-sensitive period has not yet been identified (Ohtaki, 1960; Hidaka and Ohtaki, 1963). Melanization of the cuticle in this species also depends on a hormone from the prothoracic ganglion.

The yellow/black pupal polyphenism of *Aglais urticae* is also controlled by neurohormones. As in the case of *Pieris* (and the phase polymorphism of armyworms discussed below), melanization of the cuticle and the color of the epidermis are controlled separately. Ligation experiments on animals maintained under conditions that induce the yellow form show that a melanization-inhibiting factor as well as a carotenoid (yellow)-stimulating factor are produced in the anterior portion of the prepupa (Bückmann, 1968; Maisch and Bückmann, 1987; Koch et al., 1990). The melanization-inhibiting hormone is a small polypeptide (Bückmann and Maisch, 1987) and is produced a few hours before the carotenoid-stimulating hormone. The nature of the latter hormone is not known, but by analogy to the phase polyphenism in armyworms, the carotenoid-stimulating hormone could be JH (Nijhout and Wheeler, 1982). The pupal color dimorphism of *Papilio xuthus* is controlled by a neurohormone produced in the brain and released from the first thoracic ganglion (Awiti and Hidaka, 1982). In the absence of this hormone the pupae are green, while in its presence they develop a brown pigmentation.

Seasonal Polyphenisms

Many species of insects develop distinctively different body forms or pigmentation patterns in different seasons of the year. Many species that are seasonally polyphenic can be made to develop into either of their alternative morphs by simply rearing their larvae under the temperature and photoperiod conditions characteristic of the different seasons. Seasonal polyphenism, like diapause and migration, is part of a suite of physiological and developmental adaptations to regular and predictable seasonal changes in the environment (Shapiro, 1976; Tauber et al., 1986). The developmental switches that give rise to seasonal polyphenisms are controlled by the endocrine system much in the way diapause, migration, caste, and phase polyphenism are controlled. The complex life cycles of aphids (see fig. 8.7) provide excellent examples of seasonal polyphenisms, as many of their phase transformations are triggered by seasonal changes in photoperiod, temperature, or plant senescence.

Distinctively patterned spring and summer (or summer and autumn) forms are common among the butterflies. Perhaps the best known and most extreme example of this wing pattern polyphenism is that of the European map butterfly, *Araschnia levana* (fig. 8.1). The spring and summer forms

Figure 8.1. Seasonal polyphenism in the butterfly *Araschnia levana*. (A) The black and white summer (*prorsa*) form that emerges from nondiapausing pupae. (B) The orange and brown spring form (*levana*) that emerges from diapausing pupae.

of this species are so distinctive that they were long regarded as different species. This polyphenism is under photoperiodic control and is tightly coupled with diapause. The orange and brown spring form emerges from diapausing pupae, while the black and white summer form emerges from pupae that do not diapause. Koch and Bückmann (1987) have shown that the switch to the spring form is controlled by the timing of ecdysteroid secretion during the pupal stage. When diapausing pupae are injected with 20-hydroxyecdysone so that they start development within three days after pupation, normal summer forms result; but if the injection is done more than ten days after pupation, normal spring forms develop. Injections of 20-hydroxyecdysone between three and ten days after pupation result in the development of various degrees of intermediate pattern forms. This means that the developmental switch from one form to another takes place during the first two weeks of the pupal stage. The tight correlation of the spring form with diapause results from the fact that under diapause ecdysteroid secretion is greatly delayed while in animals that develop without diapause, natural secretion of ecdysteroids occurs within two days after pupation.

A shift in the timing of ecdysteroid secretion after pupation has also been implicated in the control of spring and summer forms in *Lycaena phlaeas* (Endo and Kamata, 1985). In *Papilio xuthus* and *Polygonia c-aureum*, by contrast, seasonal polyphenism appears to be controlled by a neurohormone from the brain (Endo and Funatsu, 1985; Endo et al., 1988). In *P. xuthus*, implantation of brains from summer pupae into diapausing pupae cause them to develop into summer forms, not spring forms, even when the implant is made at the end of diapause. In *P. c-aureum* the polyphenism is not linked to diapause but is strictly controlled by photoperiod. Long day lengths induce the light-colored phenotype of the summer form, while short day lengths induce the darker and more cryptic pattern of the autumn form. If a water soluble extract from the brains of pupae reared under long day lengths is injected into short-day pupae within 48 hours after pupation, it induces development of the summer phenotype. This hormone has been called the summer-morph-producing-hormone (SMPH). The SMPH has not yet been identified. Like all neurohormones, it is presumed to be a small polypeptide. Masaki et al. (1988) have shown that SMPH has a molecular weight of approximately 4500 D, close to that of small PTTH. In addition, SMPH-active extracts from *P. c-aureum* have been shown to have PTTH activity in the pupae of *Papilio xuthus*. Given the prior demonstration in *Araschnia* (see above) that temporal shifts in ecdysteroid secretion can have dramatic effects on seasonal morph development, Masaki et al. (1988) and Endo et al. (1990) examined the possibility that SMPH might be identical to small PTTH. These investigators have shown that there are slight differences in the fluctuation patterns of SMPH and PTTH activity which indicate that the two activities are probably due to different hormones. This conclusion is strengthened by the observation that a factor with SMPH activity in *P. c-aureum* can be extracted from the brains of *Bombyx mori* (a species that does not exhibit a seasonal polyphenism), and that this factor has no PTTH activity when tested on prothoracic glands from *P. c-aureum* in vitro (Endo et al., 1990).

Gross features of morphology can also vary with the seasons. Seasonal variation in wing length is common in the Hemiptera, Homoptera, and Orthoptera (Dingle, 1985; Tauber et al., 1986). The wing diphenism of the grasshopper *Zonocerus variegatus* has been shown to be controlled by JH. Allatectomy of larvae causes many of the resulting adults to be of the long-winged form, while injection of JH causes all adults to become brachypterous (McCaffery and Page, 1978). The critical period for this hormonal effect has not yet been determined.

Castes in Ants

Ants possess the best differentiated caste system among the social Hymenoptera. The queen is the only reproductive caste in a colony, and in most species the worker caste is not only sterile but also incapable of differentiating into a reproductive caste (Wilson, 1971). The worker caste is often polyphenic as well. In many species the worker caste has a broad size range, and when the individuals near the lower and upper extremes of that range look and behave differently they are referred to as minor workers and major workers, respectively. The distribution of body sizes in the worker caste can be unimodal or bimodal, and across their size range the allometry of body parts can be monophasic (i.e., with a single coefficient of allometry across the entire size range), or diphasic (with two different coefficients in different parts of the size range; fig. 8.2). Triphasic allometry has also been

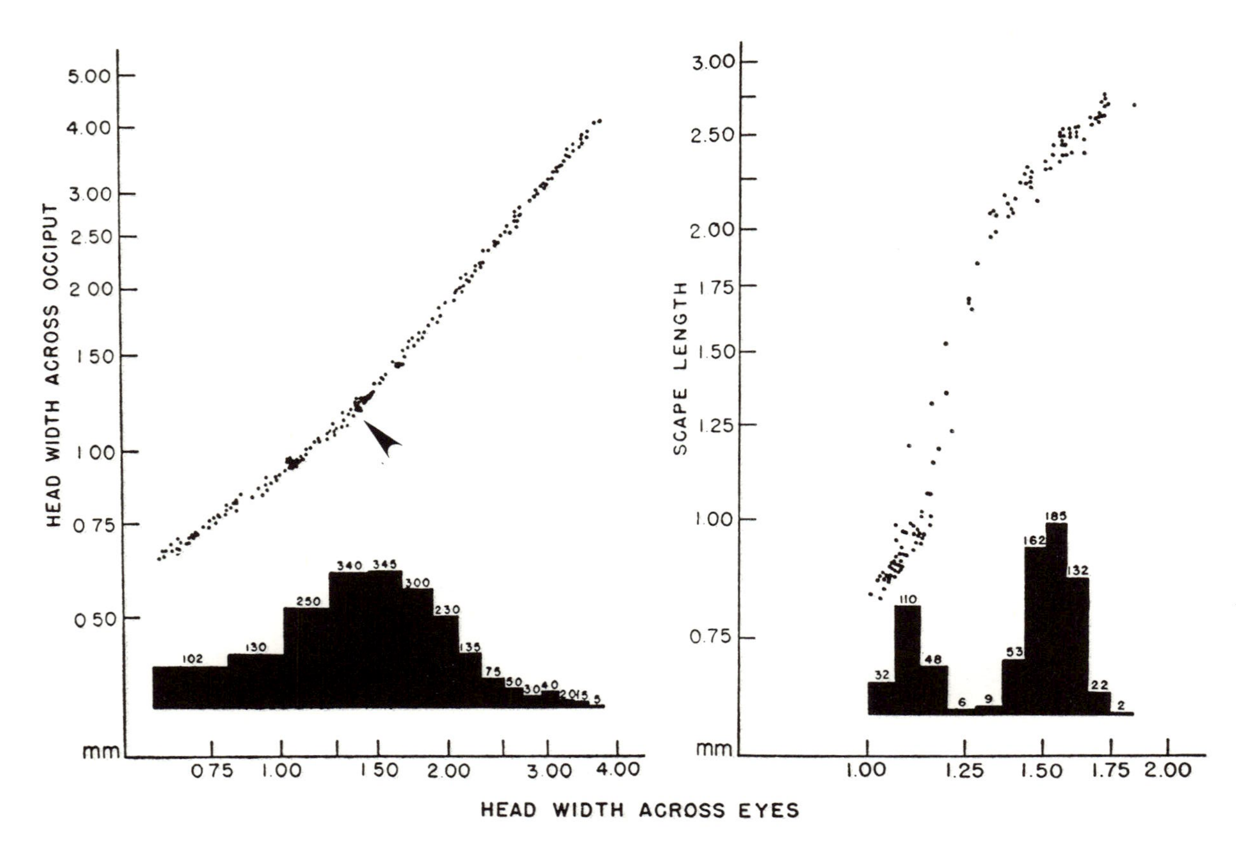

Figure 8.2. Two kinds of complex allometry in ants. *Left*: Diphasic allometry (allometry curve has two different slopes) with a unimodal size distribution in *Atta texana*. Arrow indicates the point at which the allometric curve changes slope. *Right*: Triphasic allometry (curve has three different slopes) with a bimodal size distribution in *Oecophylla smaragdina*. (Modified from Wilson, 1953. Reprinted with permission of *Quarterly Review of Biology* and the University of Chicago Press.)

described (Wilson, 1971), but appears to be rare. The consequence of a broad size distribution coupled with allometry is that the morphology of individuals at the two extremes of the size range can be dramatically different (see fig. 4.14B). Major workers usually have disproportionately large heads and mandibles compared to those of the minor workers. When allometry is diphasic, the differences between major and minor worker become accentuated, and when the size distribution of one or both characters is bimodal, the two forms become discontinuous (fig. 8.2). The large form is then usually referred to as a soldier, and the small form as a worker. But the difference between soldiers and workers is more than just a matter of allometry. In the genus *Pheidole*, soldiers and workers also have different patterns of protein expression in their cuticle (Passera, 1974), which indicates that gene switching may play a role in establishing the differences between the two castes. The combination of different size frequency distributions and different allometry patterns across the size range yields a diverse array of caste systems (Wilson, 1953). In ants with a unimodal size distribution and monophasic allometry there appear to be no special controls over the production of major and minor workers. But bimodal size distributions and biphasic allometry are the results of special regulatory steps during larval growth.

The control over worker caste polyphenism has been best studied in ants of the genus *Pheidole*. In *P. bicarinata* any larva can be induced to develop into a soldier by application of a small dose of JH. The JH-sensitive period for this reprogramming occurs during days 4 to 6 of the final larval instar. If a larva experiences no JH during this period, it pupates soon after reaching its normal critical size. But if JH is present during the JH-sensitive period, the critical size for metamorphosis is reprogrammed. Such larvae continue to grow beyond the small critical size characteristic of workers and pupate only after they reach a distinct and much larger critical size (fig. 8.3; Wheeler and Nijhout, 1981, 1983). Not only is the critical size for pupation revised during the JH-sensitive period, but there is also a developmental switch that causes the larger animal to develop characteristics not possessed by the smaller one. Soldiers have unique proteins, not possessed by the smaller workers (Passera, 1974), which means that the developmental switch involves a change in gene expression. It is possible that such a switch also causes the imaginal disks of soldiers to grow with different allometries than those of workers, and may therefore be the immediate cause of the differences in allometric coefficients between the two castes.

A bimodal size-frequency distribution is due to the reprogramming of the critical size, while a diphasic allometry is the result of a concomitant switch in the allometric growth of body parts. Species of ants differ in the degree and timing of these two developmental reprogramming events. Wheeler (1990, 1991) has shown that differences in reprogramming of

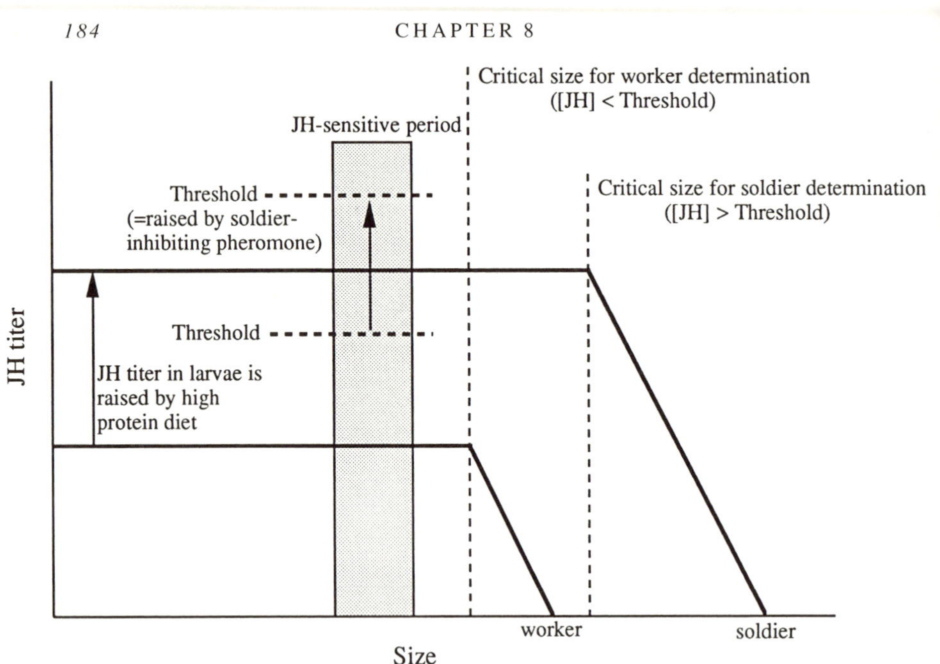

Figure 8.3. Model for the control of worker/soldier polyphenism during the last larval instar of the ant *Pheidole bicarinata*. Under a low protein diet, JH titers remain low and JH secretion ceases when larvae reach a particular (low) critical size; such larvae transform into workers. When food is high in protein, the JH titer is higher, and, if it rises above a threshold, the critical size at which JH secretion ceases is revised upward. Such larvae continue to grow for a longer period of time until the higher critical size is reached, at which time they cease to secrete JH. During this time new soldier-specific genes also become expressed, and the larva transforms into a soldier. The presence of excess soldiers in the colony raises the JH threshold for soldier induction (via an inhibitory pheromone), so that even high titers of JH remain below threshold, the critical size remains low, and larvae transform into workers. (Based on Wheeler and Nijhout, 1984.)

critical sizes and of the allometric growth of body parts can account for the entire diversity of nonnormal size-frequency distributions and nonlinear allometries found in the complex worker caste system of ants.

Ants regulate the proportion of soldiers in their colonies quite accurately. When an excess of soldiers is added to a colony, a smaller portion of larvae in the colony switches to soldier development until the original proportion of soldiers is restored. This regulatory behavior is believed to be due to a soldier-inhibiting pheromone, produced by soldiers, that prevents larvae from switching to soldier development during their JH-sensitive period (Gregg, 1942; Passera, 1974; Wheeler and Nijhout, 1984). This inhibition works in a very interesting way. Soldier development requires a high protein diet and normally only the best-fed larvae become soldiers (Passera, 1974, 1985; Wheeler and Nijhout, 1983). Only in such well-fed larvae do the corpora allata secrete enough JH to raise the hemolymph JH titer above threshold during the JH-sensitive period for soldier determination. The soldier inhibiting pheromone does not affect JH secretion but alters the sensitivity threshold to JH during the JH-sensitive period (fig. 8.3). This is demonstrated by the fact that in artificial laboratory colonies which contain an excess number of soldiers the larvae are much less sensitive to a given dose of JH than larvae in colonies with few or no soldiers (Wheeler and Nijhout, 1984). The presence of soldiers in a colony appears to raise the threshold of sensitivity to JH in final instar larvae so that fewer larvae switch to soldier development during the JH-sensitive period. The proportion of soldiers in a colony, then, depends on a balance between the mean titer of JH, which is a complex function of the availability of food and the number of foraging workers a colony has, and the threshold of sensitivity to JH, which is determined by the soldier-inhibiting pheromone produced by the soldiers already present in the colony.

Queen determination in *Pheidole* and other ants also occurs during a JH-sensitive period, but takes place much earlier in development. When JH is applied to eggs of *P. pallidula* or *Solenopsis invicta*, a large percentage of the resulting larvae develop into queens at metamorphosis. Treating the queen mother with JH has the same effect: a large portion of the eggs she lays will become queens at metamorphosis (Robeau and Vinson, 1976; Passera and Suzzoni, 1979). How queen determination is regulated during the normal life of a colony is not known, but it is reasonable to suppose that it is under maternal control, possibly via the secretion of JH into the eggs (Brian, 1974; Nijhout and Wheeler, 1982).

Castes in Termites

Termites are divided into two groups, the higher termites, which are all in the single large family Termitidae, and the lower termites, which make up the remaining families. The higher and lower termites differ considerably in their life cycle and mechanisms of caste determination (Wilson, 1971). Most of our knowledge of the physiology of caste determination comes from studies on the lower termites (Lüscher, 1960; Miller, 1969; Watson et al., 1985).

The lower termites do not have a specialized worker caste. Instead, the function of workers is assumed by mature larvae called *pseudergates* (i.e., "false workers"). Pseudergates continue to undergo infrequent stationary molts and may remain in that stage of arrested development for their entire life. Pseudergates can metamorphose into any of the three terminal castes:

winged imagos (which can become queens and kings of new colonies), soldiers, or replacement reproductives. Metamorphosis to an imago or a soldier actually requires two molts, the intervening partially differentiated forms being called a brachypterous nymph and a presoldier, respectively. Interestingly, brachypterous nymphs and presoldiers can, under certain conditions, molt back into totipotent pseudergates (Miller, 1969), constituting the only known cases of natural reversal of metamorphosis in the insects.

The control over caste differentiation of pseudergates begins with pheromones. As in the case of *Pheidole*, soldier termites produce a pheromone that inhibits differentiation of pseudergates into new soldiers. But in termites the queen also produces a pheromone that inhibits the differentiation of pseudergates into replacement reproductives (Lüscher, 1972; Miller, 1969; Bordereau, 1975). It has been shown that there are distinct windows of pheromone sensitivity during the pseudergate instar. The sensitive period during which pseudergates are most sensitive to the soldier-inhibiting pheromone occurs during the last part of the instar, while the sensitive period for the induction of the supplementary reproductive caste is during the early portion of the instar (Lüscher, 1953; Springhetti, 1972).

The inhibitory pheromones that control soldier or reproductive development appear to exert their effect by altering the pattern of JH secretion. When active corpora allata are transplanted into pseudergates of *Kalotermes*, many of the recipients molt into soldiers. If artificial JH is applied to pseudergates, a large proportion likewise metamorphose to soldiers (Lüscher, 1974). The switch to supplementary reproductives, by contrast, appears to require the complete absence of JH. The corpora allata of pseudergates that metamorphose to replacement reproductives are much diminished in size, a feature that is generally correlated with the cessation of JH secretion (Stuart, 1979). The JH-sensitive periods that control these caste determination switches are believed to coincide with the periods of pheromone sensitivity. Nijhout and Wheeler (1982) have proposed a general model for the way in which JH titers and JH-sensitive periods may control the developmental switches during a pseudergate instar, shown in figure 8.4. This model shows hypothetical JH profiles during a pseudergate instar for animals that will molt to a soldier, a replacement reproductive, and an imago, or that will remain a pseudergate. During the pseudergate instar there are three JH-sensitive periods. The first one controls the development of sexual characters (reproductive organs, genitalia), the second one controls the nonsexual characters of the imago (wings, pigmentation), and the third control characters typical of the soldier (head, mandibles, pigmentation). A high JH titer during the first and second period inhibits commitment to sexual and imaginal development. During the third JH-sensitive period a high JH titer serves to induce development of soldier

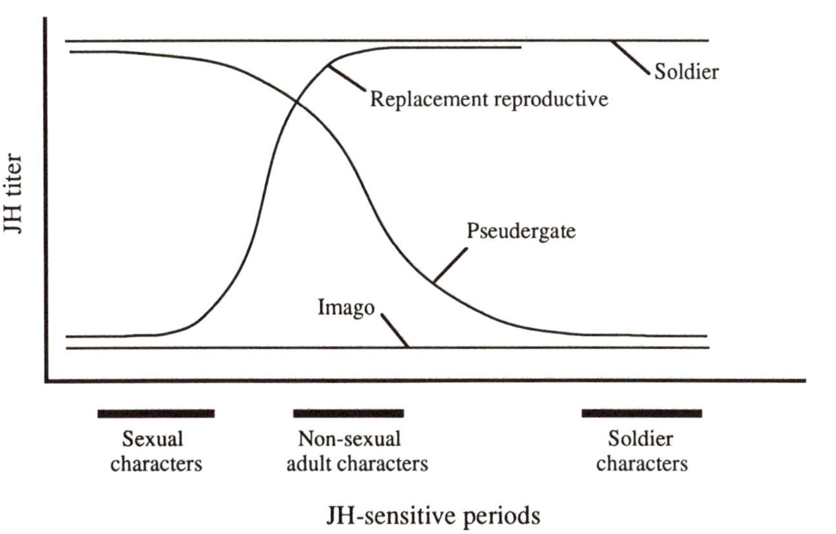

Figure 8.4. Model for the JH-mediated control of caste determination in the lower termites. Each curve represents a hypothetical JH titer profile during the course of a pseudergate instar. Bars show the timing of the JH-sensitive periods for various characters. Pseudergates destined to become soldiers or replacement reproductives molt first to an intermediately differentiated presoldier or nymphal stage, respectively. (Redrawn form Nijhout and Wheeler, 1982.)

characteristics while a low titer maintains the pseudergate status quo. To remain a pseudergate JH has to be high during the first two sensitive periods to inhibit imaginal determination, and low during the last one to prevent induction of soldier characters. Presumptive soldiers have a high JH titer throughout the instar, which inhibits imaginal development but induces soldier development. To become an imago requires that the JH titer be low during all three JH-sensitive periods. Replacement reproductives develop if the JH titer is high only during the second JH-sensitive period, so that reproductive organs but none of the other characters typical of the adult morph develop. The duration of the instar prior to a molt to a prereplacement reproductive is foreshortened, so these animals never experience the JH-sensitive period for soldier determination.

In higher termites the control of caste formation is quite different. Here we do not have a totipotent pseudergate stage, but at the end of larval life each larva metamorphoses into either a true terminal worker, a soldier, or a reproductive adult. The differentiation of soldiers and reproductives is also under control of inhibitory pheromones produced by soldiers and queens, respectively. The switch to soldier development requires a high JH titer, just as in the lower termites, but the JH-sensitive period is early in the last larval instar, instead of late (Okot-Kotber, 1980). The specific endocrine stimuli for worker and reproductive development are not yet known. Lüscher (1976) has observed that the JH concentration in the eggs of *Macrotermes subhyalinus* fluctuates considerably in the course of the year, and that low concentrations of the hormone precede the appearance of new imagos in the colony by one or two months. Lüscher suggested that queens could be determining the fate of their offspring by the amount of JH they incorporated in their eggs.

Castes in Honeybees

Honeybees, *Apis mellifera*, have only two castes, queen and worker. Queens and workers differ in size and in the morphology of their mouth parts, legs, and genital apparatus (De Wilde, 1976; Wirtz, 1973). In addition, worker bees go through a stereotyped sequence of behavioral specializations in the course of their adult life called temporal or *behavioral castes* (Wilson, 1971). Both the physical castes and temporal castes of honeybees are under endocrine control (for a discussion of behavioral castes, see chapter 9). The eggs that a queen lays are totipotent; each can develop into either a queen or a worker, depending on the conditions under which the larva grows up.

Larvae destined to become queens are reared in special cells and fed a particularly nutritious food called *royal jelly*. Any larva that is fed royal jelly during the third larval instar and beyond develops into a queen (Weaver, 1957). If a larva is not fed royal jelly until late in the second half of the third instar it develops into either a normal worker or a worker/queen intercaste. Larvae can also be induced to develop into queens by application of JH during the fourth larval instar (Wirtz, 1973). It is known that the corpora allata of honeybee larvae destined to become queens increase in size during the third larval instar and become much larger than those of worker larvae (Dogra et al., 1977). Measurements of JH titers have revealed a peak during the last portion of larval life in queen but not in worker larvae (Rembold, 1987). Rachinsky and Hartfelder (1990) have shown that the JH synthesis rate by corpora allata of queen larvae rises during the early portion of the fifth larval instar and peaks shortly before spinning, reaching a rate at least twenty-six times that found in worker larvae. This peak of JH secretion corresponds precisely with the JH-sensitive period for queen/worker switching postulated by Nijhout and Wheeler (1982). The JH-sensitive period for queen induction precedes the JH-sensitive period for pupation by about one day (fig. 8.5). By the time the JH-sensitive period for pupation comes along, the rate of JH secretion in both queen and worker larvae has declined to undetectable levels (Rachinsky and Hartfelder, 1990).

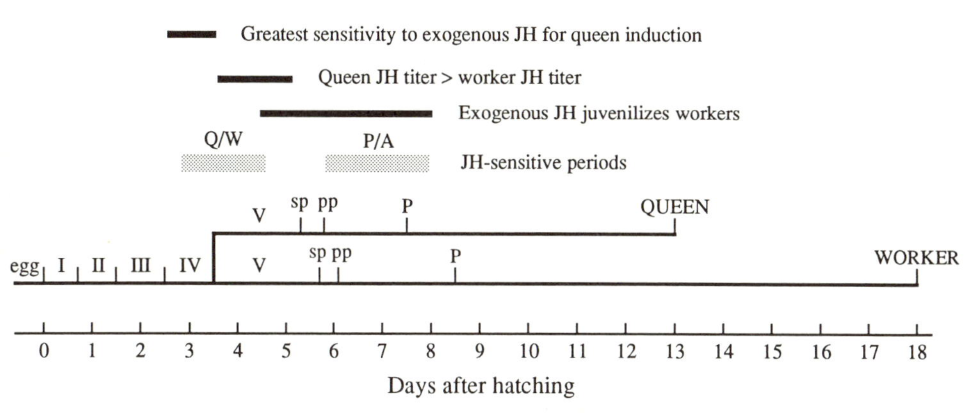

Figure 8.5. Control of queen/worker polyphenism in the honeybee *Apis mellifera*. Queen and worker developmental pathways diverge after the fourth larval instar. The pupal period of the queen is shorter than that of the worker. The gray bars indicate the JH-sensitive periods for the queen/worker (Q/W) switch and the pupal vs. adult (P/A) commitment switch. Black bars indicate the various effects of exogenous JH on these polyphenisms. (Modified from Nijhout and Wheeler, 1982; based on data from Wirtz, 1973; Zdarek and Haragsim, 1974; Dietz et al., 1979; Rachinsky and Hartfelder, 1990.)

Phase Polyphenism in Migratory Locusts

Migratory locusts have two alternative phases, solitary and gregarious, each with distinctive morphological and physiological features. Solitary-phase individuals are uniformly pale yellow, green, or brown in color with relatively short wings and large ovaries, and retain their prothoracic glands for several weeks after metamorphosis. The gregarious phase has a much darker color, often with a bright pattern of black, yellow, and orange, with long wings, much smaller ovaries, and prothoracic glands that break down immediately after metamorphosis. Behaviorally, the solitary phase is relatively sedentary while the gregarious phase is readily provoked into prolonged flight and, as its name implies, seeks out other locusts and forms massive swarms. The solitary and gregarious phases also differ in several morphometric ratios of their body parts (Faure, 1932; Joly and Joly, 1954; Uvarov, 1966; Blackith, 1972; Lauga, 1977). The physiological control over this phase polyphenism has been best studied in *Locusta migratoria* and to a somewhat lesser extent in *Schistocerca gregaria* (Staal, 1961; Pener, 1991).

The mechanism of phase determination in *Locusta* is complex and only partly understood (Pener, 1991). In the larval stages (hoppers) the control over the pigment pattern alone can be broken into three apparently independent mechanisms, as illustrated in figure 8.6. There is independent con-

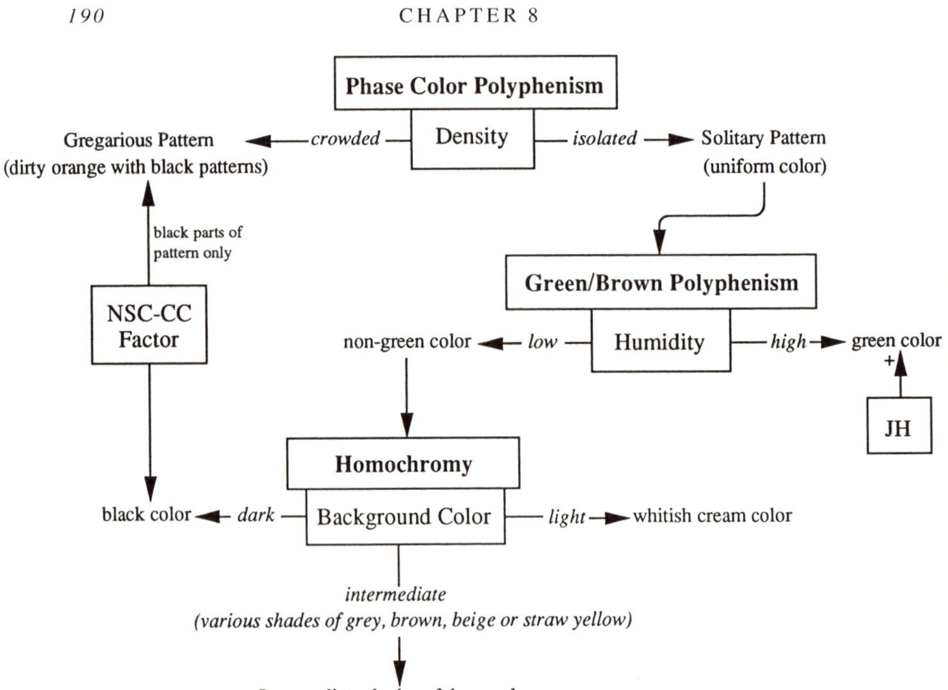

Figure 8.6. Control diagram for the regulation of the various color pattern polyphenisms in the migratory locust. Three aspects of the color pattern are controlled independently: (1) the actual pigment distribution, or pattern, which can be bold and contrasting, or homogeneous and unpatterned; (2) if unpatterned, whether the overall color is green or not; (3) if not green, whether the color is light, dark, or some intermediate color that provides a match to the background. Neurosecretory hormones and JH affect this polyphenism at two different points. (After Pener, 1991.)

trol over the gregarious and solitary phase pigmentation, which may be regulated by a neurosecretory hormone. Within the uniformly colored solitary phase there are several alternate pigmentation types, controlled by cues from the environment. Humidity appears to control whether the coloration is homogeneously green or one of several earth tones, and this effect is mediated by JH. The earth tones are variable in their relative intensity and can be modulated to adapt the hopper to its background. This adaptation is presumably mediated by the visual system.

When hoppers are reared under crowded conditions, they develop the gregarious morphology, pigmentation, and behavior. An increase of the black pigmentation characteristic of the gregarious phase can also be induced by the implantation of corpora cardiaca (Staal, 1961), and the effect appears to be due to a neurosecretory hormone from the brain (Girardie and Cazal, 1965; Bouthier, 1976). The mechanisms that control the expression of other gregarious phase characters, particularly the characteristic morphometric ratios and such features as the shape of the pronotum, are not yet understood (Pener, 1991).

Phase determination has a strong maternal component. Female locusts that are crowded tend to produce offspring of the gregarious phase, while uncrowded females tend to produce solitary-phase offspring. Females that are themselves of the solitary phase tend to have offspring that are rather difficult to switch to the gregarious phase. Several generations of crowding are usually needed to completely switch phase (Kennedy, 1961). The maternal effect appears to be mediated by JH. A solitary-green-phase mother somehow affects the pattern of JH secretion of her offspring. The JH titer in final instar larvae of the gregarious phase is low throughout the instar, but in solitary-phase larvae the JH titer rises again a few days before adult eclosion. This rise in JH coincides with a JH-sensitive period during which the green/brown coloration switch (fig. 8.6) takes place. Application of JH to gregarious larvae during this JH-sensitive period causes them to become completely or partially green adults. Implantation of extra corpora allata into gregarious crowded female adults causes them to have a mix of gregarious and solitary green-phase offspring (Cassier and Papillon, 1968), indicating that elevating the JH titer in gregarious-phase mothers affects the patterns of JH secretion in their offspring. Maternal-effect maintenance of the solitary green color may thus be due to a reprogramming of the allatotropic program of the offspring by the elevated JH titer of the mother at the time she produces eggs (see section on Delayed Effects of JH in chapter 5). The solitary green phase is thus self-reinforcing, as elevated JH titers in the mother cause elevated titers in her offspring, and so on. It appears that the JH-sensitive period for color switching comes some time after the JH-sensitive period for the larval-adult switch (De Wilde, 1975; Nijhout and Wheeler, 1982).

Pener (1991) has pointed out that the experimental work to date is insufficient to rule out additional control steps in locust phase determination. It is possible, for instance, that crowding may alter the morphological and pigmentation response to JH, suggesting an interaction with other physiological or endocrine factors. It is also possible that some studies of the action of JH on phase character determination may have been confounded by the parallel effects this hormone has on sexual maturation in adult males (see chapter 6), and on metamorphosis (chapter 5).

Phase Polyphenism in Aphids

Aphids often have extremely elaborate seasonal cycles with alternations between various adult forms at different times in the season (fig. 8.7). Two principal types of polyphenism exist among the aphids: sexual/partheno-

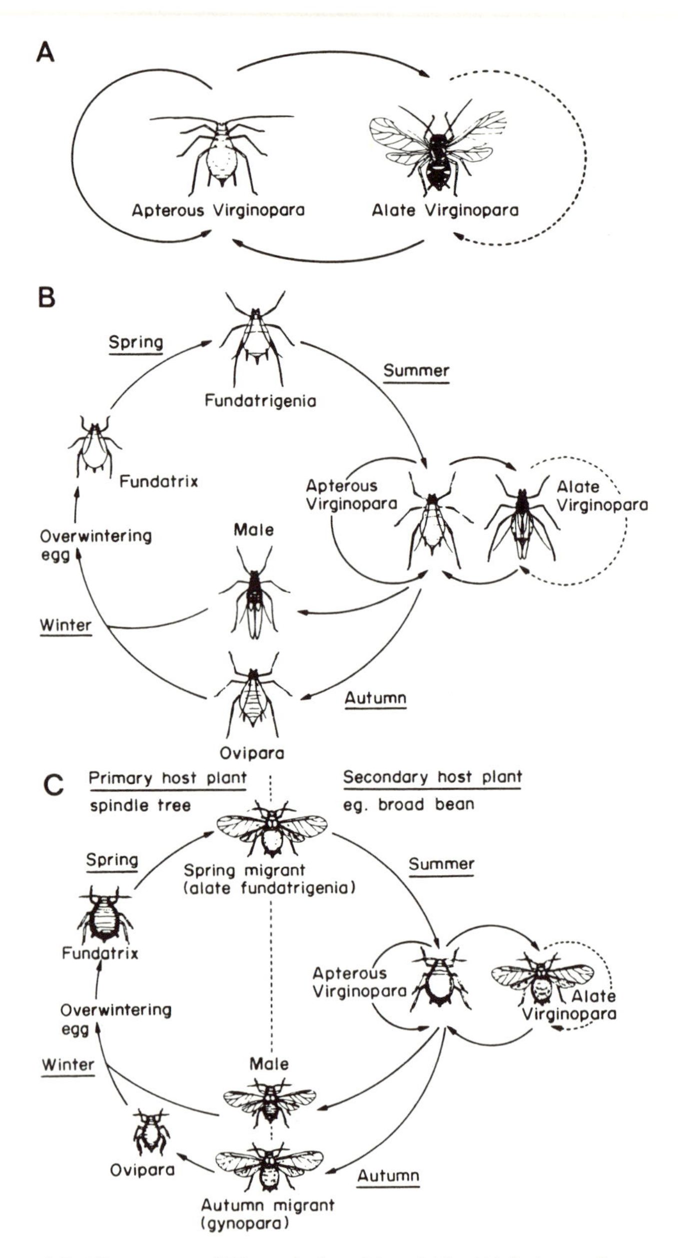

Figure 8.7. Three types of life cycle found in aphids. (A) Strict parthenogenesis: alate virginoparae usually produce only apterous progeny, but the latter may produce both alate and apterous offspring (found in some clones of *Myzus persicae*). (B) Alternation of sexual and parthenogenetic generations: sexual generations are usually produced in autumn, and eggs overwinter in diapause (found in *Megoura viciae*). (C) Alternation of sexual and parthenogenetic generations and alternation of primary and secondary hosts (found in *Aphis fabae*). (From Hardie and Lees, 1985, modified from Blackman, 1974. Reprinted with kind permission of Pergamon Press Ltd.)

genetic and winged/apterous. The sexual forms always lay eggs, while the parthenogenetic forms are viviparous. Both sexual and parthenogenetic aphids can be either winged or apterous, depending on the species. In addition, some aphids have distinct spring and summer forms. In the simplest and possibly the primitive form of an aphid life cycle, overwintering eggs give rise to apterous parthenogenetic females that produce parthenogenetic offspring during most of the spring and summer. In the late summer some winged forms are produced that disperse, and in the autumn sexual males and females are produced that mate and lay eggs that will overwinter and hatch the next spring. More complex life cycles involve an obligatory alternation of host plants in the course of the season (fig. 8.7B). The primary host is usually a tree or a root, on which the aphid overwinters, while the secondary host is often an herb and serves as host only during the summer months. Aphids that alternate host plants usually have distinctively different alate and apterous forms on each of those hosts; the alates on the primary host are different in pigmentation and body form from the alates on the secondary host, and likewise for the apterous forms. These forms are sufficiently different from one another that they have often been described as different species, and in many species of aphids it is still not known which primary and which secondary host "species" go together (Lees, 1966; Lüscher, 1976; Hardie and Lees, 1985; Moran, 1990).

The switches between the various forms are stimulated by environmental variables. In most species short day lengths induce the production of male and female sexual forms. Stimuli for the induction of the winged forms are more diverse. Depending on the species, crowding, short day lengths, low temperatures, or poor food quality (or some combination of these stimuli) induce the production of alate morphs (Hardie and Lees, 1985; Tauber et al., 1986). These environmental stimuli somehow affect the production of neurosecretory hormones by the mother, and indirectly, these hormones cause a developmental switch in her offspring. The endocrinology of form determination in aphids is still relatively poorly understood. Species differ in their response to exogenously applied hormones, and even within a species genetically distinct populations and clones may differ in their response to hormones. The analysis of hormonal effects in aphids is sometimes confounded by the fact that phase characteristics such as aptery are also larval characters, so that it is not always easy to differentiate a hormonal disturbance of metamorphosis (i.e., pathological juvenilization) from a hormonal induction of an alternate morph (i.e., true aptery).

Cautery of certain neurosecretory cells in the brain causes females of *Megoura viciae* to switch from the production of parthenogenetic offspring to the production of sexual offspring. This indicates that the production of parthenogenetic offspring requires the production of a maternal neurosecretory hormone. The hormone may be an allatotropin because topical application of JH to adult *M. viciae* and *Aphis fabae* that are producing sexual offspring causes them to switch to the production of parthenogenetic offspring. The JH-sensitive period for this transformation occurs during the embryonic life of the offspring, while they are still within their mother (Hardie 1980, 1981a,b, 1987; Hardie and Lees, 1985).

The alate/apterous polyphenism is also controlled by JH, but the JH-sensitive period here is during the early part of larval life (Hardie, 1981b). Application of JH to larvae of *Aphis fabae* early during their larval life inhibits the formation of the winged form. When JH is applied after the middle of larval life, apterous forms also result, but these are clearly juvenilized animals whose metamorphosis was disrupted so they formed a supernumerary larval instar. This suggests that the JH-sensitive period for alate-morph determination occurs shortly before the JH-sensitive period for determination of the adult. If JH is absent during both periods a winged adult is formed, whereas if JH is present only during the first of these periods an apterous adult results.

Color Phases in Armyworms

When larvae of the fall armyworm *Leucania separata* are crowded, they develop a heavily melanized cuticle. Isolated, uncrowded larvae have a yellow to reddish brown epidermis and a largely transparent cuticle, while crowded larvae develop a dark black cuticle and at the same time lose the reddish pigment from their epidermis (Ogura, 1975). Both the loss of the epidermal pigment (an ommochrome) and the melanization of the cuticle are controlled by a single hormone, the melanization-and-reddish-coloration hormone (MRCH) (Box 10). The MRCH is produced by neurosecretory cells in the subesophageal ganglion and is secreted three to six hours prior to ecdysis (Ogura, 1975). It is possible to extract MRCH activity also from whole heads of *Bombyx mori*, and this species has served as the source of the hormone for most experimental work. There are at least three molecular forms of MRCH, and they have at least a small sequence similarity to human insulinlike growth factor-II.

The fact that MRCH can be extracted in large quantities from a species such as *Bombyx*, which itself does not have a phase polyphenism, suggests that this hormone may have multiple functions, not an uncommon finding with neuropeptides. Matsumoto et al. (1990a) have, in fact, shown that the MRCH-I from *Bombyx* is exactly the same molecule as its pheromone biosynthesis activating hormone, PBAN-I (see chapter 5). Matsumoto et al. (1990b) furthermore have shown that synthetic MRCH-I induces cuticular melanization when injected into larvae of the cutworm

BOX 10

THE MELANIZATION AND REDDISH COLORATION HORMONE (MRCH)/PHEROMONE BIOSYNTHESIS ACTIVATING NEUROPEPTIDE (PBAN) FAMILY

STRUCTURE AND NOMENCLATURE: The MRCHs and PBANs form a family of structurally related peptides that share significant sequence similarity with each other and with vertebrate insulinlike growth factors. They also share structural similarity with two other groups of hormones discussed elsewhere: the diapause hormone (see chapter 7 and Box 9), and the leukopyrokinins (see chapter 3). Where tested, the hormones of this family share at least a moderate crossreactivity. The structure of MRCH-I of *Leucania separata* is identical to that of the PBAN-I of *Bombyx mori* and is given by

Leu-Ser-Glu-Asp-Met-Pro-Ala-Thr-Pro-Ala-Asp-Gln-Glu-Met-Tyr-Gln-Pro-Asp-Pro-Glu-Glu-Met-Glu-Ser-Arg-Thr-Arg-Tyr-Phe-Ser-Pro-Arg-LeuNH_2

The PBAN of *Heliothis zea* is

Leu-Ser-Asp-Asp-Met-Pro-Ala-Thr-Pro-Ala-Asp-Gln-Glu-Met-Tyr-Arg-Gln-Asp-Pro-Glu-Glen-Ile-Asp-Asp-Ser-Arg-Thr-Lys-Tyr-Phe-Ser-Pro-Arg-LeuNH_2

The structure of the PBAN of *Leucania separata* is

Lys-Leu-Ser-Tyr-Asp-Asp-Lys-Val-Phe-Glu-Asn-Val-Glu-Phe-Thr-Pro-Arg-LeuNH_2

SOURCE: The MRCH of *Leucania* (=*Pseudaletia*) *separata*, *Mamestra brassicae*, and *Spodoptera litura* is produced by neurosecretory cells in the subesophageal ganglion. The PBAN of *Heliothis zea* is also produced primarily in the subesophageal ganglion.

REFERENCES: Hiruma et al., 1984; Matsumoto et al., 1985; 1990, 1992a,b; Morita et al., 1988; Raina, 1993; Raina et al., 1987.

Spodoptera litura and stimulates sex pheromone production when injected into adults. Evidently these hormones are not species-specific, though they appear to have quite different functions in different developmental stages of a species.

Experiments on the cabbage armyworm *Mamestra brassicae* suggest that JH may also be involved in the control of larval phase polyphenism.

The larvae of *Mamestra* are pale white and yellow during most of their larval life but become very darkly pigmented at the outset of their last larval instar. This darkening is due to the development of a brown pigment (primarily xanthommatin) in the epidermis and black melanin in the cuticle, as in *Leucania*. This coloration switch is not facultative, however. Final instar larvae of *Mamestra* live on or near the ground during the daytime, and their dark coloration serves as camouflage. Hiruma et al. (1984) have shown that in *Mamestra* melanization of the cuticle is controlled solely by a MRCH from the subesophageal ganglion, but that ommochrome synthesis in the epidermis requires JH in addition to MRCH (fig. 8.8). JH appears to have a potentiating rather than an activating role in this system. JH must be present during a fairly brief critical period, 12–14 hours after head capsule slippage (the onset of apolysis in the molt to the final larval instar) to allow subsequent ommochrome synthesis to occur. If the source of JH is removed by allatectomy prior to this critical period, no ommochrome synthesis takes place. Topical application of JH restores the ability to synthesize ommochromes in allatectomized animals. The actual synthesis of ommochrome is stimulated by MRCH and does not begin until after melanin synthesis in the cuticle is nearly complete.

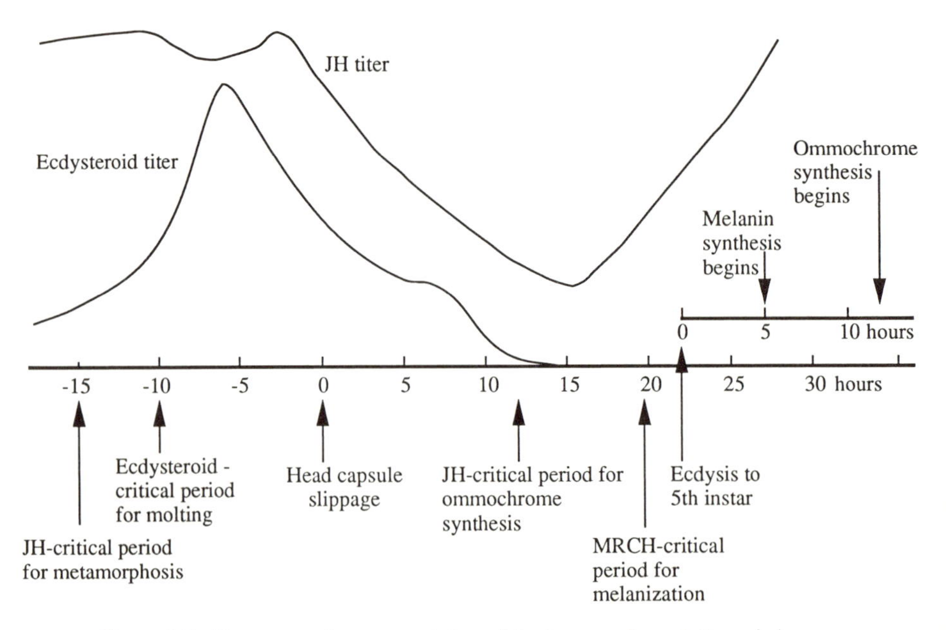

Figure 8.8. Diagrammatic representation of the hormonal regulation of pigmentation in *Mamestra brassicae*. Both MRCH from the subesophageal ganglion and JH are necessary for ommochrome synthesis, while melanin synthesis requires only MRCH. (Redrawn from Hiruma et al., 1984.)

CHAPTER 9

HORMONES AND BEHAVIOR

IN SPITE OF their relatively simple nervous systems, insects exhibit a rich diversity of behaviors. While some insect behaviors are preprogrammed fixed action patterns, the majority of insects can respond effectively and with considerable flexibility to a great variety of events and contingencies in their environment. Some insects such as honeybees, ants, sand wasps, and butterflies are even capable of learning, and must therefore possess some form of memory. As in vertebrates, a great variety of insect behaviors are stimulated, modified, or inhibited by hormones. Most of the documented cases of hormonal effects on insect behavior were discovered accidentally, by investigators working on the physiology or developmental effects of hormones. Unfortunately, relatively few of those observations have been followed up to the degree that they deserve, although there are a few exceptions, such as the hormonal control of eclosion behavior and the control of pheromone production, which have become significant areas of investigation in their own right. Beyond these, there are many scattered observations that are intriguing and still in need of fuller exploration, and a broadly synthetic insight is still some time off. Probably the most significant difficulty that besets studies of the role of hormones in insect behavior is that, with the exception of eclosion hormone and the effects of hormones on social behavior, the behaviors in question are parts of complex and integrated sets of physiological and developmental processes that are regulated by episodic releases of hormones, and it has been difficult to study the behaviors in isolation of their developmental and physiological context. Injections of hormones or removal of endocrine glands may alter behavior, but such interventions also alter a number of physiological processes, and it is not always clear whether the change in behavior is a direct response to the hormone or an indirect consequence of the altered physiology. On the positive side, the simplicity of insect nervous systems makes them attractive systems for studying the mechanism by which hormones affect the central nervous system, which should be more accessible here than in the immensely more complex nervous systems of the vertebrates (Truman and Riddiford, 1977).

It should be noted that behaviorally, an insect is really two completely different individuals at different stages in its life cycle. The larva and adult stages each have unique functions and unique behavioral capacities with which to execute those functions. The experiences of one life stage do not appear to carry over significantly to the next one. One of the reasons there is little carryover of information may be that the central nervous system undergoes extensive remodeling during metamorphosis. Many larval neurons die and new adult neurons arise from special nests of neuroblasts; larval neural pathways degenerate and new adult neural pathways are established (Truman, 1988; Truman and Riddiford, 1989). The sensory and muscular systems likewise undergo more or less extensive transformations during metamorphosis as new kinds of sensillae are added, old ones degenerate, flight musculature develops, and larval muscles degenerate. Insofar as ecdysteroids and juvenile hormones control metamorphosis, they also control this extensive remodeling of the central nervous system, and therefore they control indirectly the new patterns of behavior made possible by this restructured system. Whether these hormones act simply as general triggers for development in the central nervous system, or whether they direct the development of specific neural pathways in more detail remains to be established. In addition to controlling the development of the central nervous system, ecdysteroid and juvenile hormones also control specific behaviors associated with such diverse activities as reproduction, metamorphosis, caste differentiation, and dispersal, as will be outlined below.

Hormonal Control of Behaviors Associated with Metamorphosis

At the end of larval life, holometabolous insects stop feeding, void their gut contents, and go in search of a suitable place in which to pupate. The majority of insects dig burrows and pupate underground. The larvae of many species of Lepidoptera and Hymenoptera spin an elaborate cocoon in which they pupate. We noted in chapter 5 that this sequence of behaviors, beginning with the cessation of feeding and culminating within a complexly spun cocoon, are preprogrammed in the central nervous system and are triggered by the first peak of ecdysteroids that occurs in the absence of JH.

Prior to pupation, caterpillars of the European linden moth *Mimas tiliae* become positively geotactic and climb down from the tree in which they have been feeding. This behavior can be reversed, and the animals made negatively geotactic, by implantation of active corpora allata (Piepho et al., 1960). This experiment indicates that ecdysteroids may trigger only the enhanced locomotory activity known as "wandering," but that the sign of the geotaxis during this wandering phase is determined by the titer of JH.

Cocoon-spinning behavior is also affected by hormones. Larvae of the silkworm *Samia cynthia* spin two different types of cocoons at different times of the year. These are referred to as winter and summer cocoons: winter cocoons are spun by animals that are reared under short-day conditions and that will diapause, while summer cocoons are spun by animals reared under long-day conditions and that continue normal development without diapause (Pammer, 1966). A hormone from the brain, possibly PTTH, controls which of the two types of cocoons is constructed. Brains from short-day penultimate larval instars implanted into short-day final instar larvae caused the latter to spin summer-type cocoons in a large percentage of cases, which indicates that differences in the timing of PTTH secretion may be responsible for differences in the spinning behavior of these larvae (Nopp-Pammer and Nopp, 1968). Brains from short-day larvae, cultured in vitro for six days under long-day conditions and then implanted into short-day larvae, also caused the latter to build primarily summer-type cocoons (Nopp-Pammer and Nopp, 1968). Thus the entire photoreceptor, integration, and behavior-altering mechanism resides in the brain and can be expressed by isolated brain implants, via a humoral pathway.

Application of JH prior to cocoon spinning can delay the onset of spinning for several days, presumably until the JH decays (Riddiford, 1972). This effect is possibly similar to the modulation of geotaxis by JH, though it is not clear what the role of this response is in the normal biology of the animal. Injection of 20-hydroxyecdysone during the middle of cocoon spinning in *Antheraea pernyi* can cause a premature termination of cocoon-spinning behavior (Lounibos, 1975, 1976), presumably by premature stimulation of apolysis and pupation. Larvae of many species of moths construct complex cocoons with distinctive outer (spun first) and inner (spun last) envelopes. This sequence of cocoon construction is not fully preprogrammed. When larvae of *Ephestia kühniella* are removed from their cocoons while they are still constructing the outer envelope, they begin to construct a new cocoon from the very beginning. But if they are removed from their cocoon during construction of the inner envelope, they do not make a new outer envelope but only an inner envelope. The behavioral switch from outer to inner envelope construction appears to be regulated by ecdysteroids. When larvae of *Ephestia kühniella* are removed from their cocoon during the stage of outer envelope construction, injected with 20-hydroxyecdysone, and made to construct a new cocoon, they construct only the inner envelope, whereas control animals injected with saline always construct a new outer envelope first (Giebultowicz et al., 1980). The time at which *Ephestia* larvae normally switch from outer to inner envelope construction corresponds to the critical period for metamorphosis determined by ligation (Giebultowicz et al., 1980), which suggests that in intact animals the switch in behavior is also controlled by ecdysteroids. The difference in the responses of *Antheraea* and *Ephestia* to ecdysteroid injection may be due to differences in the normal patterns of ecdysteroid secretion prior to metamorphosis. This pattern has not yet been studied in detail in either species.

Hormonal Control of Ecdysis

The times of ecdysis and eclosion are delicate and dangerous periods in the life of an insect, because immediately after shedding the old cuticle the new cuticle is still very soft and largely unsclerotized. The soft cuticle not only offers little physical protection for the insect but it also prevents efficient and rapid locomotion as the muscles have nothing to work against. A period of up to several hours is required to fully sclerotize and harden the cuticle after an ecdysis, and during that period the insect is largely defenseless. This situation is worst, perhaps, at the time of adult eclosion, which requires an especially prolonged period of undisturbed quiescence during which to inflate, flatten, and harden the wings before the animal can take flight. It has been known for a long time that ecdysis of many insects takes place during a specific time of day, usually at twilight or at night, when they are likely to be least exposed to predators or other disturbances.

The timing of ecdysis of many insects is controlled by a biological clock and takes place during a time window of only a few hours' duration. The timing of this window is cued by the photoperiod (the length of days and nights) the insect experiences and is species-specific (Truman, 1971a; Beck, 1980). Ecdysis, and particularly the eclosion of adult insects, involves an elaborate and stereotyped series of behaviors, including air swallowing to expand and burst the old cuticle, peristalsis and wriggling movements to shed the cuticle, and finally the prolonged contraction of abdominal muscles that raises the hemolymph pressure and inflates the wings.

Eclosion Hormone

The fact that ecdysis occurs during a specific time of day indicates that a special physiological control mechanism must exist which initiates the unique behavioral repertoire associated with escape from the old cuticle and expansion of the new one. The physiological mechanism whereby the circadian biological clock controls the onset of eclosion behavior was discovered by Truman and Riddiford (1970). They showed that the biological clock, which resides in the brain, controls ecdysis behavior not directly

BOX 11

Eclosion Hormone (EH)

Structure and Nomenclature: The eclosion hormone of *Manduca sexta* is a neurosecretory polypeptide of 62 amino acids. A nearly similar eclosion hormone has been identified from *Bombyx mori*. The amino acid sequence of this hormone does not appear to have sequence similarities to any other known proteins. The *Bombyx mori* EH exists as a family of at least four nearly identical peptides. The structure of EH-IV is given below, together with the EH of *Manduca* (common sequences are underlined). EH-I and EH-III have their N-termini truncated by two positions and EH-I and EH-II terminate at residue 61.

	1 20
Manduca sexta EH:	NPAIATGYDP MEICIENCAQCKKMLGAWFEGPL
Bombyx mori EH-IV:	S PAIASS YDAMEICIENCAQCKKMFGPWFEGSL
	40 60
Manduca sexta:	CAESCIKFKGKL IPECEDFASIAPFLNKL
Bombyx mori:	CAESCIKARGKDIPECESFASISPFLNKL

Source: The eclosion hormone in larvae of *Manduca* is produced by the ventral tritocerebral neurosecretory cells. These neurosecretory cells send their axons down the length of the ventral nerve cord and have their neurohemal area along the proctodeal nerve (a nerve of the terminal abdominal ganglion). In adult *Manduca*, eclosion hormone is also produced by the ventral tritocerebral cells, but these now send new axons to the corpora cardiaca instead. Eclosion hormone acts via cGMP as a second messenger.

References: Copenhaver and Truman, 1986a; Horodyski et al., 1989; Kataoka et al., 1987; Kono et al., 1991; Marti et al., 1987; Truman, 1981; Truman and Copenhaver, 1989; Truman and Riddiford, 1970; Truman et al., 1981.

through nerves, but via a neurosecretory hormone, the eclosion hormone (EH) (Box 11). Truman and Riddiford worked on the adult eclosion behavior of giant silk moths. They showed that when the brain of a moth was removed some time prior to the expected time of eclosion, eclosion was delayed considerably and eclosion behavior was very abnormal. But when the brain was reimplanted into the abdomen, not only did eclosion occur at its normal time, but eclosion *behavior* was also quite normal. Evidently, the brain is necessary both for normal timing and behavior, but it need not be connected to the rest of the central nervous system to be effective. The role of hormones in this process was demonstrated when Truman (1971c, 1973a) showed that extracts prepared from the brains or corpora cardiaca of pharate (ready to eclose) adult moths induced premature but normal eclosion when injected into intact as well as brainless pharate adults.

Eclosion hormone is not species-specific. This was demonstrated by Truman (1971b) with reciprocal transplants of the brains of two species of silkmoths (fig. 9.1). In each case, the implanted brain caused the recipient to eclose at the time of day characteristic of the brain donor, but the eclosion behavior pattern was that of the recipient. In subsequent experiments Truman showed that when isolated abdomens of pharate adults were injected with EH, they exhibited the complete stereotyped sequence of rotational and peristaltic motions normally associated with eclosion. EH evidently acts directly on the ganglia of the ventral nerve cord to trigger a preprogrammed neural behavior pattern that results in the stereotyped sequence of motions that constitute eclosion behavior. Eclosion "behavior" can even be demonstrated in an isolated ventral nerve cord of *Hyalophora* maintained in organ culture (Truman, 1978). When recording electrodes are placed on the stumps of the motoneurones that emerge from the segmental ganglia of an isolated nerve cord, and EH is added to the culture medium, the ganglia begin to produce synchronous bursts of action potentials that alternate from their left and right sides. In an intact animal such alternating left and right stimulation of the abdominal muscles would give rise to the wiggling and rotatory motions of the abdomen characteristic of early eclosion behavior. This pattern is followed some 40 minutes later by bursts of action potentials that are synchronous on left and right sides of a given ganglion but that alternate between successive ganglia. In an intact animal such a pattern would give rise to the peristaltic motions by which the animal normally escapes from its pupal cuticle. Some half hour later, all ganglia begin to produce prolonged bursts of action potentials simultaneously from all their motor nerves, and these correspond to the steady tonic abdominal contraction necessary for inflation of the wings after ecdysis.

Thus the entire sequence of eclosion behavior is preprogrammed in the ventral ganglia. It requires no sensory input (although the timing of this sequence can be modulated by sensory input), but merely a single stimulus to get it started. The secretion of EH is controlled by a biological clock in the brain. The physiology of this photoperiodic clock for EH secretion has been studied by Truman (1971d, 1972) and Reynolds et al. (1979), who showed that EH release is restricted to a narrow "gate" whose timing depends on the species and on the exact photoperiod under which an individ-

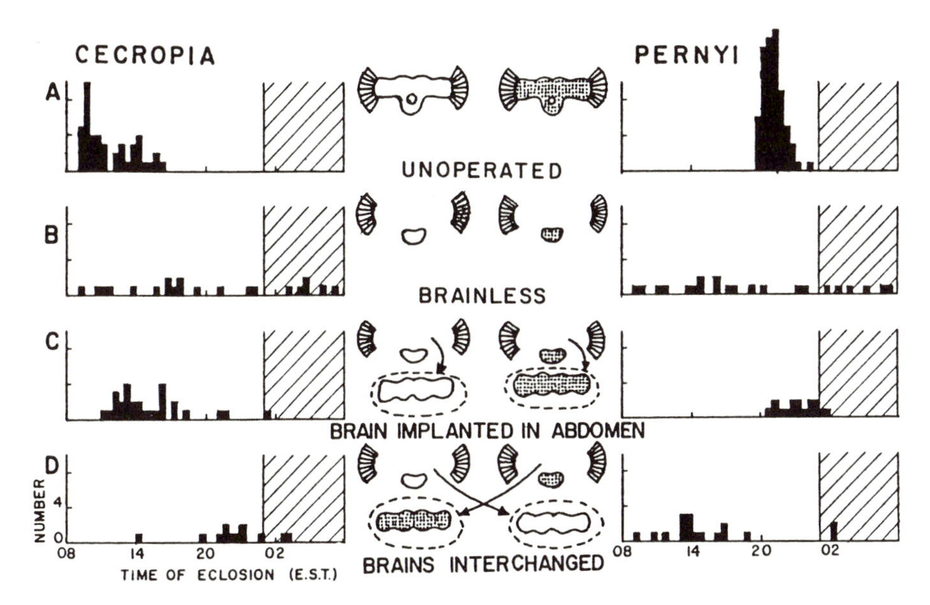

Figure 9.1. Timing of eclosion of *Hyalophora cecropia* and *Antheraea pernyi* under a 17L : 7D photoperiod regimen (daytime is white, nighttime is cross-hatched). (A) Species-specific eclosion times of intact unoperated animals. (B) Brainless animals eclose at random times during the day and night. (C) Brains reimplanted into the abdomens of brainless animals restore the normal eclosion time for each species. (D) Brains interchanged between the two species and implanted into the abdomen cause each to eclose at the time characteristic of the brain donor, although the eclosion behavior remains characteristic of the recipient. (From Truman, 1971b. Reprinted with permission of Pudoc Press.)

ual is maintained. In adult *Manduca sexta* and *Antheraea pernyi* kept under long-day conditions, the gate occurs between 20 and 23 hours after a lights-off signal, and in adult *Hyalophora cecropia* between 8 and 12 hours after lights-off. Eclosion hormone release for pupal and larval ecdyses is not gated (Truman, 1985). It is possible that a precise gating mechanism has evolved only for adult eclosion because it occurs in an exposed environment (pupation usually occurs underground or in a cocoon) and requires a much longer time than larval or pupal ecdyses. Gating allows the timing of eclosion to become adjusted to a time of day when the animal is least likely to be subject to discovery by predators.

While the precise timing of eclosion depends on the timing of EH secretion, animals also need to develop a competence to respond to the hormone. In *Manduca*, that competence develops only about four hours prior to the actual release of the EH; injections of EH prior to that time fail to provoke eclosion behavior (Reynolds et al., 1979). The onset of responsiveness is regulated by the disappearance of ecdysteroids at the end of adult development, and can be delayed by injections of exogenous ecdysteroids (Schwartz and Truman, 1983; Truman, 1985). Whether the onset of responsiveness is due to the de novo appearance or activation of hormone receptors or of some critical component of the intracellular transduction pathway is unknown (Truman, 1990).

In addition to controlling the timing of adult eclosion, EH has also been shown to control larval-larval ecdyses in *Manduca*, and it may be that the timing of all ecdyses in the Lepidoptera, and perhaps in most insects, is controlled by the secretion of eclosion hormones (Truman et al., 1981; Truman, 1985, 1990). In *Manduca*, the EH also plays an important role in the plasticization of the cuticle at ecdysis (Reynolds, 1977). This function is nearly identical to that of bursicon (see below) but occurs at a slightly different time in the molting/eclosion cycle. Truman (1985) has proposed the following roles for EH and bursicon during adult eclosion in moths. Eclosion hormone is released about two hours prior to the actual ecdysis and causes an initial softening of the cuticle which aids in escape from the old skin but leaves the new cuticle hard enough to withstand abrasion during ecdysis. After successful ecdysis, secretion of bursicon causes a further softening, particularly of the wings, which allows them to be more easily inflated. The plasticizing effect of bursicon is transient and soon after expansion bursicon-induced sclerotization of the cuticle begins.

The abdominal intersegmental muscles that are responsible for the eclosion behavior begin to degenerate soon after the adult ecloses (Lockshin, 1969). Injections of EH into isolated abdomens of *Antheraea polyphemus* induce a prompt degeneration of the intersegmental muscles (Schwartz and Truman, 1982, 1984; Truman and Schwartz, 1980). EH-stimulated cell death of the intersegmental muscles can be delayed, however, by a high level of neural input (Lockshin and Williams, 1965a,b), and this provides a safety mechanism to extend muscle life in case the insect encounters an obstacle in escaping from its pupal cuticle or cocoon, or fails to find an immediate place to rest and spread its wings (Truman, 1985).

Hormones and Migration

Immediately after eclosion many adult insects take to the wing and exhibit a prolonged period of flight activity. This behavior serves to disperse the adults away from the site where they grew up as larvae, and is presumably an adaptation to enhance outcrossing and the discovery of new habitats and food sources. This early dispersal flight is closely related to the migratory flight that some insects undertake in anticipation of unfavorable seasonal conditions (Dingle, 1985). The termination of the dispersal flight phase often coincides with the onset of reproduction (Dingle, 1972, 1985; Pener, 1985). The term "migration" refers to a more prolonged period of flight that is either seasonal in nature or is a response to specific environmental stimuli such as crowding or diminished food quality. Migrating insects are characteristically in reproductive diapause. The apparently mutually exclusive relation between reproduction and migration (and dispersal flight) leads to the suspicion that both may share at least portions of a control pathway.

It is now clear that the seasonal migratory flight of some (but not all) insects is regulated by the titer of JH. The best studied case so far is that of the large milkweed bug, *Oncopeltus fasciatus*. Migratory flight in *Oncopeltus* is stimulated by short day lengths, moderate to low temperatures, and poor food quality. These environmental conditions also inhibit ovarian development. Under long-day photoperiods, warm temperatures, and high food quality, prolonged flight is inhibited and females begin reproduction. Photoperiod, temperature, and food quality affect the level of JH in the hemolymph, and whether or not an individual is reproductive, migratory, or in deep quiescent diapause depends on the exact titer of JH. In *Oncopeltus* there are two thresholds of sensitivity to JH (fig. 9.2). When JH is higher than the upper threshold, normal reproduction takes place and migratory flight is inhibited. When the JH titer falls below the upper threshold, the ovaries regress and the animals are prone to undertake long migratory flights. When the JH titer falls below the lower threshold, migratory activity ceases, metabolism drops, and the animals enter the quiescent phase of adult diapause (Rankin and Riddiford, 1977; Rankin, 1978; Dingle, 1985). Under long-day conditions the JH titer rises immediately after adult emergence and dispersal flight takes place during the brief period that the JH titer is in the migratory range. As the JH titer rises above the upper threshold, the animals settle down and ovarian development begins. Whenever the food quality diminishes, or as the photoperiod shortens and temperature drops in the autumn, the JH titer declines gradually. It eventually falls below the upper threshold and migratory flight takes place. If migration takes the animal into an environment where food quality is better, the JH titer rises again, migratory flight ceases, and reproduction begins again. If adverse conditions persist, JH gradually falls below the lower threshold and the animal becomes quiescent (Rankin and Riddiford, 1978).

The control of migratory behavior in other species is not as well understood. In migratory locusts (see chapter 8) there is evidence that the adipokinetic hormone from the corpora cardiaca is necessary to stimulate migration (Pener, 1985), but whether it acts as a general stimulus of metabolic activity or as a specific migratory stimulus is not clear. A reduction of JH

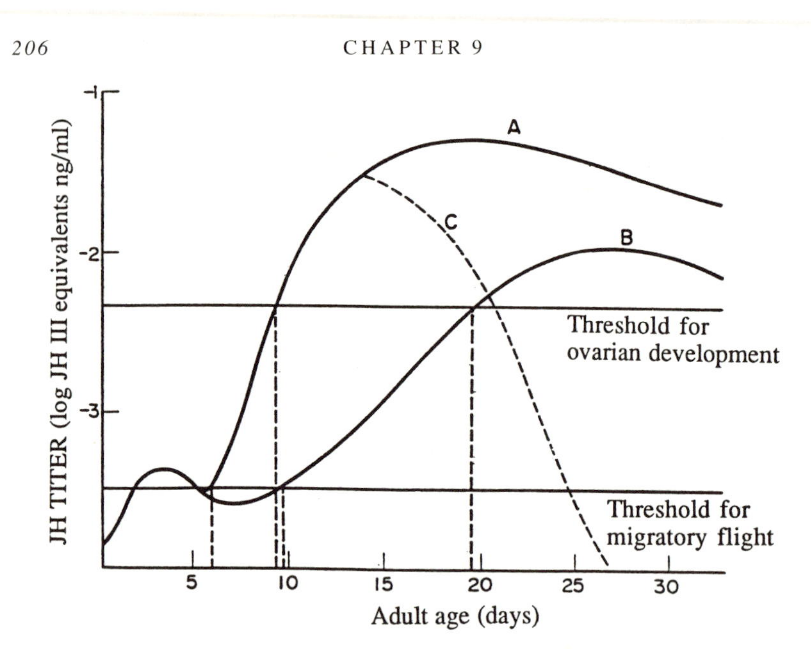

Figure 9.2. A model for the control of reproduction, migration, and diapause of adults of the milkweed bug *Oncopeltus fasciatus* under various environmental conditions. A, high temperatures and long photoperiods stimulate high JH titers and reproduction after a brief dispersal flight; B, lower temperatures and short photoperiods result in a slower rise of JH and stimulate migratory flight in response to intermediate titers of JH; C, poor food quality or starvation results in declining JH titers and flight activity, followed by cessation of flight and diapause if lack of food persists. (From Rankin and Riddiford, 1978. Reprinted with kind permission of *J. of Insect Physiol.* and Pergamon Press Ltd.)

titers and an increase in AKH titers may be correlated with migratory behavior in the monarch butterfly *Danaus plexippus* (Herman, 1973; Dallman and Herman, 1978), but an experimental demonstration of the role of these hormones in controlling migratory behavior is still lacking.

Reproductive Behavior

The onset of sexual receptivity in virgin female insects is often controlled by JH. In many insects secretion of JH begins soon after adult eclosion and begins to stimulate ovarian development. This rising JH titer is known to stimulate mating behavior as well. Removal of the corpora allata from newly ecdysed females of the grasshopper *Gomphocerus rufus* prevents them from becoming receptive to courting males (Loher and Huber, 1966). Instead, allatectomized females continue to actively reject courting males, much as freshly ecdysed females do. Injections of JH into such refractory females induce them to accept courting males. The rise of JH titer soon after adult eclosion stimulates mating behavior (as well as host-seeking behavior) in female *Aedes* mosquitoes (Lea, 1968; Hagedorn, 1985). In *Drosophila melanogaster* the implantation of active corpora allata–corpora cardiaca complexes into pharate adult females several hours before eclosion causes them to become receptive to courting males a full 24 hours before they would normally have done so (Manning, 1966). JH has been shown to be required for the onset of sexual behavior or sexual receptivity in females of various other species of Diptera, Hemiptera, and some Orthoptera (Truman and Riddiford, 1974b). Unfortunately, we have as yet little or no insight into what exactly JH does to stimulate mating behavior and receptivity to courtship and copulation.

Not all species require JH (or any other hormone for that matter) to stimulate sexual receptivity or sexual behavior. Barth and Lester (1973) have compared the taxonomic distribution and life cycles of species that do and do not use hormones to regulate their reproductive behavior, and their analysis indicates that hormonal control may be restricted to species that have relatively long life spans as adults, much like hormonal control over ovarian maturation (chapter 6). Insects with short-lived adults are apparently ready to mate at any time and require no special hormonal stimuli or conditioning. There is little evidence that hormones other than JH play a role in regulating female reproductive behavior. In *Leucophaea maderae*, cauterization of the portion of the brain containing the medial neurosecretory cells eliminates female mating behavior, which suggests that a neurohormone may be involved in the regulation of mating behavior in this species (Engelmann and Barth, 1968). Mating behavior in cauterized animals cannot be restored by implantation of active corpora allata, so the effect is probably not due to the elimination of allatotropin-producing cells, and we may, therefore, be dealing with the direct effect of a neuroendocrine factor on reproductive behavior.

There are a few cases in which male reproductive behavior has been shown to be under hormonal control. Males of *Schistocerca gregaria*, for instance, require an elevated JH titer for normal sexual behavior. Removal of the corpora allata from *Locusta* males results in the cessation of sexual activity, which can be restored by implantation of active corpora allata or by application of exogenous JH but only in crowded, gregarious males (Pener, 1967, 1991). Males of the fly *Scatophaga stertocaria* mate only when they are well fed. Starved males exhibit no sexual behavior, nor do males that are fed but have been allatectomized. Starved males can be induced to mate by implantation of active corpora allata, which suggests that under normal conditions feeding stimulates JH secretion by the corpora allata (Foster, 1967). In the bug *Pyrrhocoris apterus* mating activity can also be induced by application of JH (Zdarek and Slama, 1968). There is evidence that an as yet unidentified neurohormone is required for stimulation of courtship and sexual activity in the cockroach, *Periplaneta americana* (Milburn et al., 1960).

In addition to controlling the onset of sexual receptivity and mating behavior, there is an interesting hormonal effect in many insects that causes a change in a female's behavior once she has mated and which makes her refractory to further mating attempts by males. In some species this refractoriness to further mating lasts for the rest of the female's life, making her essentially monogamous, while in others the effect is temporary (Truman and Riddiford, 1974b; Gillott, 1988). This switch in female behavior is stimulated by peptides produced by the accessory glands of the male and transferred with the seminal fluid during copulation. In mosquitoes this factor is called matrone (Craig, 1967), while in *Drosophila* it is referred to as the sex peptide (Gillott, 1988; Kubli, 1992). Similar receptivity-inhibiting substances have been described in several other species of Diptera and in at least one lepidopteran, *Hyalophora cecropia* (Gillott, 1988). Since these peptides are transferred from one individual to another they are properly called pheromones. Their mode of action, however, is more like that of a hormone, rather than that of a pheromone, in that they appear to enter the circulatory system of the recipient female and act directly on her nervous system. Injection of matrone or sex peptide into the circulatory system of a female is as effective in altering her behavior as is transfer of the material via the customary route of copulation. Matrone and the sex peptide appear to be unrelated molecules. Sex peptide is a 36-amino acid peptide, while matrone is a protein with a molecular weight between 50,000 and 100,000 D (Gillott, 1988; Kubli, 1992). They also have quite different modes of action: matrone acts directly on the terminal abdominal ganglion of the female mosquito (Gwadz, 1972), while sex peptide appears to act on the brain of *Drosophila* (Kubli, 1992).

In addition to inhibiting receptivity to further mating, matrone and sex peptide affect several other behaviors and physiological events associated with reproduction. Both substances enhance the fecundity of females. Virgin females tend to mature eggs more slowly and in smaller numbers than females that have been mated or that have been injected with a preparation of matrone or sex peptide (Leahy and Craig, 1965; Leahy and Lowe, 1967). Fertility-enhancing effects of male accessory gland substances appear, in fact, to be very widespread among the insects (Gillott, 1988). In mosquitoes, injections of matrone preparations also enhance host-seeking and biting behavior (Judson, 1967), and the search for oviposition sites (Yeh and Klowden, 1990).

Hormonal Control of Social Behavior Patterns

A well-established endocrine effect on adult behavior is found in the control over the temporal castes in honeybees. Adult worker bees go through a stereotyped set of behavioral tasks during their lifespan. During the first two weeks after metamorphosis worker bees act primarily as nurses for growing larvae in the central portion of the hive. During the third week of life they work mainly in the peripheral region of the hive on food storage and nest maintenance, and during their fourth week and thereafter they forage outside the nest for food (Seeley, 1982; Winston, 1987). The sequence of this temporal division of labor is regulated by JH. Titers of JH gradually increase during adult life so that low titers are consistently associated with nursing behavior and high titers with foraging behavior. Application of JH to young adult worker bees causes them to shift their behavior and become precocious foragers (Huang et al., 1991; Robinson, 1987; Robinson et al., 1989, 1991). Huang et al. (1991) point out that by tying adult behavior patterns to JH titers, honeybees acquire a flexible response capacity to changing environments. Since the titer of JH may be modulated by changes in environmental conditions (Robinson et al., 1989), honeybees can use this response mechanism to accelerate, retard, or reverse their behavioral development.

Adult behavior in female paper wasps of the genus *Polistes* is also regulated by JH. Overwintering females of *Polistes* often cooperate in the founding of colonies in the spring. Eventually these females establish a linear dominance hierarchy in which the most dominant female becomes the functional queen of the colony (West-Eberhard, 1967; Wilson, 1971; Röseler, 1985). The dominant female lays most or all of the eggs for the colony. Her aggressive behavior toward the other females causes their ovaries to regress, and they take on the role of workers for the colony. The female that becomes dominant appears to be the one that has the largest ovaries at the time of colony founding. The probability that a female will become the dominant female in the colony is also strongly correlated with the size of her corpora allata (Röseler et al., 1984, 1985). Barth et al. (1975) found that treatment with JH greatly increased the aggressive behavior of females of *Polistes annularis*. Röseler et al. (1984) have shown that injection of JH-I as well as injection of 20-hydroxyecdysone increases the probability that a female of *P. gallicus* will become the dominant one in pairwise tests. The combination of large ovaries and large corpora allata suggests that dominant females have higher JH titers (produced by the corpora allata) as well as higher ecdysteroid titers (produced by the ovaries) than subordinate ones. Both hormones can evidently stimulate dominant behavior, and it remains to be determined what their relative roles are in the normal biology of the animal. The mechanism by which a dominant female suppresses hormone production in her subordinates is not yet understood.

Hormonal Control of Pheromone Production

In addition to the direct releasing or inhibitory effects on mating behavior discussed above, hormones can have an indirect effect on reproductive behavior through the control over the synthesis and secretion of sex pheromones. The corpora allata have been shown to be required for sex pheromone production in the roaches *Byrsotria fumigata* and *Pycnoscelus surinamensis* (Barth, 1962, 1965). In *Tenebrio molitor* allatectomy also inhibits sex pheromone production, and application of a JH analog restores production (Menon, 1970).

In some Lepidoptera a neurosecretory product of the brain appears to be necessary for pheromone production and release. In *Heliothis zea*, neck ligation inhibits pheromone production, and in such animals the production of pheromone is restored by injections of brain extracts behind the ligature (Raina and Klun, 1984). Two structurally related hormones, called pheromone biosynthesis activating neuropeptides (PBANs), have been isolated from homogenates of the subesophageal ganglia of *Heliothis* and of *Bombyx mori* (Raina et al., 1989; Kitamura et al., 1990). They are polypeptides of 33–34 amino acids (see Box 10), whose N-terminal sequence is similar to that of insulinlike growth factor II and the myotropic peptide of *Locusta*, and the diapause hormone of *Bombyx*. A smaller, 18-amino acid, pheromone-stimulating peptide has been isolated from the armyworm *Pseudaletia* (=*Leucania*) *separata*, which appears to differ structurally from the previous two. This latter hormone has some sequence similarity to the *Bombyx* diapause hormone, and has both a slight diapause hormone activity and high pheromonotropic activity when tested on *Bombyx mori* (Matsumoto et al., 1992a; Raina, 1993). It also has a sequence similarity to leukopyrokinins (see chapter 3) and the melanization and reddish coloration hormone (MRCH; see chapter 8), and may in fact be the active MRCH of *Pseudaletia* (Matsumoto et al., 1992b). It appears that we are dealing here with a family of closely related peptides that have been adapted for radically different regulatory functions in different species and at different stages in the life cycle.

Pheromone synthesis and release from the abdominal pheromone gland require both hormonal and nervous stimuli. It has been shown that PBAN can stimulate pheromone synthesis in the glands, but only during the dark phase of the photoperiod, that is, only at night, while during the light phase (day) PBAN has no effect. Injections of octopamine at the beginning of the dark phase of the photoperiod mimic the effects of PBAN, while injections at the beginning of the light phase have little or no effect on pheromone production (Christensen et al., 1991). PBAN also acts on the terminal abdominal ganglion to activate nervous stimulation of the pheromone glands. Artificial stimulation of the nerves that run from the terminal abdominal ganglion to the pheromone gland causes an increase in pheromone concentration in the gland, presumably by de novo synthesis, and this stimulation is effective at any time of day. Interestingly, some of the nerves that innervate the pheromone glands are neurosecretory, and the possibility exists that they are the actual sites of PBAN production and release (Christensen et al., 1991). The stimulation of pheromone production in *Heliothis* is clearly complex in that it may involve both a nervous and a neuroendocrine pathway to the pheromone gland, plus a mechanism that restricts pheromone production to nighttime (Raina, 1993).

EPILOGUE

INSECT endocrinology is, by the very nature of its subjects, a comparative science. The insects are a large, ancient, and diverse group of organisms, notable for their unparalleled ability to adapt, morphologically and physiologically, to an immense variety of natural circumstances. Metamorphosis and polyphenisms are among the most important and widespread of these adaptations. Metamorphosis allows an insect species to evolve adaptations for two or more radically different modes of life at different times in its life cycle, while alternative polyphenisms give an insect the option to differentiate into one of several alternative forms as best fits the environmental circumstances encountered. Hormones are critical components of the developmental switch mechanisms that regulate metamorphosis and polyphenisms. The titer data for PTTH, ecdysteroids, and JH published over the past two decades reveal rugged patterns of secretion with sharp peaks and valleys, not the smooth gradual rises and falls that we had initially expected for these hormones. One reason for this abrupt pattern of secretion may be that in many cases these developmental hormones act as switches (or start/stop stimuli) for developmental events. Using hormones as switching signals in essence allows central control over the onset and termination of developmental events. This produces a system through which the timing of developmental events can be flexibly regulated to meet the needs of an animal in a variable environment. A centralized switching system uncouples what would otherwise be a relatively inflexible causal chain of developmental events (with prior events stimulating subsequent ones, and so on) into a centrally controlled set of developmental modules. Even a slight developmental uncoupling of its component parts greatly increases the evolutionary potential of a complex system because it allows those components to diverge in form and function, and this may account for the unrivaled diversity of life cycles and body forms that have evolved among the insects.

It is not surprising that, as form and function evolve to meet different contingencies, the control mechanisms that enable developmental switching also evolve. Such evolution would involve shifts in the timing and number of hormonal peaks, changes in the cell types that bear receptors, and changes in the biochemical response pathways within cells, so that completely different physiological processes eventually appear in response to the same (hormonal) token stimulus. The biology of the juvenile hormones is an illustration of one of the more elaborate results of the evolution of control and response mechanisms. We have seen that these hormones are now used to control a great diversity of functions during larval and adult development and reproduction, and that responses to the hormone may range from the stimulation of massive protein synthesis (vitellogenin), to a switch in cell fate, to a change in behavior (the ancestral function of JH remains an undecided question). A similar though not quite as extreme diversity of functions is found in physiological control mechanism using adipokinetic and hypertrehalosemic hormones, whose functions can differ depending on the species and stage in the life cycle.

While we marvel at the richness of developmental and physiological mechanisms that have evolved around insect endocrine systems, this very richness suggests that it may be dangerous to rely too much on a few "model systems" for obtaining a deep understanding of the physiological and developmental endocrinology of insects. We have undeniably learned an immense amount of basic and useful information by focusing on a few such model systems. But it is also very clear that not all features of the endocrine biology of such well-studied species as *Manduca sexta*, for instance, apply to other holometabolous insects, or even to other Lepidoptera. The timing of hormonal peaks differs among species, and the detailed effects of hormones on epidermis, imaginal disks, and behavior often have a strong species-specific character. Patterns of commonality and divergence in endocrine control systems have yet to emerge. The evident diversity of mechanisms begs a comparative approach.

Advances in analytical techniques have revealed a greater diversity of hormones and greater variation in their structure than we had expected twenty years ago. The juvenile hormones and ecdysteroids occur as multiple molecular variants, though the precise origin and function of those variants remains somewhat of a mystery. The juvenile hormones have curious structural and functional similarities to substances like retinoic acid and tocopherol, which act as morphogenetic signals in vertebrates and rotifers, respectively (Gilbert, 1980; Tickle, 1991). These similarities may be superficial, though it seems quite possible that these molecules constitute a widespread but as yet unrecognized functional class of regulatory substances for developmental processes. The peptide neurohormones can be gathered into at least three major families that have structural and functional relationships: (1) the PTTHs appear to be made up of a family of related peptides, with substantial processing of the translated peptides; (2) the PBAN/myotropin/diapause hormone/MRCH cluster of hormones all share a version of the same C-terminal sequence: Phe-Thr-Pro-Arg-LeuNH$_2$ (also abbreviated FTPRLamide), and may have other commonalities in form and function; (3) the RPCH/AKH/HTH/bombyxin family share sequence similarity with vertebrate insulinlike growth factors and may play important (but yet to be fully explored) roles in the regulation of physiology and growth. It has been customary to assume that similarity of sequence in genes and peptides implies commonality of descent. But for relatively brief sequences such as we find in neuropeptides, there has been world enough and time for convergent evolution to a common structure and function, and commonality of descent is by no means granted. In any event, it would be interesting some day to map neuropeptide sequences onto independently derived cladograms to see how they may have evolved.

The last decade of research on insect endocrinology has seen a definite shift of interest from the physiology to the molecular biology of hormone action that mirrors the general shift from the organismal to the molecular we see in many other branches of the biological sciences. Molecular approaches have given us (and continue to give) powerful insights into the details of mechanisms that have greatly enriched our appreciation of the subtleties of nature, and, just as important, they have given us powerful new tools with which to manipulate the systems on which we work. But the reductionism inherent in the molecular approach is fundamentally inimical to the comparative approach and to the discovery of emergent properties that arise from interacting assemblages of parts. The regulative properties of physiological and developmental systems are emergent and must be studied at a level that integrates information from molecular, cellular, and organismal studies. Nowhere is this need more apparent than in the use of assays that test the "activity" of a hormone in a system different from the one it acts on in nature (bombyxin/PTTH and the various myotropic factors are cases in point). Such systems may be fine for monitoring steps in the purification of hormones, but can lead to great confusion when they are interpreted as revealing something about the normal role of the hormone. The recent history of insect endocrinology is teaching us not to be surprised to find that one hormone may have different functions in different species, or at different times (or on different tissues) in the same species. Comparative studies that try to tease out the patterns, causes, and consequences of those *differences* in hormonal control mechanisms are likely to yield the next great advance in our understanding of insect endocrinology.

LITERATURE CITED

Abu-Hakima, R., and K. G. Davey. 1977. The Action of juvenile hormone on the follicle cells of *Rhodnius prolixus*: The importance of volume changes. *J. Exp. Biol.* 69: 33–44.

Adams, M. E., and M. O'Shea. 1983. Peptide cotransmitter at a neuromuscular junction. *Science* 221: 286–289.

Aggarwal, S. K., and R. C. King. 1969. A comparative study of the ring glands from wild type and *1(2) gl* mutant *Drosophila melanogaster. J. Morphol.* 129: 171–200.

Agui, N., W. E. Bollenbacher, N. A. Granger, and L. I. Gilbert. 1980. Corpus allatum is release site for insect prothoracicotropic hormone. *Nature* 285: 669–670.

Agui, N., N. A. Granger, L. I. Gilbert, and W. E. Bollenbacher. 1979. Cellular localization of the insect prothoracicotropic hormone: In vitro assay of a single neurosecretory cell. *Proc. Nat. Acad. Sci. USA* 76: 5694–5698.

Albers, B., D. Bray, J. Lewis, M. Raff, K. Roberts, and J. D. Watson. 1989. *Molecular Biology of the Cell.* Garland, New York.

Allegret, P. 1964. Interrelationship of larval development, metamorphosis and age in a pyralid lepidopteran, *Galleria mellonella* (L.), under the influence of dietetic factors. *Exp. Gerontol.* 1: 49–66.

Anderson, D. T. 1972. The development of holometabolous insects. In *Developmental Systems: Insects*, ed. S. J. Counce and C. H. Waddington, vol. 1, pp. 165–142. Academic Press, New York.

Angersbach, D., and H. Kayser. 1971. Wavelength dependence of light-controlled pupal pigmentation. *Naturwiss.* 58: 571–572.

Ashburner, M., and H. D. Berendes. 1979. Puffing of polytene chromosomes. In *The Genetics and Biology of Drosophila* ed. M. Ashburner and T.R.F Wright, vol. 2b, pp. 315–395. Academic Press, London.

Ashburner, M., C. Chihara, P. Meltzer, and G. Richards. 1974. Temporal control of puffing activity in polytene chromosomes. *Cold Spring Harbor Symp. Quant. Biol.* 38: 655–662.

Atzler, M. 1930. Untersuchungen über den morphologischen und physiologischen Farbwechsel von *Dixippus* (*Carausius*) *morosus. Z. Vergl. Physiol.* 13: 505–533.

Awiti, L.R.S., and T. Hidaka. 1982. Neuroendocrine mechanism involved in pupal colour dimorphism in swallowtail *Papilio xuthus* L. *Insect. Sci. Applic.* 3: 181–191.

Baehr, J.-C., P. Porcheron, M. Papillon, and F. Dray. 1979. Haemolymph levels of juvenile hormone, ecdysteroids and protein during the last two larval instars of *Locusta migratoria. J. Insect Physiol.* 25: 415–421.

Baronio, P., and F. Sehnal. 1980. Dependence of the parasitoid *Gonia cinerascens* on the hormones of its lepidopterous host. *J. Insect Physiol.* 26: 619–626.

Barritt, G. J. 1992. *Communication within Animal Cells.* Oxford University Press, Oxford.

Barth, R. H. 1962. The endocrine control of mating behavior in the cockroach *Byrsotria fumigata* (Guérin). *Gen. Comp. Endocrinol.* 2: 53–69.

Barth, R. H. 1965. Insect mating behavior: Endocrine control of a chemical communication system. *Science* 149: 882–883.

Barth, R. H., and L. J. Lester. 1973. Neurohormonal control of sexual behavior in insects. *Ann. Rev. Entomol.* 18: 445–472.

Barth, R. H., L. J. Lester, P. Sroka, T. Kessler, and R. Hearn. 1975. Juvenile hormone promotes dominance behavior and ovarian development in social wasps (*Polistes annularis*). *Experientia* 31: 691–692.

Bean, D. W., and S. D. Beck. 1980. The role of juvenile hormone in the larval diapause of the European corn borer, *Ostrinia nubilalis*. *J. Insect Physiol.* 26: 579–584.

Beaulaton, J. 1990. Anatomy, histology, ultrastructure, and functions of the prothoracic (or ecdysial) glands in insects. In *Morphogenetic Hormones of Arthropods*, ed. A. P. Gupta, vol. 1, pp. 344–435. Rutgers University Press, New Brunswick, N.J.

Beck, S. D. 1971a. Growth and retrogression in larvae of *Trogoderma glabrum* (Coleoptera: Dermestidae). 1. Characteristics under feeding and starvation conditions. *Ann. Entomol. Soc. Amer.* 64: 149–155.

Beck, S. D. 1971b. Growth and retrogression in larvae of *Trogoderma glabrum* (Coleoptera: Dermestidae). 2. Factors influencing pupation. *Ann. Entomol. Soc. Amer.* 64: 946–949.

Beck, S. D. 1972. Growth and retrogression in larvae of *Trogoderma glabrum* (Coleoptera: Dermestidae). 3. Ecdysis and form determination. *Ann. Entomol. Soc. Amer.* 65: 1319–1324.

Beck, S. D. 1980. *Insect Photoperiodism*. 2d ed. Academic Press, New York.

Beck, S. D. 1983. Insect thermoperiodism. *Ann. Rev. Entomol.* 28: 91–108.

Beckage, N. E. 1985. Endocrine interactions between endoparasitic insects and their hosts. *Ann. Rev. Entomol.* 30: 371–413.

Beckage, N. E., and L. M. Riddiford. 1982. Effects of parasitism by *Apanteles congregatus* on the endocrine physiology of the tobacco hornworm, *Manduca sexta*. *Gen. Comp. Endocrinol.* 47: 308–322.

Beckel, W. E., and W. G. Friend. 1964. The relation of abdominal distension and nutrition to molting in *Rhodnius prolixus* (Stahl) (Hemiptera). *Can. J. Zool.* 42: 71–78.

Beckemeyer, E. F., and A. O. Lea. 1980. Induction of follicle separation in the mosquito by physiological amounts of ecdysterone. *Science* 209: 819.

Beenakkers, A.M.T. 1969. The influence of corpus allatum and corpus cardiacum on lipid metabolism in *Locusta migratoria*. *Gen. Comp. Endocrinol.* 13: 492.

Bell, R. A., F. G. Rasul, and F. G. Joachim. 1975. Photoperiodic induction of the pupal diapause in the tobacco hornworm, *Manduca sexta*. *J. Insect Physiol.* 21: 1471–1480.

Berridge, M. J. 1982. Regulation of cell secretion: The integrated action of cyclic AMP and calcium. In *Cyclic Nucleotides, part 2, Physiology and Pharmacology*, ed. J. W. Kebabian and J. A. Nathanson, pp. 227–270. Springer-Verlag, New York.

Berridge, M. J., and J. L. Oschman. 1969. A structural basis for fluid secretion by Malpighian tubules. *Tissue & Cell* 1: 247–272.

Bhaskaran, G. 1972. Inhibition of imaginal differentiation in *Sarcophaga bullata* by juvenile hormone. *J. Exp. Zool.* 182: 127–142.

Bitsch, C. 1985. Effects of ovarian maturation and insemination on the length of the intermoult in *Thermobia domestica*. *Physiol. Entomol.* 10: 15–21.

Bitsch, C., J. C. Baehr, and J. Bitsch. 1985. J.H. in the *Thermobia domestica* females: Identification and quantification during biological cycles and after precocene application. *Experientia* 41: 409–410.

Bitsch, C., and J. Bitsch. 1988. 20-Hydroxyecdysone and ovarian maturation in the firebrat *Thermobia domestica* (Thysanura: Lepismatidae). *Arch. Insect Biochem. Physiol.* 7: 281–294.

Blackith, R. E. 1972. Morphometrics in acridology: A brief survey. *Acrida* 1: 7–15.

Blackman, R. L. 1974. *Invertebrate Types: Aphids*. Ginn and Co., London.

Bogus, M. I., and K. Scheller. 1991. Activation of juvenile hormone synthesis *in vitro* by larval brains of *Galleria mellonella. Zool. Jb. Physiol.* 95: 197–208.

Bollenbacher, W. E. 1988. The interendocrine regulation of larval-pupal development in the tobacco hornworm, *Manduca sexta*: A model. *J. Insect Physiol.* 34: 941–947.

Bollenbacher, W. E., and L. I. Gilbert. 1981. Neuroendocrine control of postembryonic development in insects. The prothoracicotropic hormone. In *Neurosecretion: Molecules, Cells, Systems*, ed. D. S. Farner and K. Lederis, pp. 361–370. Plenum Press, New York.

Bollenbacher, W. E., and N. A. Granger. 1985. Endocrinology of the prothoracicotropic hormone. In *Comprehensive Insect Physiology, Biochemistry and Pharmacology*, ed. G. A. Kerkut and L. I. Gilbert, vol. 7, pp. 109–151. Pergamon, New York.

Bollenbacher, W. E., N. Agui, N. A. Granger, and L. I. Gilbert. 1979. *In vitro* activation of insect prothoracic glands by the prothoracicotropic hormone. *Proc. Nat. Acad. Sci. USA* 76: 5148–5152.

Bollenbacher, W. E., N. Agui, N. A. Granger, and L. I. Gilbert. 1980. Insect prothoracic glands in vitro: A system for studying the prothoracicotropic hormone. In *Invertebrate Systems In Vitro*, ed. E. Kurstak, K. Maramorosch, and A. Dübendorfer, pp. 253–271. Elsevier/North-Holland, Amsterdam.

Bollenbacher, W. E., E. J. Katahira, M. O'Brien, L. I. Gilbert, M. K. Thomas, N. Agui, and A. H. Baumhover. 1984. Insect prothoracicotropic hormone: Evidence for two molecular forms. *Science* 224: 1243–1245.

Bollenbacher, W. E., M. A. O'Brien, E. J. Katahira, and L. I. Gilbert. 1983. A kinetic analysis of the action of the insect prothoracicotropic hormone. *Molec. Cell. Endocrinol.* 32: 27–46.

Bollenbacher, W. E., S. L. Smith, W. Goodman, and L. I. Gilbert. 1981. Ecdysteroid titer during larval-pupal-adult development of the tobacco hornworm, *Manduca sexta. Gen. Comp. Endocrinol.* 44: 302–306.

Bordereau, C. 1975. Déterminisme des castes chez les termites supérieurs: Mise en évidence d'un contrôle royal dans la formation de la caste sexueé chez *Macrotermes bellicosus* Smeatman (Isoptera, Termitidae). *Insectes Soc.* 22: 363–374.

Borst, D. W., and J. D. O'Connor. 1972. Arthropod molting hormone: Radioimmune assay. *Science* 178: 418–419.

Borst, D. W., and J. D. O'Connor. 1974. Trace analysis of ecdysones by gas-liquid chromatography radioimmunoassay and bioassay. *Steroids* 24: 637–656.

Bouthier, A. 1976. Action des facteurs hormonaux sur le métabolisme des ommochromes chez le criquet *Locusta migratoria* L. (Orthoptères, Acrididae). *Ann. Endocrin.* 37: 539–540.

Bowen, M. F., W. E. Bollenbacher, and L. I. Gilbert. 1984a. In vitro studies on the role of the brain and prothoracic glands in the pupal duapause of *Manduca sexta. J. Exp. Biol.* 108: 9–24.

Bowen, M. F., W. E. Bollenbacher, and L. I. Gilbert. 1984b. *In vitro* reprogramming of the photoperiodic clock in an insect brain-retrocerebral complex. *Proc. Nat. Acad. Sci. USA* 81: 5881–5884.

Bowers, W. S. 1976. Discovery of insect antiallatotropins. In *The Juvenile Hormones*, ed. L. I. Gilbert, pp. 394–408. Plenum Press, New York.

Bowers, W. S. 1985. Antihormones. In *Comprehensive Insect Physiology, Biochemistry and Pharmacology*, ed. G. A. Kerkut and L. I. Gilbert, vol. 8, pp. 551–564. Pergamon, New York.

Bradley, T. J. 1985. The excretory system: Structure and Function. In *Comprehensive Insect Physiology, Biochemistry and Pharmacology*, ed. G. A. Kerkut and L. I. Gilbert, vol. 4, pp. 421–465. Pergamon, New York.

Brian, M. V. 1974. Caste differentiation in *Myrmica rubra*: The role of hormones. *J. Insect Physiol.* 20: 1351–1365.

Brown, B. E., and A. N. Starratt. 1975. Isolation of proctolin, a myotropic peptide from *Periplaneta americana*. *J. Insect Physiol.* 21: 1879–1881.

Brunet, P.C.J. 1952. The formation of the ootheca by *Periplaneta americana*. 2. The structure and functions of the left colleterial gland. *Quart. J. Microsc. Sci.* 93: 47–69.

Brunet, P.C.J., and P. W. Kent. 1955. Observations on the mechanism of a tanning reaction in *Periplaneta americana*. *Proc. Roy. Soc. London* B 144: 259–278.

Bückmann, D. 1959. Die Auslösung der Umfärbung durch das Häutungshormon bei *Cerura vinula* (Lepidoptera, Notodontidae). *J. Insect Physiol.* 3: 159–189.

Bückmann, D. 1960. Die Determination der Puppenfärbung bei *Vanessa urticae* L. *Naturwiss.* 47: 610–611.

Bückmann, D. 1968. Die Biochemie der morphologischen Farbanpassung bei Tagschmetterlingspuppen. *Zool. Anz. Suppl.* 32: 636–640.

Bückmann, D. 1974. Die hormonale Steuerung der Pigmentierung und des morphologischen Farbwechsels bei den Insekten. *Fortschr. Zool.* 22: 1–22.

Bückmann, D., and A. Maisch. 1987. Extraction and partial purification of the pupal melanization reducing factor (PMRF) from *Inachis io* (Lepidoptera). *Insect Biochem.* 17: 841–844.

Butenand, A., and P. Karlson. 1954. Über die isolierung eines Metamorphose-Hormone der Insekten in kristallierten Form. *Z. Naturforsch.* 9b: 389–391.

Carpenter, F. M. 1976. Geological history and the evolution of insects. *Proc. IV Int. Congr. Entomol.*, pp. 63–70.

Cassier, P. 1990. Morphology, histology, and ultrastructure of JH-producing glands in insects. In *Morphogenetic Hormones of Arthropods*, ed. A. P. Gupta, vol. 1, pp. 83–194. Rutgers University Press, New Brunswick, N.J.

Cassier, P., and M.-A. Fain-Maurel. 1970. Contrôle plurifactoriel de l'évolution postimaginale des glandes ventrales chez *Locusta migratoria* L. Données expérimentales et infrastructurales. *J. Insect Physiol.* 16: 301–318.

Cassier, P., and M. Papillon. 1968. Effects des implantations des corps allates sur la reproduction de femelles goupées de *Schistocerca gregaria* (Forsk.) et sur le polymorphisme de leur descendance. *C.R. Acad. Sci. Paris* 266D: 1048–1051.

Chalaye, D. 1969. La trehalosemie et son controle neuro-endocrinie chez le criquet migrateur, *Locusta migratoria migratoroides*. 2. Role des corpora cardiaca et des organes perisympathiques. *C.R. Acad. Sci. Paris* 286D: 3111–3114.

Chapman, R. F. 1985. Structure of the digestive system. In *Comprehensive Insect Physiology, Biochemistry and Pharmacology*, ed. G. A. Kerkut and L. I. Gilbert, vol. 4, pp. 165–211. Pergamon, New York.

Cheeseman, P., and G. J. Goldsworthy. 1979. The release of adipokinetic hormone during flight and starvation in *Locusta*. *Gen. Comp. Endocrinol.* 37: 35–43.

Cherbas, L., C. D. Yonge, P. Cherbas, and C. M. Williams. 1980. The morphological response of K_c-H cells to ecdysteroids: Hormonal specificity. *Wilh. Roux' Arch.* 189: 1–15.

Chiang, G. R., and K. G. Davey, 1988. A novel receptor capable of monitoring applied pressure in the abdomen of an insect. *Science* 241: 1665–1667.

Chippendale, G. M., and C.-M. Yin. 1976. Endocrine interactions controlling the larval diapause of the southwestern corn borer, *Diatraea grandiosella*. *J. Insect Physiol.* 22: 989–995.

Chippendale, G. M., and C.-M. Yin. 1979. Larval diapause of the European corn borer, *Ostrinia nubilalis*: Further experiments examining its hormonal control. *J. Insect Physiol.* 25: 53–58.

Christensen, T. A., H. Itagaki, P.E.A. Teal, R. D. Jasensky, J. H. Tumlinson, and J. G. Hildebrand. 1991. Innervation and neural regulation of the sex pheromone gland in female *Heliothis* moths. *Proc. Nat. Acad. Sci. USA* 88: 4971–4975.

Clever, U., and P. Karlson. 1960. Induktion von Puff-Veränderungen in den Speicheldrüsen Chromosomen von *Chironomus tentans* durch Ecdyson. *Exp. Cell. Res.* 20: 623–626.

Cobben, R. H. 1968. Evolutionary trends in Heteroptera. Part 1. Eggs, architecture of the shell, gross embryology and eclosion. *Meded. Landb. Hogesch. Wageningen* 151: 1–475.

Cole, B. J. 1980. Growth ratios in holometabolous and hemimetabolous insects. *Ann. Entomol. Soc. Amer.* 73: 489–491.

Copenhaver, P. F., and J. W. Truman. 1986a. Identification of the cerebral neurosecretory cells that contain eclosion hormone in the moth *Manduca sexta*. *J. Neurosci.* 6: 1738–1747.

Copenhaver, P. F., and J. W. Truman. 1986b. Metamorphosis of the cerebral neuroendocrine system in the moth, *Manduca sexta*. *J. Comp. Neurol.* 249: 186–204.

Craig, G. B. 1967. Mosquitoes: Female monogamy induced by male accessory gland substance. *Science* 156: 1499–1501.

Curtis, A. T., M. Hori, J. M. Green, W. J. Wolfgang, K. Hiruma, and L. M. Riddiford. 1984. Ecdysteroid regulation of the onset of cuticular melanization in allatectomized and *black* mutant *Manduca sexta* larvae. *J. Insect Physiol.* 30: 597–606.

Cymborowski, B. 1989. Bioassays for ecdysteroids. In *Ecdysone: From Chemistry to Mode of Action*, ed. J. Koolman, pp. 144–149. Thieme-Verlag, Stuttgart.

Cymborowski, B. 1992. *Insect Endocrinology*. Elesvier, Amsterdam; Polish Scientific Publishers, Warsaw.

Dahm, K. H., G. Bhaskaran, M. G. Peter, P. D., Shirk, K. R. Seshan, and H. Röller. 1976. On the identity of the juvenile hormone in insects. In *The Juvenile Hormones*, ed. L. I. Gilbert, pp. 19–47. Plenum Press, New York.

Dallman, S. H., and W. S. Herman. 1978. Hormonal regulation of haemolymph lipid concentration in the monarch butterfly, *Danaus plexippus*. *Gen. Comp. Endocrinol.* 36: 142–150.

Danilevskii, A. S. 1965. *Photoperiodism and Seasonal Development of Insects*. Oliver & Boyd, Edinburgh.

Darjo, A. 1976. Activité des corpora allata et controle photoperiodique de la maturation ovarienne chez *Locusta migratoria. J. Insect Physiol.* 22: 347–355.

Darnell, J., H. Lodish, and D. Baltimore. 1990. *Molecular Cell Biology*. 2nd edition. Freeman, New York.

Davey, K. G. 1967. Some consequences of copulation in *Rhodnius prolixus. J. Insect Physiol.* 13: 1629–1636.

Davey, K. G. 1985a. The male reproductive tract. In *Comprehensive Insect Physiology, Biochemistry and Pharmacology*, ed. G. A. Kerkut and L. I. Gilbert, vol. 1, pp. 1–14. Pergamon, New York.

Davey, K. G. 1985b. The female reproductive tract. In *Comprehensive Insect Physiology, Biochemistry and Pharmacology*, ed. G. A. Kerkut and L. I. Gilbert, vol. 1, pp. 15–36. Pergamon, New York.

Davey, K. G., and E. Huebner. 1974. The response of the follicle cells of *Rhodnius prolixus* to juvenile hormone and antigonadotropin *in vitro. Can. J. Zool.* 52: 1407–1412.

Davis, K. T., and A. Shearn. 1977. *In vitro* growth of imaginal disks from *Drosophila melanogaster. Science* 196: 438–439.

Dean, R. L., W. E. Bollenbacher, M. Locke, S. L. Smith, and L. I. Gilbert. 1980. Haemolymph ecdysteroid levels and cellular events in the intermoult/moult sequence of *Calpodes ethlius. J. Insect Physiol.* 26: 267–280.

Dean, R. L., M. Locke, and J. V. Collins. 1985. Structure of the fat body. In: *Comprehensive Insect Physiology, Biochemistry and Pharmacology*, ed. G. A. Kerkut and L. I. Gilbert, vol. 3, pp. 155–210. Pergamon, New York.

De Kort, C.A.D. 1981. Hormonal and metabolic regulation of adult diapause in the Colorado beetle, *Leptinotarsa decemlineata* (Coleoptera: Chrysomelidae). *Entomol. Gen.* 7: 261–271.

De Kort, C. A., and N. A. Granger. 1981. Regulation of the juvenile hormone titer. *Ann Rev. Entomol.* 26: 1–28.

Delbecque, J., and K. Slama. 1980. Ecdysteroid titers during autonomous metamorphosis in a dermestid beetle. *Z. Naturforsch.* 35c: 1066–1080.

Denlinger, D. L. 1971. Embryonic determination of pupal diapause in the flesh fly *Sarcophaga crassipalpis. J. Insect Physiol.* 17: 1815–1822.

Denlinger, D. L. 1985. Hormonal control of diapause. In *Comprehensive Insect Physiology, Biochemistry and Pharmacology*, ed. G. A. Kerkut and L. I. Gilbert, vol. 8, pp. 353–412. Pergamon, New York.

De Wilde, J. 1975. An endocrine view of metamorphosis, polymorphism, and diapause in insects. *Amer. Zool.* (suppl.) 15: 13–27.

De Wilde, J. 1976. Juvenile hormone and caste differentiation in the honey bee (*Apis mellifera* L.). In *Phase and Caste Determination in Insects: Endocrine Aspects*, ed. M. Lüscher, pp. 5–20. Pergamon, New York.

De Wilde, J., and J. A. de Boer. 1961. Physiology of diapause in the adult Colorado beetle. II. Diapause as a case of pseudoallatectomy. *J. Insect Physiol.* 6: 152–161.

De Wilde, J., and J. A. de Boer. 1969. Humoral and nervous pathways in photoperiodic induction of diapause in *Leptinotarsa decemlineata. J. Insect Physiol.* 15: 661–675.

Dietz, A., H. R. Hermann, and M. S. Blum. 1979. The role of exogenous JH-I, JH-III and anti-JH (Precocene II) on queen induction of 4.5 day-old worker honey bee larvae. *J. Insect Physiol.* 25: 503–512.

Dingle, H. 1972. Migration strategies of insects. *Science* 175: 1327–1335.

Dingle, H. 1985. Migration. In *Comprehensive Insect Physiology, Biochemistry and Pharmacology*, ed. G. A. Kerkut and L. I. Gilbert, vol. 9, pp. 375–414. Pergamon, New York.

Dingle, H., R. L. Caldwell, and J. B. Haskell. 1969. Temperature and circadian control of cuticle growth in the bug, *Oncopeltus fasciatus*. *J. Insect Physiol.* 15: 373–378.

Doane, W. W. 1973. Role of hormones in insect development. In *Developmental Systems: Insects*, ed. S. J. Counce and C. H. Waddington, vol. 2, pp. 291–497. Academic Press, London.

Dogra, G. S., G. M. Ulrich, and H. Rembold. 1977. A comparative study of the endocrine system of the honeybee larvae under normal and experimental conditions. *Z. Naturforsch.* 32c: 637–642.

Dores, R. M., S. H. Dallmann, and W. S. Herman. 1979. The regulation of post-eclosion and post-feeding diuresis in the monarch butterfly, *Danaus plexippus*. *J. Insect Physiol.* 25: 859–901.

Ejezie, G. C., and K. G. Davey. 1974. Changes in neurosecretory cells, corpus cardiacum and corpus allatum during pregnancy in *Glossina austeni* Newst. *Bull. Entomol. Res.* 64: 247–256.

Ejezie, G. C., and K. G. Davey. 1976. Some effects of allatectomy in the female tsetse, *Glossina austeni*. *J. Insect Physiol.* 22: 1743–1749.

Ejezie, G. C., and K. G. Davey. 1977. Some effects of mating in female tsetse *Glossina austeni*. *J. Exp. Zool.* 200: 303–310.

Elzinga, R. J. 1978. *Fundamentals of Entomology*. Prentice Hall, Englewood Cliffs, N.J.

Endo, K., and S. Funatsu. 1985. Hormonal control of seasonal morph determination in the swallowtail butterfly, *Papilio xuthus* (Lepidoptera: Papilionidae). *J. Insect Physiol.* 31: 669–674.

Endo, K., and Y. Kamata. 1985. Hormonal control of seasonal-morph determination in the small copper butterfly, *Lycaena phlaeas daimio* Seitz. *J. Insect Physiol.* 31: 701–706.

Endo, K., Y. Fujimoto, T. Masaki, and K. Kumagai. 1990. Stage-dependent changes in the activity of the prothoracicotropic hormone (PTTH) in the brain of the Asian comma butterfly, *Polygonia c-aureum* L. *Zool. Sci.* 7: 697–704.

Endo, K., T. Masaki, and K. Kumagai. 1988. Neuroendocrine regulation of the development of seasonal morphs in the Asian comma butterfly, *Polygonia c-aureum* L.: Difference in activity of summer-morph-producing hormone from brain extracts of the long-day and short-day pupae. *Zool. Sci.* 5: 145–152.

Engelmann, F. 1959. The control of reproduction in *Diploptera punctata* (Blattaria). *Biol. Bull.* 116: 406–419.

Engelmann, F. 1970. *The Physiology of Insect Reproduction*. Pergamon, New York.

Engelmann, F. 1983. Vitellogenesis controlled by juvenile hormone. In *Endocrinology of Insects*, ed. R.G.H. Downer H. Laufer, pp. 259–270. Alan Liss, New York.

Engelmann, F., and R. H. Barth. 1968. Endocrine control of female receptivity in *Leucophaea maderae* (Blattaria). *Ann. Entomol. Soc. Amer.* 61: 503–505.

Fain, M. J., and L. M. Riddiford. 1975. Juvenile hormone titers in the hemolymph during later larval development of the tobacco hornworm, *Manduca sexta* (L.). *Biol. Bull.* 149: 506–521.

Fallon, A. M., H. H. Hagedorn, G. R. Wyatt, and H. Laufer. 1974. Activation of vitellogenin synthesis in the mosquito *Aedes aegypti* by ecdysone. *J. Insect Physiol.* 20: 1815–1823.

Faure, J. C. 1932. The phases of locusts in South Africa. *Bull. Entomol. Res.* 23: 293–424.

Feinsod, F. M., and A. Spielman. 1980. Independently regulated juvenile hormone activity and vitellogenesis in mosquitoes. *J. Insect Physiol.* 26: 829–832.

Feir, D., and G. Winkler. 1969. Ecdysone titers in the last larva and adult stages of the milkweed bug. *J. Insect Physiol.* 15: 899–904.

Feldlaufer, M. F. 1989. Diversity of molting hormones in insects. In *Ecdysone: From Chemistry to Mode of Action*, ed. J. Koolman, pp. 308–312. Thieme-Verlag, Stuttgart.

Ferenz, H. J., and I. Diehl. 1983. Stimulation of juvenile hormone biosynthesis *in vitro* by locust allatotropin. *Z. Naturforsch.* 38c: 856–858.

Filshie, B. K., M. F. Day, and E. H. Mercer. 1975. Color and color change in the grasshopper *Kosciuscola tristis*. *J. Insect Physiol.* 21: 1763–1770.

Flanagan, T. R., and H. H. Hagedorn. 1977. Vitellogenin synthesis in the mosquito: The role of juvenile hormone in the development of responsiveness to ecdysone. *Physiol. Entomol.* 2: 173–178.

Fogal, W., and G. Fraenkel. 1969. The role of bursicon in melanization and endocuticle formation in the adult fleshfly, *Sarcophaga bullata*. *J. Insect Physiol.* 15: 1235–1247.

Foster, W. 1967. Hormone-mediated nutritional control of sexual behavior in male dung flies. *Science* 158: 1596–1597.

Fournier, B., and J. Girardie. 1988. A new function for the locust neuroparsins: Stimulation of water reabsorption. *J. Insect Physiol.* 34: 309–313.

Fox, A. M., and S. E. Reynolds. 1990. Quantification of *Manduca* adipokinetic hormone in nervous and endocrine tissue by a specific radioimmunoassay. *J. Insect Physiol.* 36: 683–689.

Fraenkel, G. 1935. Observations and experiments on the blowfly (*Calliphora erythrocephala*) on the first day after emergence. *Proc. Zool. Soc. London* (1935): 893–904.

Fraenkel, G., and C. Hsiao. 1962. Hormonal and nervous control of tanning in the fly. *Science* 138: 27–29.

Fraenkel, G., and C. Hsiao. 1965. Bursicon, a hormone which mediates tanning of the cuticle in the adult fly and other insects. *J. Insect Physiol.* 11: 513–556.

Fukuda, S. 1944. The hormonal mechanism of larval molting and metamorphosis in the silkworm. *J. Fac. Sci. Tokyo Imp. Univ.* 4: 477–532.

Fukuda, S. 1951a. Factors determining the production of non-diapause eggs in the silkworm. *Proc. Jap. Acad.* 27: 582–586.

Fukuda, S. 1951b. The production of diapause eggs by transplanting the subesophageal ganglion in the silkworm. *Proc. Imp. Acad. Japan* 27: 672–677.

Fukuda, S. 1952. Function of the pupal brain and subesophageal ganglion in the production of non-diapause and diapause eggs in the silkworm. *Annot. Zool. Jap.* 25: 149–155.

Gäde, G. 1981. Activation of fat body glycogen phosphorylase in *Locusta migratoria* by corpus cardiacum extracts and synthetic adipokinetic hormone. *J. Insect Physiol.* 27: 155–161.

Gäde, G. 1990. The adipokinetic hormone/red pigment concentrating hormone peptide family: Structures, interrelationships and functions. *J. Insect Physiol.* 36: 1–12.

Gäde, G., and D. A. Holwerda. 1976. Involvement of adenosine 3':5'-cyclic monophosphate in lipid mobilization in *Locusta migratoria. Insect Biochem.* 6: 535–540.

Gee, J. D. 1976. Active transport of sodium by the Malpighian tubules of the tsetse fly, *Glossina morsitans. J. Exp. Biol.* 64: 357–368.

Gehring, W., and R. Nothiger. 1973. The imaginal disks of *Drosophila*. In *Developmental Systems: Insects*, ed. S. J. Counce and C. H. Waddington, vol. 2, pp. 211–290. Academic Press, London.

Gersch, M., H. Birkenbeil, and J. Ude. 1975. Ultrastructure of the prothoracic gland cells of the last instar of *Galleria mellonella* in relation to the state of development. *Cell Tiss. Res.* 160: 389–397.

Ghiradella, H. 1974. Development of ultraviolet-reflecting scales: How to make an interference filter. *J. Morphol.* 142: 395–410.

Ghiradella, H. 1985. Structure and development of iridescent lepidopteran scales: The Papilionidae as a showcase Family. *Ann. Entomol. Soc. Amer.* 78: 252–264.

Giebultowicz, J. M., J. Zdarek, and U. Chroscikowska. 1980. Cocoon spinning behaviour in *Ephestia kuehniella*: Correlation with endocrine events. *J. Insect Physiol.* 26: 459–464.

Gilbert, J. J. 1980. Developmental polymorphism in the rotifer *Asplanchna sieboldi. Amer. Sci.* 68: 636–646.

Gilbert, L. I. 1989. The endocrine control of molting: The tobacco hornworm, *Manduca sexta*, as a model system. In *Ecdysone: From Chemistry to Mode of Action*, ed. J. Koolman, pp. 448–471. Thieme Verlag, Stuttgart.

Gilbert, L. I., and H. A. Schneiderman. 1959. Prothoracic gland stimulation by juvenile hormone of insects. *Nature* 184: 171–173.

Gilbert, L. I., W. L. Combest, W. A. Smith, V. H. Meller, and D. B. Rountree. 1988. Neuropeptides, second messengers and insect molting. *BioEssays* 8: 153–157.

Gillott, C. 1988. Arthropoda—Insecta. In: *Reproductive Biology of Invertebrates. Volume III. Accessory Glands*, ed. K. G. Adiyodi and R. Adiyodi, pp. 319–471. John Wiley and Sons, New York.

Girardie, A., and M. Cazal. 1965. Rôle de la pars intercerebralis et des corpora cardiaca sur la mélanisation chez *Locusta migratoria* (L.). *C.R. Acad. Sci. Paris* 261D: 4525–4527.

Girardie, A., and A. Vogel. 1966. Étude du contrôle neuro-humoral de l'activité sexuelle mâle de Locusta migratoria (L.). *C.R. Acad. Sci. Paris* 263D: 543–546.

Goldbard, G. A., J. R. Sauer, and R. R. Mills. 1970. Hormonal control of excretion in the American cockroach. II. Preliminary purification of a diuretic and an antidiuretic hormone. *Comp. Gen. Pharmacol.* 1: 82–86.

Goldsworthy, G. J. 1970. The action of hyperglycaemic factors from the corpus cardiacum of *Locusta migratoria* on glycogen phosphorylase. *Gen. Comp. Endocrinol.* 14: 78–85.

Goldsworthy, G. J., R. A. Johnson, and W. Mordue. 1972a. In vivo studies on the release of hormones from the corpora cardiaca of locusts. *J. Comp. Physiol.* 79: 85–96.

Goldsworthy, G. J., W. Mordue, and J. Guthkelch. 1972b. Studies on insect adipokinetic hormones. *Gen. Comp. Endocrinol.* 18: 545–551.

Goldsworthy, G. J., K. Mallison, C. H. Wheeler, and G. Gäde. 1986. Relative adipokinetic activities of members of the adipokinetic hormone/red pigment concentrating hormone family. *J. Insect Physiol.* 32: 433–438.

Gole, J.W.D., and R.G.H. Downer. 1979. Elevation of adenosine 3'5'-monophosphate by octopamine in fat body of the American cockroach, *Periplaneta americana. Comp. Biochem. Physiol.* 64C: 223–226.

Goodman, W. G., and E. S. Chang. 1985. Juvenile hormone cellular and hemolymph binding proteins. In *Comprehensive Insect Physiology, Biochemistry and Pharmacology*, ed. G. A. Kerkut and L. I. Gilbert, vol. 7, pp. 491–510. Pergamon, New York.

Goodwin, T. W., D.H.S. Horn, P. Karlson, J. Koolman, K. Nakanishi, W. E. Robbins, J. B. Siddall, and T. Takemoto. 1978. Ecdysteroids: A new generic term. *Nature* 272: 122.

Gould, S. J. 1966. Allometry and size in ontogeny and phylogeny. *Biol. Rev.* 41: 587–640.

Granger, N. A., and W. E. Bollenbacher. 1981. Hormonal control of insect metamorphosis. In *Metamorphosis: A Problem in Developmental Biology*, 2d ed., ed. L. I. Gilbert and E. Frieden, pp. 105–137. Plenum Press, New York.

Granger, N. A., and W. G. Goodman. 1988. Radioimmunoassay: Juvenile hormones. In *Immunological Techniques in Insect Biology*, ed. L. I. Gilbert and T. A. Miller, pp. 215–251. Springer-Verlag, New York.

Gregg, R. E. 1942. The origin of castes in ants with special reference to *Pheidole morrisi* Forel. *Ecol.* 23: 295–308.

Grimstone, A. V., A. M. Mullinger, and J. A. Ramsay. 1968. Further studies on the rectal complex of the mealworm *Tenebrio molitor*, L. (Coleoptera, Tenebrionidae). *Phil. Trans. Roy. Soc.* B 253: 343–382.

Gruetzmacher, M. C., L. I. Gilbert, N. A. Granger, W. Goodman, and W. E. Bollenbacher. 1984a. The effect of juvenile hormone on prothoracic gland function during the larval-pupal development of *Manduca sexta*: An *in situ* and *in vitro* analysis. *J. Insect Physiol.* 30: 331–340.

Gruetzmacher, M. C., L. I. Gilbert, and W. E. Bollenbacher. 1984b. Indirect stimulation of the prothoracic glands of *Manduca sexta* by juvenile hormone: Evidence for a fat body stimulatory factor. *J. Insect Physiol.* 30: 771–778.

Gwadz, R. W. 1972. Neuro-hormonal regulation of sexual receptivity in female *Aedes aegypti. J. Insect Physiol.* 18: 259–266.

Gwadz, R. W., and A. Spielman. 1973. Corpus allatum control of ovarian development in *Aedes aegypti. J. Insect Physiol.* 19: 1441–1448.

Hagedorn, H. H. 1985. The role of ecdysteroids in reproduction. In *Comprehensive Insect Physiology, Biochemistry and Pharmacology*, ed. G. A. Kerkut and L. I. Gilbert, vol. 8, pp. 205–262. Pergamon, New York.

Hagedorn, H. H. 1989. Physiological roles of hemolymph ecdysteroids in the adult insect. In *Ecdysone: From Chemistry to Mode of Action*, ed. J. Koolman, pp. 279–289. Thieme-Verlag, Stuttgart.

Hagedorn, H. H., and J. G. Kunkel. 1979. Vitellogenin and vitellin in insects. *Ann. Rev. Entomol.* 24: 475–505. Thieme-Verlag, Stuttgart.

Hagedorn, H. H., J. D. O'Connor, M. S. Fuchs, B. Sage, D. A. Schlaeger, and M. K. Bohm. 1975. The ovary as a source of α-ecdysone in an adult mosquito. *Proc. Nat. Acad. Sci. USA* 72: 3255–3259.

Hagedorn, H. H., J. P. Shapiro, and K. Hanaoka. 1979. Ovarian ecdysone secretion is controlled by a brain hormone in an adult mosquito. *Nature* 282: 92–94.

Hammock, B. D. 1985. Regulation of juvenile hormone titer: Degradation. In *Comprehensive Insect Physiology, Biochemistry and Pharmacology*, ed. G. A. Kerkut and L. I. Gilbert, vol. 7, pp. 431–472. Pergamon, New York.

Hammock, B. D., B. C. Bonning, R. D. Possee, T. N. Hanzlik, and S. Maeda. 1990. Expression and effects of the juvenile hormone esterase in a baculovirus vector. *Nature* 344: 458–460.

Hammock, B. D., S. M. Mumby, and P. W. Lee. 1977. Mechanisms of resistance to the juvenoid methoprene in the house fly *Musca domestica. Pest. Biochem. Physiol.* 7: 261–272.

Hanaoka, K., and S. Y. Takahashi. 1977. Adenylate cyclase system and the hyperglycaemic factor in the cockroach, *Periplaneta americana. Insect Biochem.* 7: 95–99.

Hannah-Alava, A. 1965. The premeiotic stages of spermatogenesis. *Adv. Genet.* 13: 157–226.

Hanton, W. K., R. D. Watson, and W. E. Bollenbacher. 1993. Ultrastructure of prothoracic glands during larval-pupal development of the tobacco hornworm, *Manduca sexta*: A reappraisal. *J. Morphol.* 216: 95–112.

Hardie, J. 1980. Juvenile hormone mimics the photoperiodic apterization of the alate gynoparae of the aphid, *Aphis fabae. Nature* 286: 602–604.

Hardie, J. 1981a. Juvenile hormone and photoperiodically controlled polymorphism in *Aphis fabae*: Prenatal effects on presumptive oviparae. *J. Insect Physiol.* 27: 257–265.

Hardie, J. 1981b. Juvenile hormone and photoperiodically controlled polymorphism in *Aphis fabae*: Postnatal effects on presumptive gynoparae. *J. Insect Physiol.* 27: 347–355.

Hardie, J. 1987. The corpus allatum, neurosecretion and photoperiodically controlled polymorphism in an aphid. *J. Insect Physiol.* 33: 201–205.

Hardie, J., and A. D. Lees. 1985. Endocrine control of polymorphism and polyphenism. In *Comprehensive Insect Physiology, Biochemistry and Pharmacology*, ed. G. A. Kerkut and L. I. Gilbert, vol. 8, pp. 441–490. Pergamon, New York.

Hasegawa, K. 1951. Studies in voltinism in the silkworm, *Bombyx mori* L., with special reference to the organs concerning determination of voltinism (a preliminary note). *Proc. Jap. Acad.* 27: 667–671.

Hasegawa, K. 1952. Studies on the voltinism of the silkworm, *Bombyx mori* L., with special reference to the organs concerning determination of voltinism. *J. Fac. Agric. Tottori Univ.* 1: 83–124.

Hasegawa, K. 1957. The diapause hormone of the silkworm, *Bombyx mori. Nature* 179: 1300–1301.

Hennig, W. 1965. Phylogenetic systematics. *Ann. Rev. Entomol.* 10:97–116.

Henrick, C. A., G. B. Staal, and J. B. Siddall. 1976. Structure activity relationships in some juvenile hormone analogs. In *The Juvenile Hormones*, ed. L. I. Gilbert, pp. 48–60. Plenum Press, New York.

Hepburn, H. R. 1985. Structure of the integument. In *Comprehensive Insect Physiology, Biochemistry and Pharmacology*, ed. G. A. Kerkut and L. I. Gilbert, vol. 3, pp. 1–58. Pergamon, New York.

Herault, J.-P., J. Girardie, and J. Proux. 1985. Separation and characteristics of antidiuretic factors from the corpora cardiaca of the migratory locust. *Int. J. Invert. Reprod. Dev.* 8: 325–335.

Herman, W. S. 1973. The endocrine basis of reproductive inactivity in monarch butterflies overwintering in Central California. *J. Insect Physiol.* 19: 1883–1887.

Herman, W. S. 1975. Juvenile hormone stimulation of tubular and accessory glands in male monarch butterflies. *Comp. Biochem. Physiol.* 51A: 507–510.

Herman, W. S., and L. I. Gilbert. 1966. The neuroendocrine system of *Hyalophora cecropia*. *Gen. Comp. Endocrinol.* 7: 275–291.

Hertel, W., and H. Penzlin. 1992. Function and modulation of the antennal heart of *Periplaneta americana* (L.). *Acta Biol. Hung.* 43: 113–125.

Hidaka, T., and T. Ohtaki. 1963. Effet de l'hormone juvenil et du farnesol sur la coloration tegumentaire de la nymphe de *Pieris rapae crucivorae* (Boisd.). *C.R. Soc. Biol. Paris* 157: 928–930.

Highnam, K. C., and L. Hill. 1977. *The Comparative Endocrinology of the Invertebrates*. University Park Press, Baltimore.

Hinton, H. E. 1958. Concealed phases in the metamorphosis of insects. *Sci. Progr.* 182: 260–275.

Hinton, H. E. 1963. The origin and function of the pupal stage. *Proc. Roy. Entomol. Soc.* A 38:77–85.

Hiruma, K. 1980. Possible roles of juvenile hormone in the prepupal stage of *Mamestra brassicae*. *Gen. Comp. Endocrinol.* 41: 392–399.

Hiruma, K., H. Shimada, and S. Yagi. 1978. Activation of the prothoracic gland by juvenile hormone and prothoracicotropic hormone in *Mamestra brassicae*. *J. Insect Physiol.* 24: 215–220.

Hiruma, K., S. Matsumoto, A. Isogai, and A. Suzuki. 1984. Control of ommochrome synthesis by both juvenile hormone and melanization hormone in the cabbage armyworm, *Mamestra brassicae*. *J. Comp. Physiol.* 154: 13–21.

Hodkova, M. 1977. Function of the neuroendocrine complex in diapausing *Pyrrhocoris apterus* females. *J. Insect Physiol.* 23: 23–28.

Hoffmann, J. A., and M. Lageux. 1985. Endocrine aspects of embryonic development in insects. In *Comprehensive Insect Physiology, Biochemistry and Pharmacology*, ed. G. A. Kerkut and L. I. Gilbert, vol. 1, pp. 435–460. Pergamon, New York.

Holman, G. M., B. J. Cook, and R. J. Nachman. 1986. Primary structure and synthesis of two additional neuropeptides from *Leucophaea maderae*: Members of a new family of cephalomyotropins. *Comp. Biochem. Physiol.* 84C: 271–276.

Holman, G. M., B. J. Cook, and R. J. Nachman. 1987. Isolation, primary structure and synthesis of leukokinins V and VI: Myotropic peptides of *Leucophaea maderae*. *Comp. Biochem. Physiol.* 85C: 27–30.

Hopkins, T. L., and K. J. Kramer. 1992. Insect cuticle sclerotization. *Ann. Rev. Entomol.* 37: 273–302.

Horn, D.H.S., and R. Bergamasco. 1985. Chemistry of ecdysteroids. In *Comprehensive Insect Physiology, Biochemistry and Pharmacology*, ed. G. A. Kerkut and L. I. Gilbert, vol. 7, pp. 185–248. Pergamon, New York.

Horodyski, F. M., L. M. Riddiford, and J. W. Truman. 1989. Isolation and expression of the eclosion hormone gene from the tobacco hornworm, *Manduca sexta*. *Proc. Nat. Acad. Sci. USA* 86: 8123–8127.

Howarth, O. W., M. J. Thompson, and H. H. Rees. 1989. Reaction of 3-dehydroecdysone with certain n.m.r. solvents. *Biochem. J.* 259: 299–302.

Huang, Z.-Y., G. E. Robinson, and D. W. Borst. 1993. Physiological correlates of division of labor among similarly aged honey bees. *J. Comp. Physiol.* A. (in press).

Huang, Z.-Y., G. E. Robinson, S. S. Tobe, K. J. Yagi, C. Strambi, A. Strambi, and B. Stay. 1991. Hormonal regulation of behavioral development in the honey bee is based on changes in the rate of juvenile hormone biosynthesis. *J. Insect Physiol.* 37: 733–741.

Hudson, A. 1966. Proteins in the haemolymph and other tissues of the developing tomato hornworm, *Protoparce quinquemaculata* Haworth. *Can J. Zool.* 44: 541–555.

Huxley, J. S. 1972. *Problems of Relative Growth*. Dover, New York.

Imai, K., T. Konno, Y. Nakazawa, T. Komiya, M. Isobe, K. Koga, T. Goto, T. Yagunima, K. Sakakibara, K. Hasegawa, and O. Yamashita. 1991. Isolation and structure of diapause hormone of the silkworm, *Bombyx mori*. *Proc. Jap. Acad.* 67B: 98–101.

Ishizaki, H., and M. Ichikawa. 1967. Purification of the brain hormone of the silkworm *Bombyx mori*. *Biol. Bull.* 133: 355–368.

Isobe, M., K. Hasegawa, I. Kubota, and T. Goto. 1976. Diapause hormone B: Its selective extraction and isolation from the silkworm *Bombyx mori*. *Agric. Biol. Chem.* 40: 1189–1199.

Iwami, M. 1990. The genes encoding bombyxin, a brain secretory peptide of *Bombyx mori*: Structure and expression. In *Molting and Metamorphosis*, ed. E. Ohnishi and H. Ishizaki, pp. 49–66. Springer-Verlag, Berlin.

Iwami, M., T. Adachi, H. Kondo, A. Kawakami, Y. Suzuki, H. Nagasawa, A. Suzuki, and H. Ishizaki. 1990. A novel family C of the genes that encode bombyxin, an insulin-related brain secretory peptide of the silkmoth *Bombyx mori*: Isolation and characterization of gene C-1. *Insect Biochem.* 20: 295–303.

Iwami, M., A. Kawakami, H. Ishizaki, S. Y. Takahashi, T. Adachi, Y. Suzuki, H. Nagasawa, and A. Suzuki. 1989. Cloning of a gene encoding bombyxin, an insulin-like brain secretory peptide of the silkmoth *Bombyx mori* with prothoracicotropic activity. *Dev. Growth Diff.* 31: 31–37.

Johnson, C. G. 1969. *Migration and Dispersal of Insects by Flight*. Methuen, London.

Joly, P. 1945. La fonction ovarienne et son contrôle humorale chez les dytiscides. *Arch. Zool. Exp. Gen.* 84: 47–164.

Joly, P. 1968. *Endocrinologie des Insectes*. Masson, Paris.

Joly, P., and L. Joly. 1954. Resultats de greffes de corpora allata chez *Locusta migratoria* L. *Ann. Sci. Nat. (Zool.)* 15: 331–345.

Jones, C. G., and R. D. Firn. 1978. The role of phytoecdysteroids in bracken fern *Pteridium aquilinum* (L.) Kühn as a defense against phytophagous insects. *J. Chem. Ecol.* 4: 117–138.

Jones, G., K. Wing, D. Jones, and B. D. Hammock. 1981. The source and action of head factors regulating juvenile hormone esterase in larvae of the cabbage looper, *Trichoplusia ni*. *J. Insect Physiol.* 27: 85–91.

Judson, C. L. 1967. Feeding and oviposition behavior in the mosquito *Aedes aegypti* (L.). I. Preliminary studies of physiological control mechanisms. *Biol. Bull.* 133: 369–377.

Jutsum, A. R., and G. J. Goldsworthy. 1977. The role of the glandular lobes of the corpora cardiaca during flight in *Locusta. Physiol. Entomol.* 2: 125–132.

Kai, H., and K. Nishi. 1976. Diapause development in *Bombyx* eggs in relation to 'esterase A' activity. *J. Insect Physiol.* 22: 1315–1320.

Kambysellis, M. P., and C. M. Williams. 1972. Spermatogenesis in cultured testes of the cynthia silkworm: Effects of ecdysone and of prothoracic glands. *Science* 175: 769–770.

Karlson, P., and M. D. Stamm-Menedez. 1956. Notiz über den Nachweis von Metamorphose-Hormon in den Imagines von *Bombyx mori. Hoppe Seyler's Z. Physiol. Chem.* 306: 109–111.

Karlson, P., H. Hoffmeister, H. Hummel, P. Hocks, and G. Spitteler. 1965. Zur Chemie des Ecdysons. VI. Reaktionen des Ecdysonsmolekuls. *Chem. Ber.* 98: 2394–2402.

Kataoka, H., A. Toschi, J. P. Li, R. L. Carney, D. A. Schooley, and S. J. Kramer. 1989. Identification of an allatotropin from adult *Manduca sexta. Science* 243: 1481–1483.

Kataoka, H., R. G. Troetschler, S. J. Kramer, B. Cesarin, and D. A. Schooley. 1987. Isolation and primary structure of the eclosion hormone of the tobacco hornworm, *Manduca sexta. Biochem. Biophys. Res. Comm.* 146: 746–750.

Kataoka, H., R. G. Troetschler, J. P. Li, S. J. Kramer, R. L. Carney, and D. A. Schooley. 1989. Isolation and identification of a diuretic hormone from the tobacco hornworm, *Manduca sexta. Proc. Nat. Acad. Sci. USA* 86: 2976–2980.

Kawakami, A., M. Iwami, H. Nagasawa, A. Suzuki, and H. Ishizaki. 1989. Structure and organization of four clustered genes that encode bombyxin, an insulin-related brain secretory peptide of the silkmoth *Bombyx mori. Proc. Nat. Acad. Sci. USA* 86: 6843–6847.

Kawakami, A., H. Kataoka, T. Ota, A. Mizoguchi, M. Kimura-Kawakami, T. Adachi, M. Iwami, H. Nagasawa, A. Suzuki, and H. Ishizaki. 1990. Molecular cloning of the *Bombyx mori* prothoracicotropic hormone. *Science* 247: 1333–1335.

Kayser-Wegmann, I. 1975. Untersuchungen zur Photobiologie und Endocrinologie der Farbmodifikationen bei Kohlweisslingspuppe *Pieris brassicae*: Zeitverlauf der sensibelen und kritischen Phasen. *J. Insect Physiol.* 21: 1065–1072.

Keeley, L. L., and T. K. Hayes. 1987. Speculations on biotechnology applications for insect neuroendocrine research. *Insect Biochem.* 17: 639–651.

Kelly, T. J., L. V. Whisenton, E. J. Katahira, M. S. Fuchs, A. B. Borkovec, and W. E. Bollenbacher. 1986. Inter-species cross-reactivity of the prothoracicotropic hormone of *Manduca sexta* and egg-development neurosecretory hormone of *Aedes aegypti. J. Insect Physiol.* 32: 757–762.

Kennedy, J. S. 1961. Continuous polymorphism in locusts. In *Insect Polymorphism*, ed. J. S. Kennedy, pp. 80–90. Royal Entomological Society, London.

Kerkut, G. A., and L. I. Gilbert. 1985. *Comprehensive Insect Physiology, Biochemistry and Pharmacology*. 13 vols. Pergamon, New York.

Kiely, M. L., and L. M. Riddiford. 1985a. Temporal programming of epidermal cell protein synthesis during the larval-pupal transformation of *Manduca sexta. Wilh. Roux' Arch.* 194: 325–335.

Kiely, M. L., and L. M. Riddiford. 1985b. Temporal patterns of protein synthesis in *Manduca* epidermis during the change to pupal commitment in vitro: Their modulation by 20-hydroxyecdysone and juvenile hormone. *Wilh. Roux'Arch.* 194: 336–343.

Kiguchi, K., and L. M. Riddiford. 1978. A role of juvenile hormone in pupal development of the tobacco hornworm, *Manduca sexta. J. Insect Physiol.* 24: 673–680.

King, D. S., and M. P. Marks. 1974. The secretion and metabolism of a-ecdysone by cockroach (*Leucophaea maderae*) tissues *in vitro. Life Sci.* 15: 147–154.

King, R. C. 1970. *Ovarian Development in Drosophila melanogaster*. Academic Press, New York.

King, R. C., S. K. Aggarwal, and D. Bodenstein. 1966. The comparative submicroscopic morphology of the ring gland of *Drosophila melanogaster* during the second and third larval instars. *Z. Zellforsch.* 73: 272–285.

Kiriishi, S., H. Nagasawa, H. Kataoka, A. Suzuki, and S. Sakurai. 1992. Comparison of the *in vivo* and *in vitro* effects of bombyxin and prothoracicotropic hormone on prothoracic glands of the silkworm, *Bombyx mori. Zool. Sci.* 9: 149–155.

Kiriishi, S., D. B. Rountree, S. Sakurai, and L. I. Gilbert. 1990. Prothoracic gland synthesis of 3-dehydoecdysone and its hemolymph 3β-reductase mediated conversion to ecdysone in representative insects. *Experientia* 46: 716–721.

Kitamura, A., H. Nagasawa, H. Kataoka, T. Ando, and A. Suzuki. 1990. Amino acid sequence of pheromone biosynthesis activating neuropeptide-II (PBAN-II) of the silkmoth, *Bombyx mori. Agric. Biol. Chem.* 54: 2495–2497.

Kitazawa, T., T. Kanda, and T. Takami. 1963. Changes in mitotic activity in the silkworm egg in relation to diapause. *Bull. Seric. Exp. Sta.* 18: 283–295.

Knobloch, C. A., and C.G.H. Steel. 1987. Effects of decapitation at the head critical period for moulting on haemolymph ecdysteroid titres in final-instar male and female *Rhodnius prolixus* (Hemiptera). *J. Insect Physiol.* 33: 967–972.

Koch, P. B., and D. Bückmann. 1984. Vergleichende Untersuchungen der Farbmuster und der Farbanpassung von Nymphalidenpuppen (Lepidoptera). *Zool. Beitr.* 28: 369–401.

Koch, P. B., and D. Bückmann. 1987. Hormonal control of seasonal morphs by the timing of ecdysteroid release in *Araschnia levana* (Nymphalidae: Lepidoptera). *J. Insect Physiol.* 33: 823–829.

Koch, P. B., G. Starnecker, and D. Bückmann. 1990. Interspecific effects of the pupal melanization reducing factor on pupal colouration in different lepidopteran families. *J. Insect Physiol.* 36: 159–164.

Koeppe, J. K., M. Fuchs, T. T. Chen, L.-M. Hunt, G. E. Kovalick, and T. Briers. 1985. The role of juvenile hormone in reproduction. In *Comprehensive Insect Physiology, Biochemistry and Pharmacology*, ed. G. A. Kerkut and L. I. Gilbert, vol. 8, pp. 165–203. Pergamon, New York.

Kono, T., H. Nagasawa, A. Isogai, H. Fugo, and A. Suzuki. 1991. Isolation and complete amino acid sequences of eclosion hormones of the silkworm *Bombyx mori. Insect Biochem.* 21: 185–195.

Koolman, J., and P. Karlson. 1985. Regulation of ecdysone titer: Degradation. In *Comprehensive Insect Physiology, Biochemistry and Pharmacology*, ed. G. A. Kerkut and L. I. Gilbert, vol. 7, pp. 343–361. Pergamon, New York.

Kopec, S. 1917. Experiments on metamorphosis of insects. *Bull. Int. Acad. Cracov* B, pp. 57–60.

Kopec, S. 1922. Studies on the necessity of the brain for the inception of insect metamorphosis. *Biol. Bull.* 142: 323–342.

Kremen, C. 1989. Patterning during pupal commitment of the epidermis in the butterfly, *Precis coenia*: The role of intercellular communication. *Dev. Biol.* 133: 336–347.

Kremen, C., and H. F. Nijhout. 1989. Juvenile hormone controls the onset of pupal commitment in the imaginal disks and epidermis of *Precis coenia* (Lepidoptera: Nymphalidae). *J. Insect Physiol.* 35: 603–612.

Kriger, F. L., and K. G. Davey. 1982. Ovarian motility in mated *Rhodnius prolixus* requires an intact neurosecretory system. *Gen. Comp. Endocrinol.* 48: 130–134.

Kubli, E. 1992. The sex peptide. *BioEssays* 14: 779–784.

Kuntze, H. 1935. Die Flügelentwicklung bei *Philosamia cynthia* Drury, mit besonderer Berücksichtigung des Gäders der Lakunen und der Tracheensysteme. *Z. Morphol. Ökol. Tiere* 30: 544–572.

Lafont, R., and D.H.S. Horn. 1989. Phytoecdysteroids: Structure and occurrence. In *Ecdysone: From Chemistry to Mode of Action*, ed. J. Koolman, pp. 39–64. Thieme-Verlag, Stuttgart.

Lageux, M., C. Hethru, F. Goltzené, C. Kappler, and J. Hoffmann. 1979. Ecdysone titre and metabolism in relation to cuticulogenesis in embryos of *Locusta migratoria*. *J. Insect Physiol.* 25: 709–723.

Lanzrein, B., V. Gentinetta, H. Abegglen, F. C. Baker, C. A. Miller, and D. A. Schooley. 1985. Titers of ecdysone, 20-hydroxyecdysone and juvenile hormone III throughout the life cycle of a hemimetabolous insect, the ovoviviparous cockroach *Nauphoeta cinerea*. *Experientia* 41: 913–917.

Lauga, J. 1977. Le problème de la mesure de la phase chez les acridiens migrateurs: Historique et définition d'echelles phasaires chez *Locusta migratoria* L. (Insecte, Orthoptère). *Arch. Zool. Exp. Gen.* 118: 247–272.

Lawrence, P. A. 1969. Cellular differentiation and pattern formation during metamorphosis of the milkweed bug *Oncopeltus*. *Dev. Biol.* 19: 12–40.

Lawrence, P. O. 1982. *Biosteres longicaudatus*: Developmental dependence on host (*Anastrepa suspensa*) physiology. *Exp. Parasitol.* 53: 396–405.

Lea, A. O. 1968. Mating without insemination in virgin *Aedes aegypti*. *J. Insect Physiol.* 14: 305–308.

Lea, A. O. 1972. Regulation of egg maturation in the mosquito by the neurosecretory system: The role of the corpus cardiacum. *Gen. Comp. Endocrinol.* (suppl.) 3: 602–608.

Lea, A. O., and M. R. Brown. 1989. Neuropeptides of mosquitoes. In *Molecular Insect Science*, ed. H. H. Hagedorn, J. G. Hildebrand, M. G. Kidwell, and J. H. Law, pp. 181–188. Plenum Press, New York.

Leahy, M. G., and G. B. Craig 1965. Male accessory gland as a stimulant for oviposition in *Aedes aegypti* and *A. albopictus*. *Mosquito News* 25: 448–452.

Leahy, M. G., and M. L. Lowe. 1967. Purification of the male factor increasing egg deposition in *D. melanogaster*. *Life Sci.* 6: 151–156.

Lees, A. D. 1966. The control of polymorphism in aphids. *Adv. Insect Physiol.* 3: 207–277.

Lehmann, M., and J. Koolman. 1989. Regulation of ecdysone metabolism. In *Ec-*

dysone: From Chemistry to Mode of Action, ed. J. Koolman, pp. 217–220. Thieme-Verlag, Stuttgart.

Lepesant, J.-A., and G. Richards. 1989. Ecdysteroid-regulated genes. In *Ecdysone: From Chemistry to Mode of Action*, ed. J. Koolman, pp. 355–367. Thieme-Verlag, Stuttgart.

Lezzi, M., and G. Richards. 1989. Salivary glands. In *Ecdysone: From Chemistry to Mode of Action*, ed. J. Koolman, pp. 393–405. Thieme-Verlag, Stuttgart.

Lipke, H., M. Sugumaran, and W. Henzel. 1983. Mechanisms of sclerotization in Dipterans. *Adv. Insect Physiol.* 17: 1–84.

Lloyd, M., and J. White. 1976. Sympatry of periodical cicadas and the hypothetical four-year acceleration. *Evolution* 30: 786–801.

Lloyd, M., G. Kritski, and C. Simon. 1983. A simple Mendelian model for 13- and 17-year life cycles of periodical cicadas, with historical evidence of hybridization between them. *Evolution* 37: 1162–1180.

Locke, M. 1958. The formation of tracheae and tracheoles in *Rhodnius prolixus. Quart. J. Microsc. Sci.* 99: 29–46.

Locke, M. 1974. The structure and formation of the integument in insects. In *The Pysiology of Insecta*, ed. M. Rockstein, vol. 6, pp.123–213. Academic Press, New York.

Locke, M. 1985. A structural analysis of postembryonic development. In *Comprehensive Insect Physiology, Biochemistry and Pharmacology*, ed. G. A. Kerkut and L. I. Gilbert, vol. 2, pp. 87–149. Pergamon, New York.

Locke, M. 1990. Epidermal cells. In *Molting and Metamorphosis*, ed. E. Ohnishi and H. Ishizaki, pp. 173–206. Springer-Verlag, Berlin.

Locke, M., and P. Huie. 1979. Apolysis and the turnover of plasma membrane plaques during cuticle formation in an insect. *Tissue & Cell* 11: 277–291.

Lockshin, R. A. 1969. Programmed cell death. Activation of lysis by a mechanism involving the synthesis of a protein. *J. Insect Physiol.* 15: 1505–1516.

Lockshin, R. A. 1985. Programmed cell death. In *Comprehensive Insect Physiology, Biochemistry and Pharmacology*, ed. G. A. Kerkut and L. I. Gilbert, vol. 2, pp. 301–317. Pergamon, New York.

Lockshin, R. A., and C. M. Williams. 1965a. Programmed cell death. III. Neural control of the breakdown of the intersegmental muscles of silkmoths. *J. Insect Physiol.* 11: 601–610.

Lockshin, R. A., and C. M. Williams. 1965b. Programmed cell death. IV. The influence of drugs on the breakdown of the intersegmental muscles of silkmoths. *J. Insect Physiol.* 11: 803–809.

Loeb, M. J., C. W. Woods, E. P. Brandt, and A. B. Borkovec. 1982. Larval testes of the tobacco budworm: A new source of insect ecdysteroid. *Science* 218: 896–898.

Loher, W. 1960. The chemical acceleration of the maturation process and its hormonal control in the male desert locust. *Proc. Roy. Soc. London* B 153: 380–397.

Loher, W., and F. Huber. 1966. Nervous and endocrine control of sexual behavior in a grasshopper (*Gomphocerus rufus* L., Acridinae). *Symp. Soc. Exp. Biol.* 20: 381–400.

Loughton, B. G., and I. Orchard. 1981. The nature of the hyperglycaemic factor from the glandular lobe of the corpus cardiacum of *Locusta migratoria. J. Insect Physiol.* 27: 383–385.

Lounibos, L. P. 1975. The cocoon spinning behavior of the chinese oak silkworm, *Antheraea pernyi. Animal Behav.* 23: 843–853.

Lounibos, L. P. 1976. Initiation and maintenance of cocoon spinning behavior by saturniid silkworms. *Physiol. Entomol.* 1: 195–206.

Lüscher, M. 1953. Kann die Determination durch eine monomolekulare Reaktion ausgelöst werden? *Rev. Suisse Zool.* 60: 524–528.

Lüscher, M. 1960. Hormonal control of caste differentiation in termites. *Ann. N.Y. Acad. Sci.* 89: 549–563.

Lüscher, M. 1972. Environmental control of juvenile hormone (JH) secretion and caste differentiation in termites. *Comp. Endocrinol.* (suppl.) 3: 509–514.

Lüscher, M. 1974. Die Kompetenz zur Soldatenbildung bei Larven (Pseudergaten) der Termite *Zootermopsis angusticollis. Rev. Suisse Zool.* 81: 710–714.

Lüscher, M. 1976. Evidence for an endocrine control of caste determination in higher termites. In *Phase and Caste Determination in Insects*, ed. M. Lüscher, pp. 91–103. Pergamon Press, Oxford.

McCaffery, A. R. 1976. Effects of electrocoagulation of cerebral neurosecretory cells and implantation of corpora allata on oocyte development in *Locusta migratoria. J. Insect Physiol.* 22: 1081–1092.

McCaffery, A. R., and W. W. Page. 1978. Factors influencing the production of long-winged *Zonocerus variegatus. J. Insect Physiol.* 24: 465–472.

McClure, J. B., and J. E. Steele. 1981. The role of extracellular calcium in hormonal activation of glycogen phosphorylase in cockroach fat body. *Insect Biochem.* 11: 605–613.

Maddrell, S.H.P. 1963. Excretion in the blood-sucking bug, *Rhodnius prolixus* Stal. I. The control of diuresis. *J. Exp. Biol.* 40: 247–256.

Maddrell, S.H.P. 1964. Excretion in the blood-sucking bug, *Rhodnius prolixus* Stal. III. The control of the release of the diuretic hormone. *J. Exp. Biol.* 41: 459–472.

Maddrell, S.H.P. 1966. The site of release of the diuretic hormone in *Rhodnius*—a new neurohaemal system in insects. *J. Exp. Biol.* 45: 499–508.

Maddrell, S.H.P. 1969. Secretion by the Malpighian tubules of *Rhodnius*. The movement of ions and water. *J. Exp. Biol.* 51: 71–97.

Maddrell, S.H.P., and J. J. Nordmann. 1979. *Neurosecretion*. Blackie, London.

Maddrell, S.H.P., and J. E. Phillips. 1975. Secretion of hypo-osmotic fluid by the lower Malpighian tubules of *Rhodnius prolixus. J. Exp. Biol.* 62: 671–683

Maddrell, S.H.P., and S. E. Reynolds. 1972. Release of hormones in insects after poisoning with insecticides. *Nature* 236: 404–406.

Maisch, A., and D. Bückmann. 1987. The control of cuticular melanin and lutein incorporation in the morphological colour adaptation of a nymphalid pupa, *Inachis io* L. *J. Insect Physiol.* 33: 393–402.

Manning, A. 1966. Corpus allatum and sexual receptivity in female *Drosophila melanogaster. Nature* 211: 1321–1322.

Marti, T., K. Takio, K. A. Walsh, G. Terzi, and J. W. Truman. 1987. Microanalysis of the amino acid sequence of the eclosion hormone of the tobacco hornworm, *Manduca sexta. FEBS Lett.* 219: 415–418.

Masaki, T., K. Endo, and K. Kumagai. 1988. Neuroendocrine regulation of the development of seasonal morphs in the Asian comma butterfly, *Polygonia c-aureum* L.: Is the factor producing summer morphs (SMPH) identical to the small prothoracicotropic hormone (4K-PTTH)? *Zool. Sci.* 5: 1051–1057.

Mason, C. A. 1973. New features of the brain-retrocerebral neuroendocrine complex of the locust *Schistocerca vaga* (Scudder). *Z. Zellforsch.* 141: 19–32.

Matsumoto, S., M. R. Brown, A. Suzuki, and A. O. Lea. 1989. Isolation and characterization of ovarian ecdysteroidogenic hormones from the mosquito, *Aedes aegypti. Insect Biochem.* 19: 651–656.

Matsumoto, S., A. Fonagy, M. Kurihara, K. Uchiumi, T. Nagamine, M. Chijimatsu, and T. Mitsui. 1992a. Isolation and primary structure of a novel pheromonotropic neuropeptide structurally related to leukopyrokinin from the armyworm larvae, *Pseudaletia separata. Biochem. Biophys. Res. Comm.* 182: 534–539.

Matsumoto, S., O. Yamashita, A. Fonagy, M. Kurihara, K. Uchiumi, T. Nagamine, and T. Mitsui. 1992b. Functional diversity of a pheromonotropic neuropeptide: Induction of cuticular melanization and embryonic diapause in lepidopteran insects by *Pseudaletia* pheromonotropin. *J. Insect Physiol.* 38: 847–851.

Matsumoto, S., A. Isogai, and A. Suzuki. 1985. N-terminal amino acid sequence of an insect neurohormone, melanization and reddish coloration hormone (MRCH): Heterogeneity and sequence homology with human insulin-like growth factor II. *FEBS Lett.* 189: 115–118.

Matsumoto, S., A. Kitamura, H. Nagasawa, H. Kataoka, C. Orikasa, T. Mitsui, and A. Suzuki. 1990. Functional diversity of a neurohormone produced by the suboesophageal ganglion: Molecular identity of melanization and reddish colouration hormone and pheromone biosynthesis activating neuropeptide. *J. Insect Physiol.* 36: 427–432.

Mayer, R. J., and D. J. Candy. 1969. Control of haemolymph lipid concentration during locust flight: an adipokinetic hormone from the corpora cardiaca. *J. Insect Physiol.* 15: 611–620.

Mead-Briggs, A. R., and J. A. Vaughan. 1969. Some requirements for mating in the rabbit flea *Spilopsyllus cuniculi* (Dale). *J. Exp. Biol.* 51: 495–511.

Meller, V. H., W. L. Combest, W. A. Smith, and L. I. Gilbert. 1988. Calmodulin-sensitive adenylate cyclase in the prothoracic glands of the tobacco hornworm, *Manduca sexta. Mol. Cell. Endocrinol.* 59: 67–76.

Menon, M. 1970. Hormone-pheromone relationships in the beetle, *Tenebrio molitor. J. Insect Physiol.* 16: 1123–1139.

Meola, R. W., and P. L. Adkisson. 1977. Release of prothoracicotropic hormone and potentiation of developmental ability during diapause in the bollworm, *Heliothis zea. J. Insect Physiol.* 23: 683–688.

Meola, R. W., and A. O. Lea. 1972. Humoral inhibition of egg development in mosquitoes. *J. Med. Entomol.* 9: 99–103.

Mercola, M., and C. D. Stiles. 1988. Growth factor superfamilies and mammalian embryogenesis. *Development* 102: 451–460.

Meurant, K., and C. Sernia. 1993. The ultrastructure of the prothoracic gland/corpus allatum/corpus cardiacum ring complex of the Australian sheep blowfly larva *Lucilia cuprina* (Wied.) (Insecta: Diptera). *Insect Biochem. Molec. Biol.* 23: 47–55.

Milburn, N., E. A. Weiant, and K. D. Roeder. 1960. The release of efferent nerve activity in the roach, *Periplaneta americana*, by extracts of the corpus cardiacum. *Biol. Bull.* 118: 111–119.

Miller, E. M. 1969. Caste differentiation in lower termites. In *Biology of Termites*, ed. K. Krishna and F. M. Weesner, pp. 283–307. Academic Press, New York.

Miller, T. A. 1980. *Neurohormonal Techniques in Insects*. Springer-Verlag, New York.

Mizoguchi, A. 1990. Immunological approach to synthesis, release, and titre fluctuation of bombyxin and prothoracicotropic hormone of *Bombyx mori*. In *Molting and Metamorphosis*, ed. E. Ohnishi and H. Ishizaki, pp. 17–32. Springer-Verlag, Berlin.

Mizoguchi, A., M. Hatta, S. Sato, H. Nagasawa, A. Suzuki, and H. Ishizaki. 1990a. Developmental changes of bombyxin content in the brain of the silkmoth *Bombyx mori*. *J. Insect Physiol* 36: 655–664.

Mizoguchi, A., H. Ishizaki, H. Nagasawa, H. Kataoka, A. Isogai, S. Tamura, A. Suzuki, M. Fujino, and C. Kitada. 1987. A monoclonal antibody against a synthetic fragment of bombyxin (4K-protoracicotropic hormone) from the silkmoth, *Bombyx mori*: Characterization and immunohistochemistry. *Molec. Cell. Endocrinol.* 51: 227–235.

Mizoguchi, A., T. Oka, H. Kataoka, H. Nagasawa, A. Suzuki, and H. Ishizaki. 1990b. Immunohistochemical localization of prothoracicotropic hormone-producing neurosecretory cells in the brain of *Bombyx mori*. *Dev. Growth & Diff.* 32: 591–598.

Moran, N. A. 1990. Aphid life cycles: Two evolutionary steps. *Amer. Nat.* 136: 135–138.

Mordue, W., and G. J. Goldsworthy. 1969. The physiological effects of corpus cardiacum extracts in locusts. *Gen. Comp. Endocrinol.* 12: 360–369.

Mordue, W., and P. J. Morgan. 1985. Chemistry of peptide hormones. In *Comprehensive Insect Physiology, Biochemistry and Pharmacology*, ed. G. A. Kerkut and L. I. Gilbert, vol. 7, pp. 153–183. Pergamon, New York.

Moreau, R., and L. Lavenseau. 1975. Rôle des organes pulsatiles thoraciques et du coeur pendant l'emergence et l'expansion des ailes des Lepidoptères. *J. Insect Physiol.* 21: 1531–1534.

Morita, M., M. Hatakoshi, and S. Tojo. 1988. Hormonal control of cuticular melanization in the common cutworm, *Spodoptera litura*. *J. Insect Physiol.* 34: 751–758.

Mosna, G. 1981. Insulin can completely replace serum in *Drosophila melanogaster* cell cultures *in vitro*. *Experientia* 37: 466–467.

Muehleisen, D. P., R. S. Gray, E. J. Katahira, M. K. Thomas, and W. E. Bollenbacher. 1993. Immunoaffinity purification of the neuropeptide prothoracicotropic hormone from *Manduca sexta*. *Peptides* 14: 531–541.

Nagasawa, H., H. Kataoka, A. Isogai, S. Tamura, H. Suzuki, H. Ishizaki, A. Mizoguchi, Y. Fujiwara, and A. Suzuki. 1984. Amino-terminal amino acid sequence of the silkworm prothoracicotropic hormone: Homology with insulin. *Science* 226: 1344–1345.

Nagasawa, H., H. Kataoka, and A. Suzuki. 1990. Chemistry of *Bombyx* prothoracicotropic hormone and bombyxin. In *Molting and Metamorphosis*, ed. E. Ohnishi and H. Ishizaki, pp. 33–48. Springer-Verlag, Berlin.

Naisse, J. 1966. Contrôle endocrinien de la différentiation sexuelle de *Lampyris noctiluca* (Coléoptère Malacoderme Lampyridae). I. Rôle androgène des testicules. *Arch. Biol. Liege* 77: 139–201.

Naisse, J. 1969. Rôle des neurohormones dans la différentiation sexuelle de *Lampyris noctiluca*. *J. Insect Physiol.* 15: 877–892.

Neville, A. C. 1967. Chitin orientation in cuticle and its control. *Adv. Insect Physiol.* 4: 213–286.

Neville, A. C. 1983. Daily cuticular growth layers and the teneral stage in adult insects. *J. Insect Physiol.* 29: 211–219.

Neville, A. C. 1984. Cuticle: Organization. In *Biology of the Integument.* vol. 1. *Invertebrates* ed. J. Bereiter-Hahn, A. G. Matoltsy, and K. S. Richards, pp. 611–625. Springer-Verlag, Berlin.

Nijhout, H. F. 1975a. The brain-retrocerebral neuroendocrine complex of *Manduca sexta* (L.) (Lepidoptera: Sphingidae). *Int. J. Insect Morphol. Embryol.* 4: 529–538.

Nijhout, H. F. 1975b. A threshold size for metamorphosis in the tobacco hornworm, *Manduca sexta. Biol. Bull.* 149: 214–225.

Nijhout, H. F. 1976. The rôle of ecdysone in pupation of *Manduca sexta. J. Insect Physiol.* 22: 453–463.

Nijhout, H. F. 1979. Stretch-induced moulting in *Oncopeltus fasciatus. J. Insect. Physiol.* 25: 277–281.

Nijhout, H. F. 1981. Physiological control of molting in insects. *Amer. Zool.* 21: 631–640.

Nijhout, H. F. 1983. Definition of a juvenile hormone-sensitive period in *Rhodnius prolixus. J. Insect Physiol.* 29: 669–677.

Nijhout, H. F. 1984. Abdominal stretch reception in *Dipetalogaster maximus* (Hemiptera: Reduviidae). *J. Insect Physiol.* 30: 629–633.

Nijhout, H. F. 1985. The developmental physiology of color patterns in Lepidoptera. *Adv. Insect Physiol.* 18: 181–247.

Nijhout, H. F. 1991. *The Development and Evolution of Butterfly Wing Patterns.* Smithsonian Institution Press, Washington, D.C.

Nijhout, H. F., and G. M. Carrow. 1978. Diuresis after a bloodmeal in female *Anopheles freeborni. J. Insect Physiol.* 24: 293–298.

Nijhout, H. F., and D. E. Wheeler. 1982. Juvenile hormone and the physiological basis of insect polymorphisms. *Quart. Rev. Biol.* 57: 109–133.

Nijhout, H. F., and C. M. Williams. 1974a. Control of moulting and metamorphosis in the tobacco hornworm, *Manduca sexta* (L.): Growth of the last instar larva and the decision to pupate. *J. Exp. Biol.* 61: 481–491.

Nijhout, H. F., and C. M. Williams. 1974b. Control of moulting and metamorphosis in the tobacco hornworm, *Manduca sexta* (L.): Cessation of juvenile hormone secretion as a trigger for pupation. *J. Exp. Biol.* 61: 493–501.

Nopp-Pammer, E., and H. Nopp. 1968. Gehirnhormon und Spinnverhalten bei *Philosamia cynthia* Dru. *Verh. Deutsch. Zool. Gesellsch.* 33: 508–519.

Normann, T. C. 1975. Neurosecretory cells in insect brain and production of hypoglycaemic hormone. *Nature* 354: 259–261.

Norris, M. J. 1954. Sexual maturation in the desert locust (*Schistocerca gregaria* Forsk.) with special reference to the effects of grouping. *Anti-Locust Bull.* 18: 1–44.

Nöthiger, R. 1972. The larval development of imaginal disks. In *The Biology of Imaginal Disks*, ed. H. Ursprung and R. Nöthiger, pp. 1–34. Springer-Verlag, New York.

Novak, V.J.A. 1975. *Insect Hormones.* Chapman and Hall, London.

Oberlander, H. 1985. The imaginal disks. In *Comprehensive Insect Physiology, Biochemistry and Pharmacology*, ed. G. A. Kerkut and L. I. Gilbert, vol. 2, pp. 151–182. Pergamon, New York.

O'Brien, M. A., E. J. Katahira, T. R. Flanagan, L. W. Arnold, G. Haughton, and W. E. Bollenbacher. 1988. A monoclonal antibody to the insect prothoracicotropic hormone. *J. Neurosci.* 8: 3247–3257.

O'Connor, J. D. 1985. Ecdysteroid action at the molecular level. In *Comprehensive Insect Physiology, Biochemistry and Pharmacology*, ed. G. A. Kerkut and L. I. Gilbert, vol. 8, pp. 85–98. Pergamon, New York.

Ogura, N. 1975. Hormonal control of larval coloration in the armyworm, *Leucania separata. J. Insect Physiol.* 21: 559–576.

Ohnishi, E. 1990. Ecdysteroids in insect ovaries. In *Molting and Metamorphosis*, ed. E. Ohnishi and H. Ishizaki, pp. 121–129. Springer-Verlag, Berlin.

Ohtaki, T. 1960. Humoral control of pupal coloration in the cabbage white butterfly, *Pieris rapae crucivorae. Annot. Zool. Jap.* 33: 97–103.

Okajima, A., and K. Kumagai. 1989. The inhibitory control of prothoracic gland activity by the neurosecretory neurones in a moth, *Mamestra brassicae. Zool. Sci.* 6: 851–858.

Okot-Kotber, B. M. 1980. Competence of *Macrotermes michaelseni* (Isoptera: Macrotermitinae) larvae to differentiate into soldiers under the influence of juvenile hormone analogue (ZR-515, methoprene). *J. Insect Physiol.* 26: 655–659.

Orchard, I. 1987. Adipokinetic hormones—an update. *J. Insect Physiol.* 33: 451–463.

Orchard, I., and B. G. Loughton. 1981. The neural control of release of hyperlipaemic hormone from the corpus cardiacum of *Locusta migratoria. Comp. Biochem. Physiol.* 68A: 25–30.

Orchard, I., and B. G. Loughton. 1985. Neurosecretion. In *Comprehensive Insect Physiology, Biochemistry and Pharmacology*, ed. G. A. Kerkut and L. I. Gilbert, vol. 7, pp. 61–107. Pergamon, New York.

Orchard, I., B. G. Loughton, and R. A. Webb. 1981. Octopamine and short-term hyperlipaemia in the locust. *Gen. Comp. Endocrinol.* 45: 175–180.

O'Shea, M., and M. E. Adams. 1986. Proctolin: From "gut factor" to model neuropeptide. *Adv. Insect Physiol.* 19: 1–28.

Overton, J. 1966. Microtubules and microfibrils in morphogenesis of the scale cells of *Ephestia kühniella* Zeller. *Zeitschr. Zellforsch.* 63: 840–870.

Ozeki, K. 1968. Experimental studies on the regression of the ventral glands of the earwig, *Anisolabis maritima*, during metamorphosis. *Univ. Tokyo Coll. Gen. Educ. Papers* 20: 143–155.

Palli, S. R., L. M. Riddiford, and K. Hiruma. 1991. Juvenile hormone and "retinoic acid" receptors in *Manduca* epidermis. *Insect Biochem.* 21: 7–15.

Pammer, E. 1966. Auslösung und Steuerung des Spinnverhaltens und der Diapause bei *Philosamia cynthia* Dru. (Saturniidae, Lep.). *Z. Vergl. Physiol.* 53: 99–113.

Panov, A. A. 1980. Demonstration of neurosecretory cells in the insect central nervous system. In *Neuroanatomical Techniques*, ed. N. J. Strausfeld and T. A. Miller, pp. 26–51. Springer-Verlag, New York.

Panov, A. A., and O. K. Bassurmanova. 1970. Fine structure of the gland cells in inactive and active corpus allatum of the bug, *Eurygaster integriceps. J. Insect Physiol.* 16: 1265–1281.

Pass, G., G. Sperk, H. Agricola, E. Baumann, and H. Penzlin. 1988. Octopamine in a neurohaemal area within the antennal heart of the American cockroach. *J. Exp. Biol.* 135: 495–498.

Passera, L. 1974. Differenciation des soldats chez la fourmi *Pheidole pallidula* (Nyl.) (Formicidae Myrmicinae). *Insect. Soc.* 21: 71–86.

Passera, L. 1985. Soldier determination in ants of the genus *Pheidole*. In *Caste Differ-*

entiation in Social Insects ed. J.A.L. Watson, B. M. Okot-Kotber, and Ch. Noirot, pp. 331–346. Pergamon, Oxford.

Passera, L., and J.-P. Suzzoni. 1979. Le rôle de la reine de *Pheidole pallidula* (Nyl.) (Hymenoptera: Formicidae) dans la sexualization du couvain après traitement par l'hormone juvénile. *Insect. Soc.* 26: 343–353.

Pener, M. P. 1967. Effects of allatectomy and sectioning of the nerves of the corpora allata on oocyte growth, male sexual behaviour, and colour change in adults of *Schistocerca gregaria. J. Insect Physiol.* 13: 665–684.

Pener, M. P. 1985. Hormonal effects on flight and migration. In *Comprehensive Insect Physiology, Biochemistry and Pharmacology*, ed. G. A. Kerkut and L. I. Gilbert, vol. 8, pp. 491–550. Pergamon, New York.

Pener, M. P. 1991. Locust phase polymorphism and its endocrine relations. *Adv. Insect Physiol.* 23: 1–79.

Pener, M. P., L. Orshan, and J. de Wilde. 1978. Precocene II causes atrophy of corpora allata in *Locusta migratoria. Nature* 272: 350–353.

Phillips, J. E. 1981. Comparative physiology of insect renal function. *Amer. J. Physiol.* 241: R241–R257.

Phillips, J. E. 1983. Endocrine control of salt and water balance: Excretion. In *Endocrinology of Insects*, ed. R.G.H Downer and H. Laufer, pp. 411–425. Alan Liss, New York.

Phillips, J. E., J. Hanrahan, M. Chamberlin, and B. Thomson. 1986. Mechanisms and control of reabsorption in insect hindgut. *Adv. Insect Physiol.* 19: 329–422.

Phillips, J. E., J. Meredith, J. Spring, and M. Chamberlin. 1982. Control of ion reabsorption in locust rectum: Implications for fluid transport. *J. Exp. Zool.* 222: 297–308.

Piepho, H. 1942. Untersuchungen zur Entwicklungsphysiologie der Insekten-Metamorphose. Über Puppenhautung der Wachsmotte *Galleria mellonella. Wilh. Roux' Arch.* 141: 500–583.

Piepho, H. 1946. Versuche über die Rolle von Wirkstoffen in der Metamorphose der Schmetterlinge. *Biol. Zblt.* 65: 141–148.

Piepho, H. 1950. Über die Hemmung der Falterhäutung durch Corpora allata. *Biol. Zblt.* 69: 261–271.

Piepho, H. 1951. Über die Lenkung der Insektenmetamorphose durch Hormone. *Verh. Dtsch. Zool. Gesell.*, 62–76.

Piepho, H., E. Boden, and I. Holtz. 1960. Über die Hormonabhängigkeit des Verhaltens von Schwärmerraupen vor den Häutungen. *Z. Tierpsychol.* 17: 261–269.

Pitman, R. M., C. D. Tweedle, and M. J. Cohen. 1972. Branching of central neurons: Intracellular cobalt injection for light and electron microscopy. *Science* 176: 647–650.

Plapp, F. W., and S. B. Vinson. 1973. Juvenile hormone analogues: Toxicity and cross-resistance in the housefly. *Pestic. Biochem. Physiol.* 3: 131–136.

Postlethwait, J. H. 1974. Juvenile hormone and the adult development of *Drosophila. Biol. Bull.* 147: 119–135.

Pratt, G. E., D. E. Farnsworth, N. R. Siegel, K. F. Fok, and R. Feyereisen. 1989. Identification of an allatostatin from adult *Diploptera punctata. Biochem. Biophys. Res. Comm.* 163: 1243–1247.

Pratt, G. E., D. E. Farnsworth, and R. Feyereisen. 1990. Changes in the sensitivity of adult cockroach corpora allata to the brain allatostatin. *Molec. Cell. Endocrinol.* 70: 185–195.

Pratt, G. E., D. E. Farnsworth, K. F. Fok, N. R. Siegel, A. L. McCormack, J. Shabanowitz, D. F. Hunt, and R. Feyereisen. 1991. Identity of a second type of allatostatin from cockroach brains: An octadecapeptide amide with a tyrosine-rich address sequence. *Proc. Nat. Acad. Sci. USA* 88: 2412–2416.

Proux, J. P., C. A. Miller, J. P. Li, R. L. Carney, A. Girardie, M. Delaage, and D. A. Schooley. 1987. Identification of an arginine vasopressin-like diuretic hormone from *Locusta migratoria. Biochem. Biophys. Res. Comm.* 149: 180–186.

Raabe, M. 1966. Recherches sur la neurosécrétion dans la chaine nerveuse ventrale du phasme *Carausius morosus*: Liaison entre l'activité des cellules B1 et la pigmentation. *C.R. Acad. Sci. Paris* 263: 408–411.

Raabe, M. 1971. Neurosécretion dans la châine nerveuse ventrale des insectes et organes neurohémaux métameriques. *Arch. Zool. Exp. Gén.* 112: 679–694.

Raabe, M. 1983. The neurosecretory-neurohaemal system of insects: Anatomical, structural and physiological data. *Adv. Insect Physiol.* 17: 205–303.

Raabe, M. 1986. Insect reproduction: Regulation of successive steps. *Adv. Insect Physiol.* 19: 29–154.

Raabe, M., N. Baudry, J. P. Grillot, and A. Provansal. 1974. The perisympathetic organs of insects. In *Neurosecretion—The Final Neuroendocrine Pathway. VI Intern. Symp. Neurosecr. London, 1973*, pp. 59–71. Springer-Verlag, Berlin.

Rachinsky, A., and K. Hartfelder. 1990. Corpora allata activity, a prime regulating element for caste-specific juvenile hormone titre in honey bee larvae (*Apis mellifera carnica*). *J. Insect Physiol.* 36: 189–194.

Raina, A. K. 1993. Neuroendocrine control of sex pheromone biosynthesis in Lepidoptera. *Ann. Rev. Entomol.* 38: 329–349.

Raina, A. K., and J. A. Klun. 1984. Brain factor control of sex pheromone production in the female corn earworm moth. *Science* 225: 531–533.

Raina, A. K., H. Jaffe, J. A. Klun, R. L. Ridgway, and D. K. Hayes. 1987. Characteristics of a neurohormone that controls sex pheromone in *Heliothis zea. J. Insect Physiol.* 33: 809–814.

Raina, A. K., H. Jaffe, T. G. Kempe, P. Keim, R. W. Blacher, H. M. Fales, C. T. Riley, J. A. Klun, R. L. Ridgway, and D. K. Hayes. 1989. Identification of a neuropeptide hormone that regulates sex pheromone production in female moths. *Science* 244: 796–798.

Ramalingam, S., and G. B. Craig. 1977. The effects of a JH mimic and cauterization of the corpus allatum complex on the male accessory glands of *Aedes aegypti* (Diptera: Culicidae). *Can. Entomol.* 109: 897–906.

Ramsay, J. A. 1964. The rectal complex of the mealworm *Tenebrio molitor* L. (Coleoptera, Tenebrionidae). *Phil. Trans. Roy. Soc. London* B 262: 251–160.

Rankin, M. A. 1978. Hormonal control of insect migratory behavior. In *Evolution of Insect Migration and Diapause*, ed. H. Dingle, pp. 5–32. Springer-Verlag, New York.

Rankin. M. A., and L. M. Riddiford. 1977. Hormonal control of migratory flight in *Oncopeltus fasciatus*: The effects of corpus cardiacum, corpus allatum and starvation on migration and reproduction. *Gen. Comp. Endocrinol.* 33: 309–321.

Rankin, M. A., and L. M. Riddiford. 1978. Significance of haemolymph juvenile hormone titer changes in timing of migration and reproduction in adult *Oncopeltus fasciatus. J. Insect Physiol.* 24L: 31–38.

Rankin, S. M., and B. Stay. 1987. Distribution of allatostatin in the adult cockroach, *Diploptera punctata*, and effects of corpora allata *in vitro*. *J. Insect Physiol.* 33: 551–558.

Reagan, J. D., W. H. Miller, and S. J. Kramer. 1992. Allatotropin-induced formation of inositol phosphates in the corpora allata of the moth, *Manduca sexta*. *Arch. Ins. Biochem. Physiol.* 20: 145–155.

Redfern, C.P.F. 1984. Evidence for the presence of makisterone A in *Drosophila* larvae and the secretion of 20-deoxymakisterone A by the ring gland. *Proc. Nat. Acad. Sci. USA* 81: 5643–5647.

Rees, H. H. 1985. Biosynthesis of ecdysone. In *Comprehensive Insect Physiology, Biochemistry and Pharmacology*, ed. G. A. Kerkut and L. I. Gilbert, vol. 7, pp. 249–293. Pergamon, New York.

Rees, H. H. 1989. Zooecdysteroids: Structure and occurrence. In *Ecdysone: From Chemistry to Mode of Action*, ed. J. Koolman, pp. 28–38. Thieme-Verlag, Stuttgart.

Rembold, H. 1987. Caste specific modulation of juvenile hormone titers in *Apis mellifera*. *Insect Biochem.* 17: 1003–1007.

Retnakaran, A., J. Granett, and T. Ennis. 1985. Insect growth regulators. In *Comprehensive Insect Physiology, Biochemistry and Pharmacology*, ed. G. A. Kerkut and L. I. Gilbert, vol. 12, pp. 529–601. Pergamon, New York.

Reum, L., and J. Koolman. 1989. Radioimmune assays of ecdysteroids. In *Ecdysone: From Chemistry to Mode of Action*, ed. J. Koolman, pp. 131–143. Thieme-Verlag, Stuttgart.

Reynolds, S. E. 1974. Pharmacological induction of plasticization in the abdominal cuticle of *Rhodnius*. *J. Exp. Biol.* 61: 705–718.

Reynolds, S. E. 1975. The mechanical properties of the abdominal cuticle of *Rhodnius* larvae. *J. Exp. Biol.* 62: 69–80.

Reynolds, S. E. 1977. Control of cuticle extensibility in the wings of adult *Manduca* at the time of eclosion: Effects of eclosion hormone and bursicon. *J. Exp. Biol.* 70: 27–39.

Reynolds, S. E. 1983. Bursicon. In *Endocrinology of Insects*, ed. R.G.H. Downer and H. Laufer, pp. 235–248. Alan Liss, New York.

Reynolds, S. E. 1985. Hormonal control of cuticle mechanical properties. In *Comprehensive Insect Physiology, Biochemistry and Pharmacology*, ed. G. A. Kerkut and L. I. Gilbert, vol. 8, pp. 335–351. Pergamon, New York.

Reynolds, S. E., P. E. Taghert, and J. W. Truman. 1979. Eclosion hormone and bursicon titres and the onset of hormonal responsiveness during the last day of adult development in *Manduca sexta*. *J. Exp. Biol.* 78: 77–86.

Riddiford, L. M. 1970. Prevention of metamorphosis by exposure of insect eggs to juvenile hormone analogs. *Science* 167: 287–288.

Riddiford, L. M. 1972. Juvenile hormone in relation to the larval-pupal transformation of the Cecropia silkworm. *Biol. Bull.* 142: 310–325.

Riddiford, L. M. 1978. Ecdysone-induced change in cellular commitment of the epidermis of the tobacco hornworm, *Manduca sexta*, at the initiation of metamorphosis. *Gen. Comp. Endocrinol.* 34: 438–446.

Riddiford, L. M. 1981. Hormonal control of epidermal cell development. *Amer. Zool.* 21: 751–762.

Riddiford, L. M. 1982. Changes in translatable mRNAs during the larval-pupal transformation of the epidermis of the tobacco hornworm. *Dev. Biol.* 92: 330–342.

Riddiford, L. M. 1985. Hormone action at the cellular level. In *Comprehensive Insect Physiology, Biochemistry and Pharmacology*, ed. G. A. Kerkut and L. I. Gilbert, vol. 8, pp. 37–84. Pergamon, New York.

Riddiford, L. M. 1989. The epidermis as a model system for ecdysteroid action. In *Ecdysone: From Chemistry to Mode of Action*, ed. J. Koolman, pp. 407–413. Thieme-Verlag, Stuttgart.

Riddiford, L. M. 1992. Molecular approaches to insect endocrinology. In *Insect Molecular Science*, ed. J. M. Cramton and P. Eggleston, pp. 226–240. Academic Press, London.

Riddiford, L. M., and K. Hiruma. 1990. Hormonal control of sequential gene expression in lepidopteran epidermis. In *Molting and Metamorphosis*, ed. E. Ohnishi and H. Ishizaki, pp. 207–222. Springer-Verlag, Berlin.

Riddiford, L. M., and M. L. Kiely. 1981. The hormonal control of commitment in the insect epidermis—cellular and molecular aspects. In *Regulation of Insect Development and Behaviour*, ed. F. Sehnal, A. Zabza, J. J. Menn, and B. Cymborowski, pp. 485–496. Wroclaw Technical University Press, Wroclaw.

Riddiford, L. M., and J. W. Truman. 1972. Delayed effects of juvenile hormone on insect metamorphosis are mediated by the corpus allatum. *Nature* 237: 458.

Riddiford, L. M., and J. W. Truman. 1978. Biochemistry of insect hormones and insect growth regulators. In *Biochemistry of Insects*, ed. M. Rockstein, pp. 307–357. Academic Press, New York.

Riddiford, L. M., and J. W. Truman. 1993. Hormone receptors and the regulation of insect metamorphosis. *Amer. Zool.* 33: 340–347.

Riddiford, L. M., and C. M. Williams. 1967. The effects of juvenile hormone analogues on the embryonic development of silkworms. *Proc. Nat. Acad. Sci. USA* 57: 595–601.

Robeau, R. M., and S. B. Vinson. 1976. Effects of juvenile hormone analogues on caste differentiation in the imported fire ant, *Solenopsis invicta*. *J. Georgia Entomol. Soc.* 11: 198–203.

Robinson, G. E. 1987. Regulation of honey bee age polyethism by juvenile hormone. *Behav. Ecol. Sociobiol.* 120: 329–338.

Robinson, G. E., R. E. Page, C. Strambi, and A. Strambi. 1989. Hormonal and genetic control of behavioral integration in honey bee colonies. *Science* 246: 109–112.

Robinson, G. E., C. Strambi, A. Strambi, and M. F. Feldlaufer. 1991. Comparison of juvenile hormone and ecdysteroid haemolymph titers in adult worker and queen honey bees (*Apis mellifera*). *J. Insect Physiol.* 37: 929–935.

Rohdendorf, E. B., and J.A.L. Watson. 1969. The control of reproductive cycles in the female firebrat, *Lepismodes inquilinus*. *J. Insect Physiol.* 15: 2085–2101.

Röller, H., K. H. Dahm, C. C. Sweeley, and B. M. Trost. 1967. Die Struktur des Juvenilhormon. *Angew. Chem.* 79: 190–191.

Romer, F. 1971. Die Prothorakaldruse der Larve von *Tenebrio molitor* L. (Tenebrionidae, Coleoptera) und ihre Veränderungen während eines Hautungszyklus. *Z. Zellforsch. Mikrosk. Anat.* 122: 425–455.

Romer, F., H. Emmerich, and J. Nowock. 1974. Biosynthesis of ecdysones in isolated prothoracic glands and oenocytes of *Tenebrio molitor in vitro*. *J. Insect Physiol.* 20: 1975–1987.

Röseler, P.-F. 1985. Endocrine basis of dominance and reproduction in polistine paper wasps. *Forthschr. Zool.* 31: 260–272.

Röseler, P.-F., I. Röseler, and A. Strambi. 1985. Role of ovaries and ecdysteroids in dominance hierarchy establishment among foundresses of the primitively social wasp, *Polistes gallicus*. *Behav. Ecol. Sociobiol.* 18: 9–13.

Röseler, P.-F., I. Röseler, A. Strambi, and R. Augier. 1984. Influence of insect hormones on the establishment of dominance hierarchies among foundresses of the paper wasp, *Polistes gallicus*. *Behav. Ecol. Sociobiol.* 15: 133–142.

Roth, L. M., and B. Stay. 1961. Oocyte development in *Diploptera punctata* (Eschscholtz) (Blattaria). *J. Insect Physiol.* 7: 186–202.

Rothschild, M. 1975. Recent advances in our knowledge of the Order Siphonaptera. *Ann. Rev. Entomol.* 20: 241–259.

Rothschild, M., and B. Ford. 1973. Factors influencing the breeding of the rabbit flea (*Spilopsyllus cuniculi*): A springtime accelerator and a kairomone in nestling rabbit urine, with notes on *Cediopsylla simplex*, another "hormone bound" species. *J. Zool.* 170: 87–137.

Rothschild, M., B. Ford, and M. Hughes. 1970. Maturation of the male rabbit flea (*Spilopsyllus cuniculi*) and the Oriental rat flea (*Xenopsylla cheopis*): Some effects of mammalian hormones on development and impregnation. *Trans. Zool. Soc. London* 32: 105–188.

Rountree, D. B., and W. E. Bollenbacher. 1986. The release of the prothoracicotropic hormone in the tobacco hornworm, *Manduca sexta*, is controlled intrinsically by juvenile hormone. *J. Exp. Biol.* 120: 41–58.

Roussel, J. P. 1978. Reprise de la ponte chez femelles allatectomisées de *Locusta migratoria* après injection d'hormone juvénile. *C.R. Hebd. Séanc. Acad. Sci. Paris* 286: 485–488.

Rousset, A., C. Bitsch, and J. Bitsch. 1987. Vitellogenins in the firebrat, *Thermobia domestica* (Packard): Immunological quantification during reproductive cycles and in relation to insemination (Thysanura; Lepismatidae). *J. Insect Physiol.* 33: 593–601.

Rowell, C.H.F. 1967. Corpus allatum implantation and green/brown polymorphism in three African grasshoppers. *J. Insect Physiol.* 13: 1401–1412.

Rowell, C.H.F. 1971. The variable colouration of the acridoid grasshoppers. *Adv. Insect Physiol.* 8: 145–198.

Safranek, L., B. Cymborowski, and C. M. Williams. 1980. Effects of juvenile hormone on ecdysone-dependent development in the tobacco hornworm, *Manduca sexta*. *Biol. Bull.* 158: 248–256.

Sakurai, S., J. T. Warren, and L. I. Gilbert. 1989. Mediation of ecdysone synthesis in *Manduca sexta* by a hemolymph enzyme. *Arch. Insect Biochem. Physiol.* 10: 179–197.

Saunders, D. S. 1980. Some effects of constant temperature and photoperiod on the diapause response of the fleshfly, *Sarcophaga argyrostoma*. *Physiol. Entomol.* 5: 191–198.

Scarborough, R. M., G. C. Jamieson, F. Kalish, S. J. Kramer, G. A. McEnroe, C. A.

Miller, and D. A. Schooley. 1984. Isolation and primary structure of two peptides with cardioacceleratory and hyperglycaemic activity from the corpora cardiaca of *Periplaneta americana. Proc. Nat. Acad. Sci. USA* 81: 5575–5579.

Schaller, F. 1952. Effets d'une ligature postcéphalique sur le développement de larves agées d'*Apis mellifica* L. *Bull. Soc. Zool. France* 77: 195–204.

Scharrer, B. 1964. The fine structure of Blattarian prothoracic glands. *Z. Zellforsch. Mikrosk. Anat.* 64: 301–326.

Scharrer, B. 1966. Ultrastructural study of the regressing prothoracic glands of blattarian insects. *Z. Zellforsch. Mikrosk. Anat.* 69: 1–21.

Scharrer, B., and M. Von Harnack. 1958 . Histophysiological studies on the corpus allatum of *Leucophaea maderae*. I. Normal life cycle in male and female adults. *Biol. Bull.* 115: 508–520.

Schmidt-Nielsen, K. 1984. *Scaling: Why is Animal Size so Important?* Cambridge University Press, Cambridge, U.K.

Schneiderman, H. A., and L. I. Gilbert. 1964. Control of growth and development in insects. *Science* 143: 325–333.

Schoofs, L., G. M. Holman, T. K. Hayes, R. J. Nachman, and A. De Loof. 1990. Isolation, identification, and synthesis of locustamyotropin II, an additional neuropeptide of *Locusta migratoria*: Member of the cephalomyotropic peptide family. *Insect Biochem.* 20: 479–484.

Schooley, D. A., and F. C. Baker. 1985. Juvenile hormone biosynthesis. In *Comprehensive Insect Physiology, Biochemistry and Pharmacology*, ed. G. A. Kerkut and L. I. Gilbert, vol. 7, pp. 363–389. Pergamon, New York.

Schooley, D. A., C. A. Miller, and J. P. Proux. 1987. Isolation of two arginine vasopressin-like factors from ganglia of *Locusta migratoria. Arch. Insect Biochem. Physiol.* 5: 157–166.

Schooneveld, H., S. J. Kramer, H. Privee, and A. van Huis. 1979. Evidence of controlled corpus allatum activity in the adult Colorado beetle. *J. Insect Physiol.* 25: 449–453.

Schooneveld, H., A. O. Sanchez, and J. de Wilde. 1977. Juvenile hormone-induced break and termination of diapause in the Colorado potato beetle. *J. Insect Physiol.* 23: 689–696.

Schwartz, L. M., and J. W. Truman. 1982. Peptide and steroid regulation of muscle degeneration in an insect. *Science* 215: 1420–1421.

Schwartz, L. M., and J. W. Truman. 1983. Hormonal control of rates of metamorphic development in the tobacco hornworm, *Manduca sexta. Dev. Biol.* 99: 103–114.

Schwartz, L. M., and J. W. Truman. 1984. Hormonal control of muscle atrophy and degeneration in the moth *Antheraea polyphemus. J. Exp. Biol.* 111: 13–30

Sedlak, B. J. 1985. Structure of endocrine glands. In *Comprehensive Insect Physiology, Biochemistry and Pharmacology*, ed. G. A. Kerkut and L. I. Gilbert, vol. 7, pp. 25–60. Pergamon, New York.

Seeley, T. D. 1982. Adaptive significance of the age polyethism schedule in honeybee colonies, *Behav. Ecol. Sociobiol.* 11: 287–293.

Sehnal, F. 1972. The action of ecdysone on ligated larvae of *Galleria melonella* L. (Lepidoptera): Induction of development. *Acta Entomol. Bohemoslov.* 68: 143–155.

Sehnal, F. 1983. Juvenile hormone analogues. In *Endocrinology of Insects*, ed. R.G.H. Downer and H. Laufer, pp. 657–672. Alan Liss, New York.

Sehnal, F. 1985. Growth and life cycles. In *Comprehensive Insect Physiology, Biochemistry and Pharmacology*, ed. G. A. Kerkut and L. I. Gilbert, vol. 2, pp. 1–86. Pergamon, New York.

Seligman, I. M., and F. A. Doy. 1973. Hormonal regulation of disaggregation of cellular fragments in the haemolymph of *Lucillia cuprina. J. Insect Physiol.* 19: 125–135.

Sevala, V. L., and K. G. Davey. 1989. Action of juvenile hormone on the follicle cells: Evidence for a novel regulatory mechanism involving protein kinase c. *Experientia* 45: 355–356.

Shapiro, A. M. 1976. Seasonal polyphenism. *Evol. Biol.* 9: 259–333.

Shapiro, J. P., and H. H. Hagedorn. 1982. Juvenile hormone and the development of ovarian responsiveness to a brain hormone in the mosquito *Aedes aegypti. Gen. Comp. Endocrinol.* 46: 176–183.

Shapiro, A. B., G. D. Wheelock, H. H. Hagedorn, F. C. Baker, L. W. Tsai, and D. A. Schooley. 1986. Juvenile hormone and juvenile hormone esterase in adult females of the mosquito, *Aedes aegypti. J. Insect Physiol.* 32: 867–877.

Siegert, K., and R. Ziegler. 1983. A hormone from the corpora cardiaca controls fat body glycogen phosphorylase during starvation in tobacco hornworm larvae. *Nature* 301: 526–527.

Sivasubramanian, P., S. Friedman, and G. Fraenkel. 1974. Nature and role of of proteinaceous hormonal factors acting during puparium formation in flies. *Biol. Bull.* 147: 163–185.

Slama, K. 1975. Some old concepts and new findings on hormonal control of insect morphogenesis. *J. Insect Physiol.* 21: 921–955.

Slama, K., and C. M. Williams. 1966. The juvenile hormone. V. The sensitivity of the bug, *Pyrrhocoris apterus*, to a hormonally active factor in American paper pulp. *Biol. Bull.* 130: 235–246.

Smith, S. L. 1985. Regulation of ecdysteroid titer: Synthesis. In *Comprehensive Insect Physiology, Biochemistry and Pharmacology*, ed. G. A. Kerkut and L. I. Gilbert, vol. 7, pp. 295–341. Pergamon, New York.

Smith, W. A., and W. L. Combest. 1985. Role of cyclic nucleotides in hormone action. In *Comprehensive Insect Physiology, Biochemistry and Pharmacology*, ed. G. A. Kerkut and L. I. Gilbert, vol. 8, pp. 263–299. Pergamon, New York.

Smith, W. A., and H. F. Nijhout. 1982. Ultrastructural changes accompanying secretion and cell death in the molting glands of an insect (*Oncopeltus*). *Tissue Cell* 14: 243–252.

Smith, W. A., and H. F. Nijhout. 1983. In vitro stimulation of cell death in the moulting glands of *Oncopeltus fasciatus* by 20-hydroxyecdysone. *J. Insect Physiol.* 29: 169–176.

Smith, W. A., and T. J. Pasquarello. 1989. Developmental changes in phosphodiesterase activity and hormonal response in the prothoracic glands of *Manduca sexta. Molec. Cell. Endocrinol.* 63: 239–246.

Smith, W. A., M. F. Bowen, W. E. Bollenbacher, and L. I. Gilbert. 1986. Cellular changes in the prothoracic glands of diapausing pupae of *Manduca sexta. J. Exp. Biol.* 120: 131–142.

Snodgrass, R. E. 1935. *Principles of Insect Morphology*. McGraw-Hill, New York.

Sonobe, H., and E. Ohnishi. 1970. Accumulation of 3-hydroxykynurenine in ovarian follicles in relation to diapause in the silkworm, *Bombyx mori*. *Dev. Growth Diff.* 12: 41–52.

Sparagana, S. P., G. Bhaskaran, and P. Bartrera. 1985. Juvenile hormone acid methyl transferase activity in imaginal disks of Manduca sexta prepupae. *Arch. Ins. Biochem. Physiol.* 2: 191–202.

Spencer, I. M., and D. J. Candy. 1974. The effect of flight on the concentrations and composition of haemolymph diacyl glycerols in the desert locust. *Biochem. Soc. Trans.* 2: 1093–1096.

Spring, J. H. 1990. Endocrine regulation of diuresis in insects. *J. Insect Physiol.* 36: 13–22.

Springhetti, A. 1972. The competence of *Kalotermes flavicollis* Fabr. (Isoptera) pseudergates to differentiate into soldiers. *Monit. Zool. Ital.* 6: 97–111.

Staal, G. B. 1961. Studies on the physiology of phase induction in *Locusta migratoria migratorioides* R. & F. *Publ. Fond. Landb. Bur.* 40: 1–127.

Starratt, A. N., and B. E. Brown. 1975. Structure of the pentapeptide proctolin, a proposed neurotransmitter in insects. *Life Sci.* 17: 1253–1256.

Stay, B., K. K. Chan, and A. P. Woodhead. 1992. Allatostatin-immunoreactive neurons projecting to the corpora allata of adult *Diploptera punctata*. *Cell Tissue Res.* 270: 15–23.

Stay, B., T. Friedel, S. S. Tobe, and E. C. Mundall. 1980. Feedback control of juvenile hormone synthesis in cockroaches: Possible role for ecdysterone. *Science* 207: 898–900.

Steel, C.G.H., and K. G. Davey. 1985. Integration in the insect endocrine system. In *Comprehensive Insect Physiology, Biochemistry and Pharmacology*, ed. G. A. Kerkut and L. I. Gilbert, vol. 8, pp. 1–35. Pergamon, New York.

Steel, C.G.H., W. E. Bollenbacher, S. L. Smith, and L. I. Gilbert. 1982. Haemolymph ecdysteroid titres during larval-adult development in *Rhodnius prolixus*: Correlations with moulting hormone action and brain neurosecretory cell activity. *J. Insect Physiol.* 28: 519–525.

Steele, J. E. 1961. Occurrence of a hyperglycaemic factor in the corpus cardiacum of an insect. *Nature* 192: 680–681.

Steele, J. E. 1980. Hormonal modulation of carbohydrate and lipid metabolism in the fat body. In *Insect Biology in the Future*, ed. M. Locke and D. S. Smith, pp. 253–271. Academic Press, New York.

Steele, J. E. 1982. Glycogen phosphorylase in insects. *Insect Biochem.* 12: 131–147.

Steele, J. E. 1985. Control of metabolic processes. In *Comprehensive Insect Physiology, Biochemistry and Pharmacology*, ed. G. A. Kerkut and L. I. Gilbert, vol. 8, pp. 99–145. Pergamon, New York.

Stuart, A. M. 1979. The determination and regulation of the neotenic reproductive caste in the lower termites: With special reference to the genus *Zootermopsis*. *Sociobiol.* 4: 223–237.

Sugumaran, M. 1988. Molecular mechanisms of cuticular sclerotization. *Adv. Insect Physiol.* 21: 179–231.

Sweet, S. S. 1980. Allometric inference in morphology. *Amer. Zool.* 20: 643–652.

Szibbo, C. M., and S. S. Tobe. 1981. Cellular and volumetric changes in relation to the activity cycle in the corpora allata of *Diploptera punctata*. *J. Insect Physiol.* 27: 655–665.

Takami, T. 1959. Induced growth of diapausing silkworm embryos *in vitro. Science* 130: 98–99.

Takeda, S., Y. Kono, and Y. Kameda. 1988. Induction of nondiapause eggs in *Bombyx mori* by a trehalase inhibitor. *Entomol. Exp. Appl.* 46: 291–294.

Tauber, M. J., C. A. Tauber, and S. Masaki. 1986. *Seasonal Adaptations of Insects*. Oxford University Press, New York.

Tickle, C. 1991. Retinoic acid and chick limb bud development. *Development*, suppl. 1 (1991): 113–121.

Tobe, S. S., and B. Stay. 1985. Structure and regulation of the corpus allatum. *Adv. Insect Physiol.* 18: 305–432.

Tobe, S. S., J. Girardie, and A. Girardie. 1982. Enhancement of juvenile hormone biosynthesis in locusts following electrostimulation of cerebral neurosecretory cells. *J. Insect Physiol.* 23: 867–871.

Truman, J. W. 1971a. The role of the brain in the ecdysis rhythm of silkmoths: Comparison with the photoperiodic termination of diapause. In *Biochronometry*, ed. M. Menaker, pp. 483–504. National Academy of Sciences Press, Washington, D.C.

Truman, J. W. 1971b. Circadian rhythms and physiology with special reference to neuroendocrine processes in insects. In *Proceedings of the International Symposium on Circadian Rhythmicity*, Wageningen, The Netherlands, pp. 111–135. Pudoc Press, Wageningen.

Truman, J. W. 1971c. Physiology of insect ecdysis. I. The eclosion behaviour of saturniid moths and its hormonal release. *J. Exp. Biol.* 54: 805–814.

Truman, J. W. 1971d. Hour-glass behavior of the circadian clock controlling eclosion of the silkmoth *Antheraea pernyi. Proc. Nat. Acad. Sci. USA* 68: 595–599.

Truman, J. W. 1972. Physiology of insect rhythms. I. Circadian organization of the endocrine events underlying the molting cycle of larval tobacco hornworms. *J. Exp. Biol.* 57: 805–820.

Truman, J. W. 1973a. Physiology of insect ecdysis. II. The assay and occurrence of the eclosion hormone in the Chinese Oak silkmoth, *Antheraea pernyi. Biol. Bull.* 144: 200–211.

Truman, J. W. 1973b. Physiology of insect ecdysis. III. Relationship between the hormonal control of eclosion and of tanning in the tobacco hornworm, *Manduca sexta. J. Exp. Biol.* 58: 821–829.

Truman, J. W. 1978. Hormonal release of stereotyped motor programmes from the isolated nervous system of the Cecropia silkmoth. *J. Exp. Biol.* 74: 151–174.

Truman, J. W. 1981. Interaction between ecdysteroids, eclosion hormone, and bursicon titers in *Manduca sexta. Amer. Zool.* 21: 655–661.

Truman, J. W. 1985. Hormonal control of ecdysis. In *Comprehensive Insect Physiology, Biochemistry and Pharmacology*, ed. G. A. Kerkut and L. I. Gilbert, vol. 8, pp. 413–440. Pergamon, New York.

Truman, J. W. 1988. Hormonal approaches for studying nervous system development. *Adv. Insect. Physiol.* 21:1–34

Truman, J. W. 1990. Neuroendocrine control of ecdysis. In *Molting and Metamorphosis*, ed. E. Ohnishi and H. Ishizaki, pp. 67–82. Springer-Verlag, Berlin.

Truman, J. W., and P. F. Copenhaver. 1989. The larval eclosion hormone neurones in *Manduca sexta*: Identification of the brain-proctodeal neurosecretory system. *J. Exp. Biol.* 147: 457–470.

Truman, J. W., and L. M. Riddiford. 1970. Neuroendocrine control of ecdysis in silkmoths. *Science* 167: 1624–1626.

Truman, J. W., and L. M. Riddiford. 1974a. Physiology of insect rhythms. III. The temporal organization of endocrine events underlying pupation of the tobacco hornworm. *J. Exp. Biol.* 60: 371–382.

Truman, J. W., and L. M. Riddiford. 1974b. Hormonal mechanisms underlying insect behavior. *Adv. Insect Physiol.* 10: 297–350.

Truman, J. W., and L. M. Riddiford. 1977. Invertebrate systems for the study of hormonal effects on behavior. *Vitam. Horm.* 35: 283–315.

Truman, J. W., and L. M. Riddiford. 1989. Development of the insect neuroendocrine system. In *Development, Maturation, and Senescence of Neuroendocrine Systems: A Comparative Approach*. Academic Press, New York.

Truman, J. W., and L. M. Schwartz. 1980. Peptide hormone regulation of programmed death of neurons and muscle in an insect. In *Peptides: Integrators of Cell and Tissue Function*, ed. F. E. Bloom, pp. 55–67. Raven Press, New York.

Truman, J. W., and L. M. Schwartz. 1982. Programmed death in the nervous system of a moth. *Trends Neurosci.* 5: 270–273.

Truman, J. W., S. M. Mumby, and S. K. Welch. 1979. Involvement of cyclic GMP in the release of stereotyped behavior patterns in moths by a peptide hormone. *J. Exp. Biol.* 84: 201–212.

Truman, J. W., L. M. Riddiford, and L. Safranek. 1973. Hormonal control of cuticle coloration in the tobacco hornworm, *Manduca sexta*: Basis of an ultrasensitive bioassay for juvenile hormone. *J. Insect Physiol.* 19: 195–203.

Truman, J. W., L. M. Riddiford, and L. Safranek. 1974. Temporal patterns of response to ecdysone and juvenile hormone in the epidermis of the tobacco hornworm, *Manduca sexta*. *Dev. Biol.* 39: 247–262.

Truman, J. W., D. B. Rountree, S. E. Reiss, and L. M. Schwartz. 1983. Ecdysteroids regulate the release and action of eclosion hormone in the tobacco hornworm, *Manduca sexta* (L.). *J. Insect Physiol.* 29: 895–900.

Truman, J. W., P. H. Taghert, P. F. Copenhaver, N. J. Tublitz, and L. M. Schwartz. 1981. Eclosion hormone may control all ecdyses in insects. *Nature* 291: 70–71.

Tsuchida, K., M. Nagata, and A. Suzuki. 1987. Hormonal control of ovarian development in the silkworm *Bombyx mori*. *Arch. Insect Biochem. Physiol.* 5: 167–178.

Tublitz, N. J. 1989. Insect cardioactive peptides: Neurohormonal regulation of cardiac activity by two cardioacceleratory peptides during flight in the tobacco hawkmoth, *Manduca sexta*. *J. Exp. Biol.* 142: 31–48.

Tublitz, N. J., and J. W. Truman. 1981. An insect cardioactive peptide modulates heart rate during development. *Soc. Neurosci. Abstr.* 7: 253.

Tublitz, N. J., and J. W. Truman. 1985. Insect cardioactive peptides. I. Distribution and molecular characteristics of two cardioacceleratory peptides in the tobacco hawkmoth, *Manduca sexta*. *J. Exp. Biol.* 114: 365–379.

Tublitz, N. J., A. T. Allen, C. C. Cheung, K. K. Edwards, D. P. Kimble, P. K. Loi, and A. W. Sylwester. 1992. Insect cardioactive peptides: Regulation of hindgut activity by cardioacceleratory peptide 2 (CAP_2) during wandering behavior in *Manduca sexta* larvae. *J. Exp. Biol.* 165: 241–264.

Tublitz, N. J., D. Brink, K. S. Broadie, P. K. Loi, and A. W. Sylwester. 1991. From behavior to molecules: An integrated approach to the study of neuropeptides. *Trends in Neurosci.* 14: 254–259.

Tublitz, N. J., C. C. Cheung, K. K. Edwards, A. W. Sylwester, and S. E. Reynolds. 1992. Insect cardioactive peptides in *Manduca sexta*: A comparison of the biochemical and molecular characteristics of cardioactive peptides in larvae and adults. *J. Exp. Biol.* 165: 265–272.

Uvarov, B. P. 1966. *Grasshoppers and Locusts.* Vol. 1. Cambridge University Press, Cambridge, England.

Van der Horst, D. J., J. M. Van Doorn, and A. M. Beenakkers. 1979. Effects of the adipokinetic hormone on the release and turnover of haemolymph diglycerides and on the formation of the diglyceride-transporting lipoprotein system during locust flight. *Insect Biochem.* 9: 627–635.

Van Der Kloot, W. G., and C. M. Williams. 1954. Cocoon construction by the Cecropia silkworm. III. The alteration of spinning behavior by chemical and surgical techniques. *Behaviour* 5: 233–255.

Vaught, G. L., and K. W. Stewart. 1974. The life history and ecology of the stonefly *Neoperla clymene* (Newman) (Plecoptera: Perlidae). *Ann. Entomol. Soc. Amer.* 67: 167–178.

Veenstra, J. A. 1989a. Do insects really have a homeostatic hypotrehalosaemic hormone? *Biol. Rev.* 64: 305–316

Veenstra, J. 1989b. Isolation and structure of two gastrin CCK-like neuropeptides from the American cockroach homologous to the leucosulfakinins. *Neuropeptides* 14: 145–149.

Veenstra, J. 1989c. Isolation and structure of corazonin, a cardioactive peptide from the American cockroach. *FEBS Lett.* 250: 231–234.

Veron, J.E.N. 1973. Physiological control of the chromatophores of *Austrolestes annulosus* (Odonata). *J. Insect Physiol.* 19: 1689–1703.

Veron, J.E.N., A. F. O'Farrell, and B. Dixon. 1974. Fine structure of Odonata chromatophores. *Tissue & Cell* 6: 613–626.

Vincent, J.F.V. 1975a. How does a female locust dig her oviposition hole? *J. Entomol.* 50: 175–181.

Vincent, J.F.V. 1975b. Locust oviposition: Stress softening of the extensible intersegmental membranes. *Proc. Roy. Soc. London* B 188: 189–201.

Vincent, J.F.V. 1981. Morphology and design of the extensible intersegmental membrane of the female migratory locus. *Tissue & Cell* 13: 831–853.

Warren, J. T., and L. I. Gilbert. 1986. Ecdysone metabolism and distribution during the pupal-adult development of *Manduca sexta. Insect Biochem.* 16: 62–82.

Warren, J. T., and L. I. Gilbert. 1988. Radioimmunoassay: Ecdysteroids. In *Immunological Techniques in Insect Biology*, ed. L. I. Gilbert and T. A. Miller, pp. 181–214. Springer-Verlag, New York.

Warren, J. T., S. Sakurai, D. B. Rountree, and L. I. Gilbert. 1988. Synthesis and secretion of ecdysteroids by the prothoracic glands of *Manduca sexta. J. Insect Physiol.* 34: 561–576.

Warren, J. T., W. A. Smith, and L. I. Gilbert. 1984. Simplification of the ecdysteroid radioimmunoassay by the use of protein A from *Staphylococcus aureus. Experientia* 40: 393–394.

Wasserthal, L. T. 1975. Herzschlag-Umkehr bei Insekten und die Entwicklung der imaginales Herzrhytmik. *Verh. Dtsch. Zool. Ges.* 1974: 95–99.

Watanabe, K. 1935. On the hatching of silkworm eggs by acid treatment. *Tech. Bull. Seric. Exp. Sta.* 47: 1–38.

Watson, J.A.L. 1964. Moulting and reproduction in the adult firebrat, *Thermobia domestica* (Packard) (Thysanura, Lepismatidae). II. The reproductive cycles. *J. Insect Physiol.* 10: 399–408.
Watson, J.A.L., B. M. Okot-Kotber, and Ch. Noirot. 1985. *Caste Differentiation in Social Insects*. Pergamon, Oxford.
Watson, R. D., and W. E. Bollenbacher. 1988. Juvenile hormone regulates the steroidogenic competence of *Manduca sexta* prothoracic glands. *Molec. Cell. Endocrinol.* 57: 251–259.
Watson, R. D., N. Agui, M. E. Haire, and W. E. Bollenbacher. 1987. Juvenile hormone coordinates the regulation of the hemolymph ecdysteroid titer during pupal commitment in *Manduca sexta. Insect Biochem.* 17: 955–959.
Watson, R. D., M. K. Thomas, and W. E. Bollenbacher. 1989. Regulation of ecdysteroidogenesis in prothoracic glands of the tobacco hornworm *Manduca sexta. J. Exp. Zool.* 252: 255–263.
Weaver, N. 1957. Effects of larval age on dimorphic differentiation of the female honey bee. *Ann. Entomol. Soc. Amer.* 50: 283–294.
Weber, H. 1954. *Grundriss der Insektenkunde*. Gustav-Fischer-Verlag, Stuttgart.
Weis-Fogh, T. 1952. Fat combustion and metabolic rate of flying locusts (*Schistocerca gregaria* Forstral). *Phil. Trans. Roy. Soc. London* B 237: 1–36.
Weis-Fogh, T. 1964. Diffusion in insect wing muscle, the most active tissue known. *J. Exp. Biol.* 41: 229–256.
Westbrook, A. L., and W. E. Bollenbacher. 1990. The prothoracicotropic hormone neuroendocrine axis in *Manduca sexta*: Development and function. In *Molting and Metamorphosis*, ed. E. Ohnishi and H. Ishizaki, pp. 3–16. Springer-Verlag, Berlin.
Westbrook, A. L., S. L. Regan, and W. E. Bollenbacher. 1993. Developmental expression of the prothoracicotropic hormone in the CNS of the tobacco hornworm *Manduca sexta. J. Comp. Neurol.* 327: 1–16.
West-Eberhard, M. J. 1967. Foundress associations in polistine wasps: Dominance hierarchies and the evolution of social behavior. *Science* 157: 1584–1585.
Wheeler, C. H., and G. M. Coast. 1990. Assay and characterization of diuretic factors in insects. *J. Insect Physiol.* 36: 23–34.
Wheeler, D. E. 1990. The developmental basis of worker polymorphism in fire ants. *J. Insect Physiol.* 36: 315–322.
Wheeler, D. E. 1991. The developmental basis of worker caste polymorphism in ants. *Amer. Nat.* 138: 1218–1238.
Wheeler, D. E., and H. F. Nijhout. 1981. Imaginal wing disks in larvae of the soldier caste of *Pheidole bicarinata vinelandica. Int. J. Insect Morphol. Embryol.* 10: 131–139.
Wheeler, D. E., and H. F. Nijhout. 1983. Soldier determination in *Pheidole bicarinata*: Effect of methoprene on caste and size within castes. *J. Insect Physiol.* 29: 847–854.
Wheeler, D. E., and H. F. Nijhout. 1984. Soldier determination in the ant *Pheidole bicarinata*: Inhibition by adult soldiers. *J. Insect Physiol.* 30: 127–135.
Wheeler, W. M. 1910. *Ants: Their Structure and Behavior*. Columbia University Press, New York.
Whiting, P. W., R. J. Greb, and B. Speicher. 1934. A new type of sex-intergrade. *Biol. Bull.* 66: 152–165.

Whitten, J. 1968. Metamorphic changes in insects. In *Metamorphosis*, ed. W. Etkin and L. I. Gilbert, pp. 43–105. Appleton-Century-Crofts, New York.

Wiens, A. W., and L. I. Gilbert. 1967. Regulation of carbohydrate mobilization and utilization by the corpus cardiacum *in vitro. Science* 150: 614–616.

Wigglesworth, V. B. 1934. The physiology of ecdysis in *Rhodnius prolixus* (Hemiptera). II. Factors controlling moulting and 'metamorphosis'. *Quart. J. Microsc. Sci.* 77: 191–222.

Wigglesworth, V. B. 1936. The functions of the corpus allatum in the growth and reproduction of *Rhodnius prolixus* (Hemiptera). *Quart. J. Microsc. Sci.* 79: 91–121.

Wigglesworth, V. B. 1940. The determination of characters at metamorphosis in *Rhodnius prolixus* (Hemiptera). *J. Exp. Biol.* 17: 201–222.

Wigglesworth, V. B. 1952. The thoracic gland in *Rhodnius prolixus* (Hemiptera) and its role in moulting. *J. Exp. Biol.* 29: 561–570.

Wigglesworth, V. B. 1955. The breakdown of the thoracic glands in the adult insect, *Rhodnius prolixus. J. Exp. Biol.* 32: 485–491.

Wigglesworth, V. B. 1957. The action of growth hormones in insects. *Symp. Soc. Exp. Biol.* 11: 204–227.

Wigglesworth, V. B. 1959. *The Control of Growth and Form: A Study of the Epidermal Cell in an Insect.* Cornell University Press, Ithaca, NY.

Wigglesworth, V. B. 1965. *The Principles of Insect Physiology*. Methuen, London.

Wigglesworth, V. B. 1970. *Insect Hormones*. Freeman and Co., San Francisco.

Wigglesworth, V. B. 1972. *The Principles of Insect Physiology*. 7th ed. Chapman and Hall, London.

Wiley, E. O. 1981. *Phylogenetics*. Wiley, New York.

Williams, C. M. 1947. Physiology of insect diapause. II. Interaction between the pupal brain and prothoracic glands in the metamorphosis of the giant silkworm, *Platysamia cecropia. Biol. Bull.* 93: 89–98.

Williams, C. M. 1948a. Physiology of insect diapause. III. The prothoracic glands in the cecropia silkworm, with special reference to their significance in embryonic and postembryonic development. *Biol. Bull.* 94: 60–65.

Williams, C. M. 1948b. Extrinsic control of morphogenesis as illustrated in the metamorphosis of insects. *Growth Symp.* 12: 61–74.

Williams, C. M. 1952a. Physiology of insect diapause. IV. The brain and prothoracic glands as an endocrine system in the cecropia silkworm. *Biol. Bull.* 103: 120–138.

Williams, C. M. 1952b. Morphogenesis and the metamorphosis of insects. *Harvey Lectures* 47: 126–155.

Williams, C. M. 1956. The juvenile hormone of insects. *Nature* 178: 212–213.

Williams, C. M. 1959. The juvenile hormone. I. Endocrine activity of the corpora allata of the adult cecropia silkworm. *Biol. Bull.* 116: 323–338.

Williams, C. M. 1961. The juvenile hormone. II. Its role in the endocrine control of molting, pupation, and adult development in the cecropia silkworm. *Biol. Bull.* 121: 572–585.

Williams, C. M. 1963. The juvenile hormone. III. Its accumulation and storage in the abdomens of certain male moths. *Biol. Bull.* 124: 355–367.

Williams, C. M. 1967. Third generation pesticides. *Sci. Amer.* 217: 13–17.

Williams, C. M. 1980. Growth in insects. In *Insect Biology in the Future*, ed. M. Locke and D. S. Smith, pp. 369–383. Academic Press, New York.

Williams, C. M., and K. Slama. 1966. The juvenile hormone. VI. Effects of the "paper factor" on the growth and metamorphosis of the bug, *Pyrrhocoris apterus. Biol. Bull.* 130: 247–253.

Williams, J. C., and K. W. Beyenbach. 1983. Differential effects of secretagogues on Na and K secretion in the malpighian tubules of *Aedes aegypti. J. Comp. Physiol.* 149: 511–517.

Willis, J. H. 1969. The programming of differentiation and its control by juvenile hormone in saturniids. *J. Embryol. Exp. Morphol.* 22: 27–44.

Willis, J. H. 1986. The paradigm of stage specific gene sets in insect metamorphosis: Time for revision! *Arch. Insect Biochem. Physiol.* (suppl.) 1: 47–57.

Willis, J. H., and P. A. Lawrence. 1970. Deferred action of juvenile hormone. *Nature* 225: 81–83.

Willis, J. H., R. Rezaur, and F. Sehnal. 1982. Juvenoids cause some insects to form composite cuticles. *J. Embryol. Exp. Morphol.* 71: 25–40.

Wilson, E. O. 1953. The origin and evolution of polymorphism in ants. *Quart. Rev. Biol.* 28: 136–156.

Wilson, E. O. 1971. *The Insect Societies*. Harvard University Press, Cambridge, Mass.

Winston, M. L. 1987. *The Biology of the Honey Bee*. Harvard University Press, Cambridge, Mass.

Wirtz, P. 1973. Differentiation in the honeybee larva. *Meded. Landb. Hogesch. Wageningen*, 73–75: 1–66.

Wolfgang, W. J., and L. M. Riddiford. 1981. Cuticular morphogenesis during continuous growth of the final instar larva of a moth. *Tissue & Cell* 13: 757–772.

Woodhead, A. P., B. Stay, S. L. Seidel, M. A. Khan, and S. S. Tobe. 1989. Primary structure of four allatostatins: Neuropeptide inhibitors of juvenile hormone synthesis. *Proc. Nat. Acad. Sci. USA* 86: 5997–6001.

Yagi, S., and N. Akaike. 1976. Regulation of larval diapause by juvenile hormone in the European corn borer, *Ostrinia nubilalis. J. Insect Physiol.* 22: 389–392.

Yagi, S., and M. Fukaya. 1974. Juvenile hormone as a key factor in regulating larval diapause of the rice stem borer *Chilo suppressalis* (Lepidoptera: Pyralidae). *Appl. Entomol. Zool.* 9: 247–255.

Yamamoto, K., A. Chadarevian, and M. Pellegrini. 1988. Juvenile hormone action mediated in the male accessory glands of *Drosophila* by calcium and kinase c. *Science* 239: 916–919.

Yamashita, O., and K. Hasegawa. 1964. Studies on the mode of action of diapause hormone in the silkworm, *Bombyx mori.* III. Effects of diapause hormone extracts on 3-hydroxykynurenine content in ovaries of silkworm pupae. *J. Seric. Sci.* 33: 115–123.

Yamashita, O., and K. Hasegawa. 1970. Oocyte age sensitive to the diapause hormone from the standpoint of glycogen synthesis in the silkworm, *Bombyx mori. J. Insect. Physiol.* 16: 2377–2383.

Yamashita, O., and T. Yaginuma. 1991. Silkworm eggs at low temperatures: Implications for sericulture. In *Insects al Low Temperature*, ed. R. E. Lee and D. L. Denlinger, pp. 424–445. Chapman and Hall, New York.

Yamazaki, M., and M. Kobayashi. 1969. Purification of the proteinic brain hormone of the silkworm, *Bombyx mori. J. Insect Physiol.* 15: 1981–1990.

Yeh, C., and M. J. Klowden. 1990. Effects of male accessory gland substances on the pre-oviposition behavior of *Aedes aegypti* mosquitoes. *J. Insect Physiol.* 36: 799–803.

Yin, C.-M., and G. M. Chippendale. 1974. Juvenile hormone and the induction of larval polymorphism and diapause of the southwestern corn borer, *Diatraea grandiosella. J. Insect Physiol.* 20: 1833–1847.

Yin, C.-M., and G. M. Chippendale. 1976. Hormonal control of larval diapause and metamorphosis in the southwestern corn borer, *Diatraea grandiosella. J. Exp. Biol.* 64: 303–310.

Yin, C.-M., and G. M. Chippendale. 1979. Diapause of the southwestern corn borer, *Diatraea grandiosella*: Further evidence showing juvenile hormone to be the regulation. *J. Insect Physiol.* 25: 513–523.

Zdarek, J. 1985. Regulation of pupariation in flies. In *Comprehensive Insect Physiology, Biochemistry and Pharmacology*, ed. G. A. Kerkut and L. I. Gilbert, vol. 8, pp. 301–333. Pergamon, New York.

Zdarek, J., and G. Fraenkel. 1969. Correlated effects of ecdysone and neurosecretion in puparium formation (pupariation) in flies. *Proc. Nat. Acad. Sci. USA* 64: 565–572.

Zdarek, J., and O. Haragsim. 1974. Action of juvenoids on metamorphosis of the honey-bee, *Apis mellifera. J. Insect Physiol.* 20: 209–221.

Zdarek, J., and K. Slama. 1968. Mating activity in adultoids or supernumerary larvae induced by agents with high juvenile hormone activity. *J. Insect Physiol.* 14: 563–567.

Ziegler, R. 1979. Hyperglycaemic factor from the corpora cardiaca of *Manduca sexta* (L.) (Lepidoptera: Sphingidae). *Gen. Comp. Endocrinol.* 39: 350–357.

Ziegler, R., and M. Schulz. 1986. Regulation of lipid metabolism during flight in *Manduca sexta. J. Insect Physiol.* 32: 903–908.

Ziegler, R., K. Eckart, and J. H. Law. 1990. Adipokinetic hormone controls lipid metabolism in adults and carbohydrate metabolism in larvae of *Manduca sexta. Peptides* 11: 1037–1040.

Ziegler, R., K. Eckart, H. Schwartz, and R. Keller. 1985. Amino acid sequence of *Manduca sexta* adipokinetic hormone elucidated by combined fast atom bombardment (FAB)/tandem mass spectrometry. *Biochem. Biophys. Res. Comm.* 133: 337–342.

AUTHOR INDEX

SUBJECT INDEX